INTEGRATED AIRCRAFT NAVIGATION

INTEGRATED AIRCRAFT NAVIGATION

JAMES L. FARRELL

Westinghouse Systems Development Division
Baltimore, Maryland

ACADEMIC PRESS New York San Francisco London 1976
A Subsidiary of Harcourt Brace Jovanovich, Publishers

ACADEMIC PRESS, INC.
111 Fifth Avenue, New York, New York 10003

United Kingdom Edition published by
ACADEMIC PRESS, INC. (LONDON) LTD.
24/28 Oval Road, London NW1

Library of Congress Cataloging in Publication Data

Farrell, James L
Integrated aircraft navigation.

Bibliography: p.
Includes index.
1. Inertial navigation (Aeronautics)
2. Estimation theory. I. Title.
TL588.5.F34 75-32025
ISBN 0-12-249750-3

PRINTED IN THE UNITED STATES OF AMERICA

To Terri Ann

Contents

Preface

Modern aircraft navigation systems are characterized by a multifaceted, computer-oriented approach, including various branches of theoretical dynamics, inertial measurements, radar, radio navaids, celestial observations, and widely used statistical estimation techniques. Each pertinent field entails much technological development that is *not* essential for applied systems analysis. Thus, a book devoted to inertial navigation might naturally contain instrument mechanization and design principles, while rigorous mathematics may well appear in a textbook on estimation theory. The student or practicing engineer, however, often wants only a valid analytical characterization, using the simplest theoretical concepts permissible while omitting specialized mechanization details. To satisfy that need, this book presents the pertinent information extracted from a broad range of topics, expressed in terms of Newtonian physics and matrix–vector mathematics.

Commonality and modular design in avionics can be greatly enhanced by recognizing the underlying *functional* similarity of (1) aircraft navigation and tracking, (2) spheroidal earth navigation and designation of a surface point, (3) primary and secondary (backup) sensor usage, (4) gimballed platform and strapdown inertial navigation, (5) space stable, geographic, and wander azimuth coordinate references, (6) damped and undamped inertial navigation systems, (7) initialization, in-flight update, and sensor calibration, (8) updating with radio, radar, or optical measurements, and (9) block or recursive estimation algorithms for updating. Since these items involve different engineering disciplines, an integrated presentation defining the necessary interrelationships was the prime motivation for this book. A readily understandable tutorial exposition is the overriding objective, without sacrifice of generality or interpretation. This goal was approached here through the following aspects of the presentation:

(1) Material was organized for maximum study flexibility. Chapters 1, 2, 3, 5, and 7, as well as Appendices II and III, are largely self-contained. The same is true of many tutorial appendices at the end of individual chapters. Cross references and supplementary material (e.g., text-related exercises, a few sections involving Laplace and Fourier transforms, and other special topics) are included but compartmentalized for ease of omission. The exercises are designed to add considerable insight for those willing to tackle them.

(2) Notation is unified throughout all chapters, and consistent with accepted conventions of each separate discipline. Appendix I summarizes the major notation and provides applicable units, array dimensions, subscripts, superscripts, and abbreviations.

(3) Appendix II contains enough matrix theory for many common engineering problems; thus the exposition proceeds without digression in mathematical developments.

(4) Mathematical analyses are interpreted by accompanying physical explanations.

(5) Various topics are presented with closed-form solutions to restrictive examples (e.g., regular precession, gyrodynamics, equivalence between final estimates from recursion and block estimation) which, though well known, are not commonly found in full detail elsewhere.

(6) Restrictive examples are generalized to practical situations in which linearity and other idealizations are violated.

(7) Convergence enhancement techniques are provided for practical realization of sequential estimation.

(8) Succinct descriptions are provided for most common navigation sensors.

(9) Navigation modes are defined in varying degrees of complexity, with alternatives for coordinate references, with and without inertial and backup sensors.

(10) Existing mechanizations are used for examples only, without constraining the applied techniques to any specific past or present system configuration.

In assembling the text material, an effort was made to cover all important topics while avoiding excess length. The first chapter defines basic navigation quantities and functions, and introduces various subjects as an aid to subsequent developments. Chapter 2 presents the kinematics to be used throughout (definition of coordinate frames and transformations between them; analysis of mathematical relationships for Euler angle, direction cosine, and four-parameter sets; applications to illustrative problems that arise in later sections). Chapter 3 presents a unified inertial navigation analysis applicable to both gimballed and strapdown systems (basic functional description of inertial instruments; fundamental approach to platform implementation and navigation equations; propagation of bias errors through the system). Chapter 4 derives the physics of inertial measurements, with physical explanations for several error sources, and propagates combined error expressions through a general error analysis. Chapter 5 analyzes navigation with multiple sensors and, after investigation of estimators under ideal conditions (linearity, known model characterization), extends the theory to practical navigation problems. Chapter 6 develops the dynamic equations in proper form for all nav modes to be considered (geographic position and velocity determination over a spheroidal earth, with or without INS; short-term, intermediate duration, and long-term estimation; adaptation to point navigation over an essentially flat earth for short duration; extension for air-to-air tracking). Chapter 7 describes functional relationships and practical considerations for the various navaid sensors in common usage. Chapter 8 illustrates several system applications. While the presentation here is slanted toward aircraft navigation, much of the material can be applied over a far broader scope.

Acknowledgments

I gratefully acknowledge the treasure of information contributed to this writing by individuals in education, publishing, editing, advanced mathematics, applied systems analysis, and aviation. The presentation has benefited from constructive conversations with authors of some of the references (such as Dr. B. Danik of Singer-Kearfott Division, Dr. W. S. Widnall of Intermetrics, and Dr. B. A. Kriegsman of C. S. Draper Laboratory), and with Messrs. E. C. Quesinberry, B. Van Hook, R. Wancowicz, E. H. Thompson, and T. P. Haney of Westinghouse. Portions of the manuscript were improved through the suggestions of Messrs. J. H. Mims, R. C. Webber, and T. D. Merrell, and Dr. J. C. Stuelpnagel of Westinghouse. Management support was provided by Dr. P. M. Pan and Messrs. J. C. Murtha, J. G. Gregory, and G. Shapiro. A practical contribution of major importance, obtained through airborne estimator mechanization and flight test verification, was made possible by the outstanding system and software team at Westinghouse, including Messrs. L. C. Schafer, R. C. Leedom, W. L. Weigle, R. I. Heller, and R. E. Kahn. The excellent staff at Academic Press provided considerable help on notation and other editorial advice. Valuable assistance was also received from Westinghouse secretaries Norine Eitel, Dot Michalski, and Ning Wist, and from G. Madoo, who coordinated the illustrations.

The chapter on estimation theory received constructive suggestions from Messrs. G. S. Axelby of Westinghouse, T. S. Englar of B.T.S., Inc., and E. J. Lefferts of NASA–GSFC. A special note of appreciation is also directed to Mr. William H. Pleasants of Westinghouse, for his insight as both systems analyst and pilot, to Professor Arthur E. Bryson of Stanford University, who showed extraordinary generosity in reviewing practically the entire first draft, and to Dr. Bernard Friedland of Singer-Kearfott Division, for innumerable suggestions regarding organization and presentation.

Chapter 1

Introduction

1.0

The aim of avionics systems design is to balance multiple (and potentially conflicting) requirements, such as

low cost,
high accuracy,
reliability,
small size and weight,
low power consumption,
simplicity: ease of operation and maintenance,
flexibility: ease of modification and growth.

Advances in digital technology have provided a design approach whereby several or all of these goals can be achieved simultaneously. A key feature of this approach is efficient usage of all available data, through centralized processing in an integrated system of computing and data transmission. In a fully integrated system, separate instruments would not be needed for any one function; every subsystem could have access to, or benefit from, all sensor information.

As applied to navigation, system integration can provide improved accuracy for any given instrumenting configuration or minimize the sensor requirements for any specified system accuracy. In either case, the objective is to process all navigation input data so that maximum information is extracted. While this may appear self-evident, conventional navigation methods have *not* generally used all available inputs to their full potential. For instance, partial LORAN fixes (from a master plus one slave station) and air data unaccompanied by wind velocity indication have often been considered useless for accurate navigation; conventional doppler-damped inertial systems may use excess data rates and/or exhibit error transients in turns; barometric "damping" of vertical channel acceleration frequently loses accuracy during altitude changes. These limitations do not follow from inherent deficiencies in information, but rather from data *usage* by methods widely followed before digitization.

While classical approaches have served well in the past, departures from previously accepted procedures are being forced by advanced avionics system requirements. Concurrent with breakthroughs in computation, there has been a substantial broadening of analytical techniques, notably in estimation (Ref. 1-1). This combination of analytical and digital technology advances has given further impetus to the development of improved algorithms (e.g., for high resistance to numerical degradation) and improved simulation techniques. The influence of all these changes on aircraft navigation is readily apparent from the open literature, wherein various studies and flight programs are reported. Of course, successes already achieved can be realized repeatedly, *provided* that practical measures are adopted as described in subsequent chapters.

As an aid for *applied systems analysis*, this book avoids both rigorous mathematical proofs and detailed hardware mechanization. Between pure theory and practice lies a great deal of engineering, essential for the development of operational navigation (nav) systems. The material presented here will prepare the reader for such tasks as specifying a nav sensor selection with an observation schedule, conducting an extensive simulation, or defining required avionics computations. Most of the pertinent information is independent of changing technology, but some descriptions are expressed in terms of existing equipment for ease of familiarization. Preliminary analyses in pertinent subject areas, intentionally simplified for introductory purposes, are given in this chapter.

1.1 BASIC MOTION PATTERNS

Rigid body dynamics can involve rotation of an object *about* its mass center and/or translation *of* that mass center. The motion can be absolute or relative (i.e., defined with respect to a fixed or changing reference), and constraints may or may not restrict the set of possible excursions. Present discussion will begin with translation along a predetermined path.

VELOCITY AND POSITION ON A TRACK

Consider translation of a point C at constant speed V_C along the fixed straight line in Fig. 1-1, so that the instantaneous distance from the fixed reference point is

$$X_C = X_{C(t_0)} + V_C(t - t_0) \tag{1-1}$$

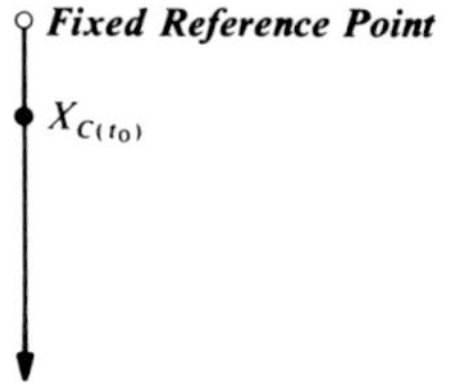

FIG. 1-1. Translation along a straight line.

As a first generalization, the speed may vary so that

$$X_C = X_{C(t_0)} + \int_{t_0}^{t} V_C \, d\tau \tag{1-2}$$

This same relation can be used for translation restricted to curved line paths, in which X_C and V_C represent arc length and tangential speed, respectively. When (1-2) is applied to Fig. 1-2 with known velocity time

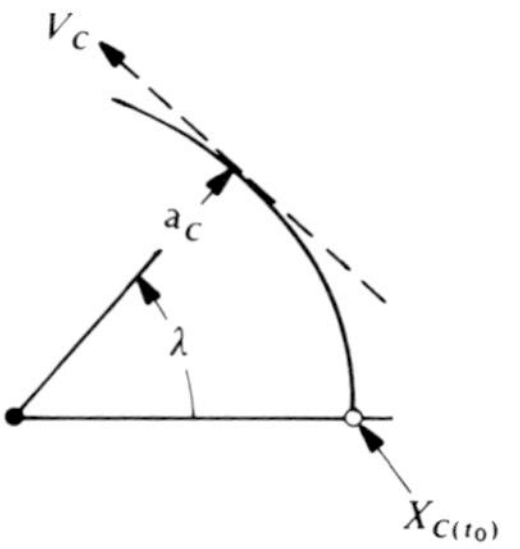

FIG. 1-2. Translation along a circular arc (instantaneous velocity V_C is tangent to the arc).

history, for example, only one reference quantity $X_{C(t_0)}$ is needed to determine both polar coordinates for position in the plane at any time t; the radius is constant, and the angle traversed is

$$\lambda = (X_C - X_{C(t_0)})/a_C \tag{1-3}$$

Extensions can also be made to various three-dimensional curves, in which displacements in two directions are specified functions of one coordinate. Attention is now directed, however, to a more general set of conditions.

MOTION WITH MULTIPLE INDEPENDENT COORDINATES

With (1-2) applied to two displacements X_{C1}, X_{C2},

$$X_{C1} = X_{C1(t_0)} + \int_{t_0}^{t} V_{C1}\, d\tau, \qquad X_{C2} = X_{C2(t_0)} + \int_{t_0}^{t} V_{C2}\, d\tau \tag{1-4}$$

For positions measured along perpendicular axes of fixed direction (Fig. 1-3) this relation could characterize arbitrary excursions of a surface vehicle on flat terrain. In fact, (1-4) suffices for excursions on any surface for which

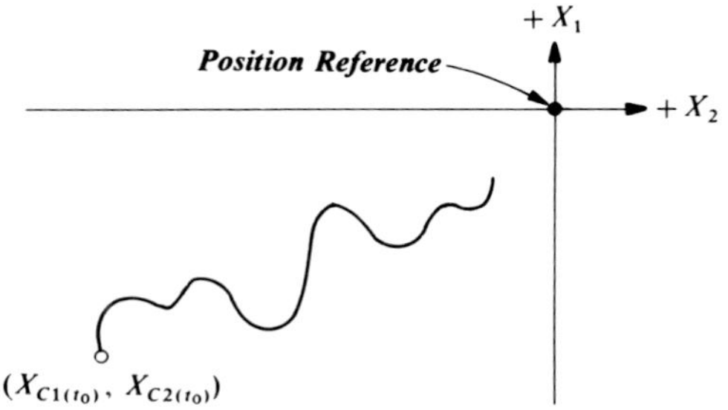

FIG. 1-3. Translation with two degrees of freedom.

the third dimension is either (1) a known "profile function" of position on the surface, or (2) unimportant for any reason (e.g., small in comparison with surface path lengths). Note that a constraint such as a road would effectively reduce the governing relations to (1-2).

Figure 1-3 can also represent the *ground track* (i.e., projection of a flight path on a level plane) for an aircraft with instantaneous horizontal position conforming to (1-4). By adding another coordinate X_{C3} for vertical position, location in all three directions can be expressed in terms of the vectors (with components again measured along fixed axes)

$$\mathbf{R}_C = \begin{bmatrix} X_{C1} \\ X_{C2} \\ X_{C3} \end{bmatrix}, \qquad \mathbf{V}_C = \begin{bmatrix} V_{C1} \\ V_{C2} \\ V_{C3} \end{bmatrix} \tag{1-5}$$

so that

$$\mathbf{R}_C = \mathbf{R}_{C(t_0)} + \int_{t_0}^{t} \mathbf{V}_C\, d\tau \tag{1-6}$$

This relation, representing a general altitude profile accompanying an arbitrary ground track, can also be written

$$\mathbf{R}_C = \mathbf{R}_{C(t_0)} + \mathbf{V}_{C(t_0)}(t - t_0) + \int_{t_0}^{t} \int_{t_0}^{\mathrm{r}} (\mathbf{A}_C + \mathbf{G}_C)\, d\mathrm{r}'\, d\mathrm{r} \tag{1-7}$$

where $\mathbf{G}_C$, $\mathbf{A}_C$ denote total gravitational and nongravitational aircraft accelerations, respectively.

The development thus far has not considered motion of the chosen position reference point. When the velocity $\boldsymbol{V}_\mathrm{T}$ of a tracked object (at instantaneous position $\boldsymbol{R}_\mathrm{T}$) is substituted into (1-6) and relative position is formed through subtraction, the result is

$$\mathbf{R}_\mathrm{r} = \mathbf{R}_{\mathrm{r}(t_0)} + \int_{t_0}^{t} \mathbf{V}_\mathrm{r}\, d\mathrm{r} \tag{1-8}$$

where $\mathbf{R}_\mathrm{r}$, $\mathbf{V}_\mathrm{r}$ denote relative position and velocity, respectively, and are given by

$$\mathbf{R}_\mathrm{r} = \mathbf{R}_C - \boldsymbol{R}_\mathrm{T}, \qquad \mathbf{V}_\mathrm{r} = \mathbf{V}_C - \boldsymbol{V}_\mathrm{T} \tag{1-9}$$

Alternatively, the total acceleration $\boldsymbol{Z}_\mathrm{T}$ of the tracked object can be inserted:

$$\mathbf{R}_\mathrm{r} = \mathbf{R}_{\mathrm{r}(t_0)} + \mathbf{V}_{\mathrm{r}(t_0)}(t - t_0) + \int_{t_0}^{t} \int_{t_0}^{\mathrm{r}} (\mathbf{A}_C + \mathbf{G}_C - \boldsymbol{Z}_\mathrm{T})\, d\mathrm{r}'\, d\mathrm{r} \tag{1-10}$$

These last three expressions are often associated with the tracking of surface or airborne vehicles.

MOTION OVER A SPHERICAL OR SPHEROIDAL SOLID

The polar coordinate representation of Fig. 1-2 will now be extended to include more dimensions. First, consider motion on a sphere of radius

$$\mathscr{R} = R_\mathrm{E} + h \tag{1-11}$$

where the altitude h above a sphere of radius R_E is temporarily restricted to a constant value. Through tentative identification of the sphere with the earth, position can be characterized by latitude λ and east longitude φ, as illustrated in Fig. 1-4, where $\mathbf{K}_\mathrm{E}$ defines the axis of earth rotation. The equatorial plane is normal to $\mathbf{K}_\mathrm{E}$, and each *meridian* is the intersection of the sphere with a plane containing $\mathbf{K}_\mathrm{E}$. Thus the equatorial plane intersects at right angles with the plane containing any meridian; the line of intersection defines the reference direction $\mathbf{I}_\mathrm{E}$ for the zero longitude meridian passing through Greenwich, England. Latitude corresponds to arc length measured along a meridian from the equator. The locus of points at any fixed latitude forms a "latitude circle" or "parallel," by intersection of the sphere with a plane parallel to the equator. It is seen that meridians

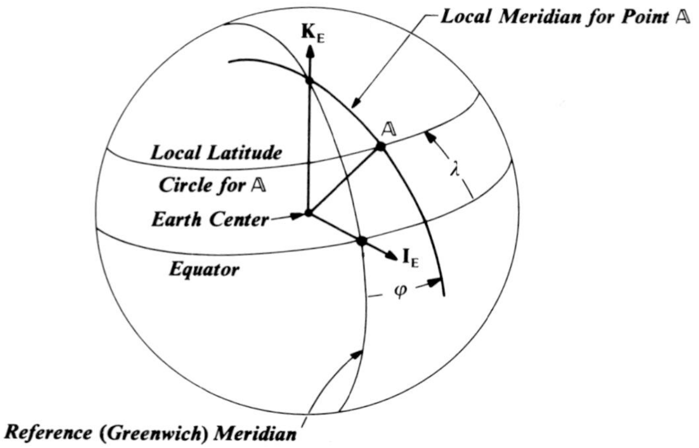

FIG. 1-4. Earth coordinate conventions.

are *great circles* (i.e., they have radius equal to that of the sphere) and the equator lies in a great circle plane, but the radius of a latitude circle is proportional to cos λ.

Tangential velocity along a meridian is simply the north component V_N of the velocity vector. Thus (1-2) and (1-3) applied here produce the relation

$$\lambda = \lambda_{(t_0)} + \int_{t_0}^{t} (V_N/\mathscr{R})\, d\tau \tag{1-12}$$

Similarly, the longitude rate is proportional to the eastward velocity V_E with respect to a point rotating with the sphere, at the location of the aircraft. In view of the above remark regarding latitude circle radii,

$$\varphi = \varphi_{(t_0)} + \int_{t_0}^{t} \left(\frac{V_E}{\mathscr{R}\cos\lambda}\right) d\tau \tag{1-13}$$

Both this equation and Fig. 1-4 show that longitude loses its meaning at $\lambda = \pm\pi/2$.

For earth navigation, extensions of this spherical example are necessary. The question of surface irregularities is easily resolved; latitude and longitude can be associated with variable radial distance or, more simply, mean sea level can be adopted as the altitude reference. A more pronounced departure from sphericity arises from oblateness. Whereas gravitational force would be everywhere normal to a nonrotating homogeneous sphere, the earth's gravity is approximated as a force everywhere normal to an ellipsoid of revolution about $\mathbf{K}_E$. This same ellipsoid is used to define *geodetic* latitude and longitude. Radius parameters for the meridians and parallels, denoted

R_M and R_P, respectively, are slowly varying functions of latitude.* In (1-12) and (1-13), $\mathscr{R}$ is simply replaced by $(R_M + h)$ and $(R_P + h)$, respectively. With these substitutions and a further generalization of variable altitude, a full three-dimensional formulation is achieved for earth navigation:

$$h = h_{(t_0)} - \int_{t_0}^{t} V_V \, d\tau \tag{1-14}$$

where V_V represents velocity *downward*, i.e., normal to the ellipsoid. In general, this local vertical line does not quite pass through the earth center, but rather conforms to the direction that a plumb-bob would assume if suspended above the ellipsoid. This convention is appropriate for terrestrial navigation over great distances, particularly with inertial instruments. For certain analytical developments an earth center may still be used as a convenient conceptual position reference, but the foregoing geographic formulation avoids the need for a precise mechanized center point computation.

MOTION DESCRIPTION CONSIDERATIONS

By elementary methods the preceding development has introduced three ways of defining observed vehicle motion:

(1) *Earth Navigation* Typically an aircraft navigation system is initialized with respect to a latitude/longitude and local vertical/north reference, and this mode is used for general flight purposes en route to a destination.

(2) *Point Navigation and Associated Operations* In the vicinity of a specified point of interest (e.g., designated "target" point; landing area; surface vehicle on a road, being tracked by police helicopter) the aircraft is operated in accordance with its relation to that point, irrespective of latitude or longitude (and, in the targeting case, irrespective of north direction).

(3) *Air-to-Air Mode* In this mode, the aircraft is relatively independent of the earth below, and all operations are dictated by its relation to a specified airborne vehicle (not necessarily restricted to air combat maneuver but including in-flight refueling and all other functions for which another airborne vehicle exerts dominating influence).

Various relations applicable to these modes are clear from the foregoing investigation, from which it might seem that navigation equations in general are self-evident. On the contrary, terrestrial navigation calls for the systematic

* Specific relations for the radii, as well as the physical basis for and local deviations from the ellipsoid, the distinction between gravity and gravitation, and translation of the earth itself are covered in Chapters 3 and 4.

development in succeeding chapters. Underlying conditions applicable to the vector $\mathbf{R}_C$,

$$\mathbf{A}_C + \mathbf{G}_C = \frac{d\mathbf{V}_C}{dt} = \dot{\mathbf{V}}_C = \begin{bmatrix} \dot{V}_{C1} \\ \dot{V}_{C2} \\ \dot{V}_{C3} \end{bmatrix} = \frac{d^2\mathbf{R}_C}{dt^2} = \ddot{\mathbf{R}}_C = \begin{bmatrix} \ddot{X}_{C1} \\ \ddot{X}_{C2} \\ \ddot{X}_{C3} \end{bmatrix} \tag{1-15}$$

owe their simplicity to the definition of (1-5) in terms of fixed axes,

$$\mathbf{R}_C = \mathbf{I}_\mathrm{I} X_{C1} + \mathbf{J}_\mathrm{I} X_{C2} + \mathbf{K}_\mathrm{I} X_{C3}, \qquad \mathbf{V}_C = \mathbf{I}_\mathrm{I} V_{C1} + \mathbf{J}_\mathrm{I} V_{C2} + \mathbf{K}_\mathrm{I} V_{C3} \tag{1-16}$$

where $(\mathbf{I}_\mathrm{I}, \mathbf{J}_\mathrm{I}, \mathbf{K}_\mathrm{I})$ are stationary unit vectors whose time derivatives are zero. There are important departures from these conditions, even in the simple geographic formulation already presented. Consider three unit vectors $(\mathbf{I}_\mathrm{G}, \mathbf{J}_\mathrm{G}, \mathbf{K}_\mathrm{G})$ pointing along local north, east, and downward directions (i.e., tangent to the local meridian, tangent to the local latitude circle, and normal to the ellipsoid), respectively. The velocity components contained in (1-12)–(1-14) comprise a vector

$$\mathbf{V} = \begin{bmatrix} V_\mathrm{N} \\ V_\mathrm{E} \\ V_\mathrm{V} \end{bmatrix} = \mathbf{I}_\mathrm{G} V_\mathrm{N} + \mathbf{J}_\mathrm{G} V_\mathrm{E} + \mathbf{K}_\mathrm{G} V_\mathrm{V} \tag{1-17}$$

as expressed in the *geographic coordinate frame*. The next section draws further attention to rotation of this frame and to the definition of V_E as "ground east" velocity, i.e., relative to a point fixed over the earth at the current aircraft position. More broadly, conventions widely accepted for navigation are systematically explained.

1.2 NAVIGATION COORDINATE FRAMES

Of three convenient "x axis" directional references $(\mathbf{I}_\mathrm{I}, \mathbf{I}_\mathrm{E}, \mathbf{I}_\mathrm{G})$ already defined, only $\mathbf{I}_\mathrm{I}$ permits immediate identification of derivatives as in (1-15). By way of contrast, the time derivative $\dot{V}_\mathrm{N}$ of north velocity would *not* be the total acceleration along the direction $\mathbf{I}_\mathrm{G}$ during typical flight over the ellipsoid. Motivation for investigating the difference is easily found: to provide readily interpretable earth-based output information, terrestrial inertial navigation systems often employ a locally level reference, and the accelerometers are physically mounted in a rotating frame. Merely equating the integrated outputs of level accelerometers to horizontal velocity components, even in the slowly rotating geographic frame, could produce errors on the order of a mile after only five minutes, or four miles after ten minutes. Thus an in-depth navigation study includes kinematics of greater complexity than (1-15).

The full range of navigation problems can best be addressed with the use of specific coordinate frames, each consisting of three mutually orthogonal unit vectors (**I**, **J**, **K**) chosen according to right-hand conventions:

$$\mathbf{I} = \mathbf{J} \times \mathbf{K}, \qquad \mathbf{J} = \mathbf{K} \times \mathbf{I}, \qquad \mathbf{K} = \mathbf{I} \times \mathbf{J}$$
$$\mathbf{I} \cdot \mathbf{I} = \mathbf{J} \cdot \mathbf{J} = \mathbf{K} \cdot \mathbf{K} = 1 \tag{1-18}$$

Of the numerous triads to be considered in this book, three important frames have already been introduced:

(1) the nonrotating frame ($\mathbf{I}_I$, $\mathbf{J}_I$, $\mathbf{K}_I$),
(2) the frame ($\mathbf{I}_E$, $\mathbf{J}_E$, $\mathbf{K}_E$) rigidly attached to the earth,
(3) the geographic frame ($\mathbf{I}_G$, $\mathbf{J}_G$, $\mathbf{K}_G$).

Additional triads include a frame ($\mathbf{I}_A$, $\mathbf{J}_A$, $\mathbf{K}_A$) rigidly attached to the aircraft, a platform reference ($\mathbf{I}_P$, $\mathbf{J}_P$, $\mathbf{K}_P$), and other frames having specified azimuth separations from geographic axes. Although a full exposition must await the next chapter, the following material satisfies the needs of an ensuing introductory nav system analysis.

1.2.1 Relation of Earth Frame to Fixed Axes

The orientation of $\mathbf{K}_E$ in Fig. 1-4 can, for purposes of this book, be considered fixed. A similar statement cannot apply to $\mathbf{I}_E$, in view of its association with a specific meridian (the zero-longitude, Greenwich meridian). Earth rotation changes the direction of $\mathbf{I}_E$ and causes translation of points on the surface. A cursory glance at a point on the equator might suggest a peripheral speed dictated by an equatorial radius of approximately 3444 nautical miles (nm) and a 24-hr period; thus

$$3444 \text{ nm} \times 2\pi/24 \text{ hr} > 900 \text{ kt} \tag{1-19}$$

but certain concepts based on this *mean solar* day are not firmly anchored for nav applications. The 24-hr time convention was adopted as an average period of earth rotation *with respect to the earth–sun line.* Since that line experiences one revolution in approximately 365.25 days, the total number of cycles that could be counted per year, from a position at the center of the earth, would essentially conform to

$$\omega_s \doteq \frac{1 + 365.25 \text{ cycles}}{365.25 \times 24 \text{ hr}} \times \frac{2\pi \text{ rad/cycle}}{3600 \text{ sec/hr}} \doteq \frac{1.0027379(2\pi)}{24(3600)} \text{ rad/sec} \tag{1-20}$$

This *sidereal rate* corresponds to a *sidereal period* ($2\pi/\omega_s$), slightly shorter than a mean solar day, which defines an accurate and consistent interval between successive pointings of $\mathbf{I}_E$ in any fixed direction (Ex. 1-1). In

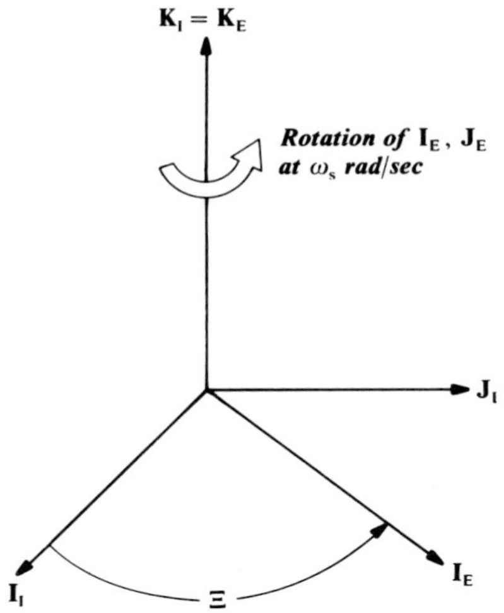

FIG. 1-5. Rotation of earth frame with respect to fixed axes ($\mathbf{I}_I$ points along the vernal equinox; $\mathbf{I}_I$, $\mathbf{J}_I$, and $\mathbf{K}_I$ have fixed direction).

particular, the choice of $\mathbf{K}_I \equiv \mathbf{K}_E$ ensures that once every sidereal day $\mathbf{I}_E$ coincides with $\mathbf{I}_I$, defined as follows:

The $\mathbf{I}_I$-$\mathbf{J}_I$ plane of Fig. 1-5 is equatorial. Its intersection with the *ecliptic* (i.e., the plane of the earth's orbit around the sun) is known as the *line of equinoxes.** The *hour angle of the vernal equinox* Ξ increases at the rate indicated in (1-20); an initial value Ξ_0 can be taken from Ref. 1-2 at zero hours Universal Time (UT; Greenwich mean solar time) on January 1, 1975, when the mean hour angle was ($6^h 40^m 5.156^s$). This hour–minute–second convention is converted to radians for Ξ_0 and, at N_D full days plus N_S seconds after 0^h UT on January 1, 1975,

$$\Xi = \Xi_0 + 2\pi\left(0.0027379\,\mathsf{N}_D + \frac{1.0027379}{86{,}400}\,\mathsf{N}_S\right)\ \text{rad} \tag{1-21}$$

while the relative orientation of ($\mathbf{I}_E$, $\mathbf{J}_E$, $\mathbf{K}_E$) and ($\mathbf{I}_I$, $\mathbf{J}_I$, $\mathbf{K}_I$) conforms to (Ex. 1-2)

$$\begin{bmatrix} \mathbf{I}_E \\ \mathbf{J}_E \\ \mathbf{K}_E \end{bmatrix} = \begin{bmatrix} \cos\Xi & \sin\Xi & 0 \\ -\sin\Xi & \cos\Xi & 0 \\ 0 & 0 & 1 \end{bmatrix} \begin{bmatrix} \mathbf{I}_I \\ \mathbf{J}_I \\ \mathbf{K}_I \end{bmatrix} \tag{1-22}$$

* The line of equinoxes defines solar position in a geocentric (i.e., earth-centered) frame near the start of spring (vernal equinox) and autumn (autumnal equinox). No significant discrepancy will arise here from the assumption of a fixed plane for the earth's orbit or of a constant sidereal rate vector.

Benefits of the above formulation include (1) elimination of time ambiguities or dependence on zones, (2) establishment of a link connecting rotating frames to an absolute reference, and (3) conformity to celestial navigation standards; a star sightline $\mathbf{c}_I$ is defined in *inertial coordinates* by two fixed angles; thus,

$$\mathbf{c}_I = \mathbf{I}_I \cos(\mathrm{DEC})\cos(\mathrm{SHA}) - \mathbf{J}_I \cos(\mathrm{DEC})\sin(\mathrm{SHA}) + \mathbf{K}_I \sin(\mathrm{DEC})$$
$$= \begin{bmatrix} \cos(\mathrm{DEC})\cos(\mathrm{SHA}) \\ -\cos(\mathrm{DEC})\sin(\mathrm{SHA}) \\ \sin(\mathrm{DEC}) \end{bmatrix} \tag{1-23}$$

in which (DEC) and (SHA) represent *declination* and *sidereal hour angle*, respectively (Ref. 1-3). An analogy can be drawn to the latitude–longitude convention, but (SHA) is defined as a westward displacement. Of interest for satellite navaids, this $(\mathbf{I}_I, \mathbf{J}_I, \mathbf{K}_I)$ frame is also used to express ephemeris data.

Although these conventions prepare celestial fix data for introduction into nav systems, the connection between $(\mathbf{I}_E, \mathbf{J}_E, \mathbf{K}_E)$ and $(\mathbf{I}_G, \mathbf{J}_G, \mathbf{K}_G)$ calls for further elaboration.

1.2.2 Relation of Geographic Axes to Earth Coordinates

The present objective is to define a transformation between the earth and geographic triads analogous to (1-22). One clue to the desired expression is the standard spherical coordinate representation for position [compare with (1-23)],

$$\mathscr{R}(\mathbf{I}_E \cos\lambda\cos\varphi + \mathbf{J}_E \cos\lambda\sin\varphi + \mathbf{K}_E \sin\lambda) = \mathscr{R}\begin{bmatrix} \cos\lambda\cos\varphi \\ \cos\lambda\sin\varphi \\ \sin\lambda \end{bmatrix} \tag{1-24}$$

while, from the definition of $\mathbf{K}_G$, the same vector expressed in geographic coordinates must be

$$\mathbf{K}_G(-\mathscr{R}) = \begin{bmatrix} 0 \\ 0 \\ -\mathscr{R} \end{bmatrix} \tag{1-25}$$

Therefore,

$$\mathbf{K}_G = -\cos\lambda\cos\varphi\,\mathbf{I}_E - \cos\lambda\sin\varphi\,\mathbf{J}_E - \sin\lambda\,\mathbf{K}_E \tag{1-26}$$

From Fig. 1-4 it is also seen that the tangent to the local latitude circle has a direction

$$\mathbf{J}_G = \mathbf{J}_E \cos\varphi - \mathbf{I}_E \sin\varphi \tag{1-27}$$

By combining (1-18) with the last two expressions (Ex. 1-3),

$$\begin{bmatrix} \mathbf{I}_G \\ \mathbf{J}_G \\ \mathbf{K}_G \end{bmatrix} = \begin{bmatrix} -\sin\lambda\cos\varphi & -\sin\lambda\sin\varphi & \cos\lambda \\ -\sin\varphi & \cos\varphi & 0 \\ -\cos\lambda\cos\varphi & -\cos\lambda\sin\varphi & -\sin\lambda \end{bmatrix} \begin{bmatrix} \mathbf{I}_E \\ \mathbf{J}_E \\ \mathbf{K}_E \end{bmatrix} \tag{1-28}$$

Through (1-22) and (1-28), vectors in earth and geographic coordinates can be reexpressed in inertial coordinates, in which differentiation via (1-15) is permissible. Exercise 1-4 determines the velocity of a point on the earth's surface directly below the aircraft, due to the sidereal rate [compare with (1-19)],

$$\omega_s R_E \cos\lambda(\mathbf{J}_E \cos\varphi - \mathbf{I}_E \sin\varphi) = \mathbf{J}_G \omega_s R_E \cos\lambda \tag{1-29}$$

and Ex. 1-5 uses the bottom row of (1-28) to obtain (1-17).

There is a more direct method for differentiating vectors in rotating frames. Instead of algebraic substitutions involving triad axes, vectors expressed in any frame are regarded as 3 × 1 matrices that can be transformed through premultiplication by other 3 × 3 matrices. Derivatives of vector components are then obtained through product differentiation. The procedure is described in the next chapter and used thereafter wherever needed.

Preparation is not yet complete, even for the ensuing simplified nav system discussions, until after the introduction of the following two additional coordinate frames.

1.2.3 Aircraft and Platform Axes

For aircraft employing inertial navigators, attention is restricted in this chapter to locally level "north-slaved" gimballed mechanizations. A frame ($\mathbf{I}_P$, $\mathbf{J}_P$, $\mathbf{K}_P$) rigidly attached to the platform is then continuously driven in an attempt to maintain coincidence with the geographic axes. Departures

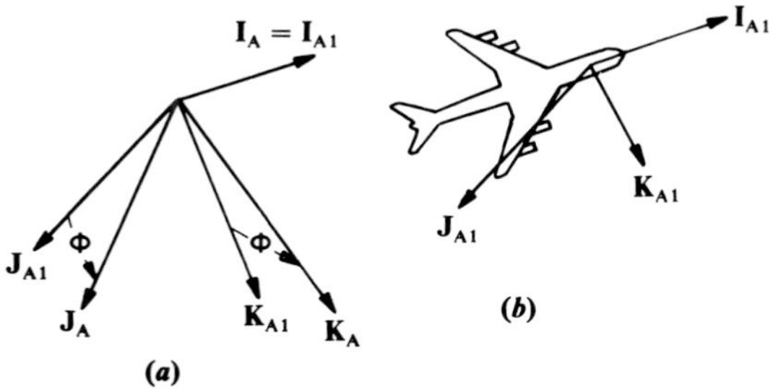

FIG. 1-6. Roll angle transformation: (a) aircraft axes and displaced axes, (b) displaced axes and aircraft orientation. $\mathbf{J}_{A1}$ is horizontal.

between the triads are expressed in terms of small angles,

$$\psi_N \doteq \mathbf{K}_G \cdot \mathbf{J}_P, \qquad \psi_E \doteq \mathbf{I}_G \cdot \mathbf{K}_P, \qquad \psi_A \doteq \mathbf{J}_G \cdot \mathbf{I}_P \tag{1-30}$$

representing orientation errors about the north, east, and azimuth axes, respectively.

Aircraft orientation as indicated by platform gimbals is described as follows: Let $(\mathbf{I}_A, \mathbf{J}_A, \mathbf{K}_A)$ point along the forward fuselage ("roll") axis, outward along the right wing ("pitch" axis), and toward the floor of the aircraft ("yaw" axis), respectively. Figure 1-6a shows these axes with another triad $(\mathbf{I}_{A1}, \mathbf{J}_{A1}, \mathbf{K}_{A1})$, obtained by rotating the aircraft about $\mathbf{I}_A$ until the wings are level as depicted in Fig. 1-6b. It is seen that

$$\begin{bmatrix} \mathbf{I}_A \\ \mathbf{J}_A \\ \mathbf{K}_A \end{bmatrix} = \begin{bmatrix} 1 & 0 & 0 \\ 0 & \cos\Phi & \sin\Phi \\ 0 & -\sin\Phi & \cos\Phi \end{bmatrix} \begin{bmatrix} \mathbf{I}_{A1} \\ \mathbf{J}_{A1} \\ \mathbf{K}_{A1} \end{bmatrix} \tag{1-31}$$

where Φ denotes the *roll* angle shown in Fig. 1-6a. When $\mathbf{I}_{A1}$ and $\mathbf{K}_{A1}$ are rotated about $\mathbf{J}_{A1}$ until the fuselage is level, another intermediate set

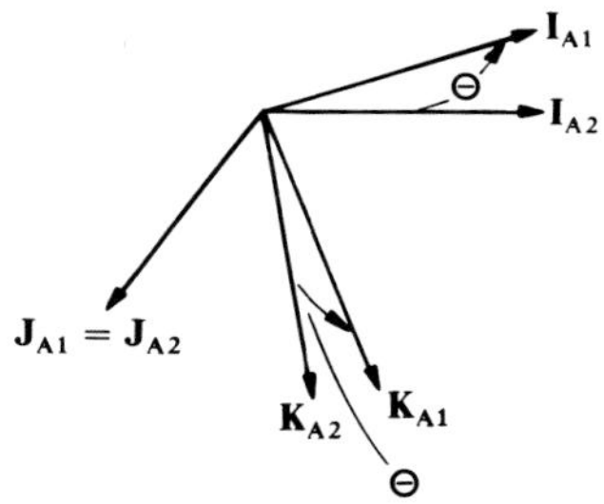

FIG. 1-7. Pitch angle transformation ($\mathbf{I}_{A2}$ and $\mathbf{J}_{A2}$ are horizontal).

of axes $(\mathbf{I}_{A2}, \mathbf{J}_{A2}, \mathbf{K}_{A2})$ appears as shown in Fig. 1-7. The *pitch* angle Θ characterizes the relation

$$\begin{bmatrix} \mathbf{I}_{A1} \\ \mathbf{J}_{A1} \\ \mathbf{K}_{A1} \end{bmatrix} = \begin{bmatrix} \cos\Theta & 0 & -\sin\Theta \\ 0 & 1 & 0 \\ \sin\Theta & 0 & \cos\Theta \end{bmatrix} \begin{bmatrix} \mathbf{I}_{A2} \\ \mathbf{J}_{A2} \\ \mathbf{K}_{A2} \end{bmatrix} \tag{1-32}$$

which is connected to the platform frame via the *heading* (Ψ) transformation in Fig. 1-8:

$$\begin{bmatrix} \mathbf{I}_{A2} \\ \mathbf{J}_{A2} \\ \mathbf{K}_{A2} \end{bmatrix} = \begin{bmatrix} \cos\Psi & \sin\Psi & 0 \\ -\sin\Psi & \cos\Psi & 0 \\ 0 & 0 & 1 \end{bmatrix} \begin{bmatrix} \mathbf{I}_P \\ \mathbf{J}_P \\ \mathbf{K}_P \end{bmatrix} \tag{1-33}$$

In the absence of a physical gimballed platform, aircraft attitude and heading

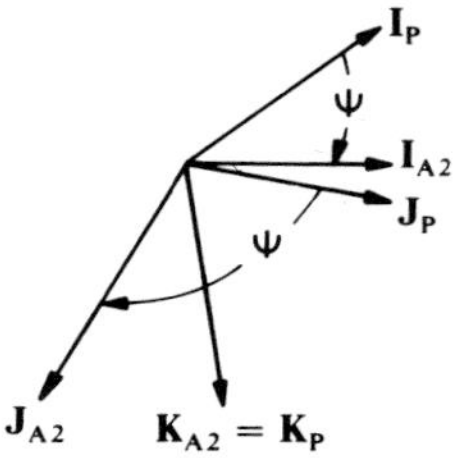

FIG. 1-8. Heading angle transformation.

can be defined by other means such as gyro-fed computations or vertical and directional gyros. In any case, these transformations serve as a basis for system descriptions to follow.

1.3 NAVIGATION TECHNIQUES AND REQUIREMENTS

Early concepts of aircraft navigation involved guidance of a sort. According to that scheme, a route is divided into a series of course legs with extremities at preselected reference points. As the aircraft nears one of these points, a position deviation is noted and used to compensate the flight plan (e.g., by altering the heading) for the next segment. Aircraft position between these *fixes* is determined by extrapolation based on a constant velocity estimated for each course leg.

Influence of early nav concepts remains evident in modern systems, but with certain modifications. Whereas the traditional meaning of airborne "navigation" implied some form of guidance, present emphasis falls on the *determination* of aircraft position, velocity, and orientation; all aspects of flight control are considered here only from the standpoint of data requirements.* Other departures from tradition involve more than a change in perspective; *area navigation* (R-nav) eliminates restrictions to corridors, thus allowing more general flight paths. Also, *time-varying* velocity information for extrapolation between fixes may be obtained from inertial instruments, doppler radar, or air data plus estimated wind. This information, with appropriate computing and updating provisions, allows considerable improvement over traditional navigation techniques. The general (constrained or unconstrained path) situation is addressed here, with a general

* This does not imply that functions such as steering and arrival time scheduling are less important, but that they are not emphasized in this book. Once the nav information is obtained, it can be used at various data rates to perform a broad class of functions, as discussed in Appendix III.

array of nav sensors. The following introductory description begins with rudimentary practices and subsequently builds the case for a more advanced approach.

1.3.1 Elementary Updating Procedures

Consider a course leg from a known starting location $\mathbb{S}$ to a known checkpoint $\mathbb{C}$, with altitude ignored, as depicted in Fig. 1-9. An uncertainty in velocity direction (e.g., due to initially unknown crosswind; compare

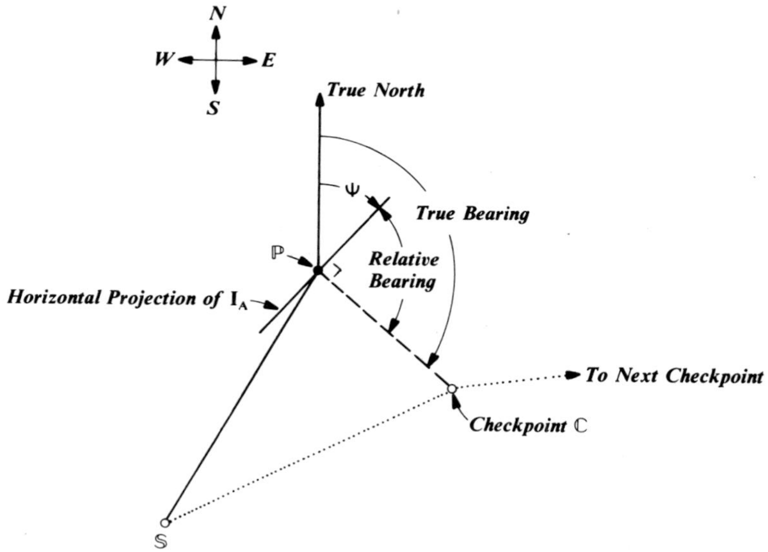

FIG. 1-9. Plan view of dead reckoning course leg. Dotted line shows preflight planned route.

orientation of $\mathbf{I}_A$ with course direction) could cause a displacement error $\overline{\mathbb{PC}}$, shown with an exaggerated scale. When the sightline to the checkpoint lies outward along a wing (90° *relative bearing*), a pilot might note time elapsed since departure from $\mathbb{S}$, and make a visual estimate of the cross-course distance $\overline{\mathbb{PC}}$. The elapsed time, in combination with this estimated displacement and with the known distance $\overline{\mathbb{SC}} \doteq \overline{\mathbb{SP}}$, provides a cross-course velocity correction and a *groundspeed* estimate, respectively. This velocity information, combined with the next checkpoint position, influences the heading desired for the next course leg.

This simple procedure of *dead reckoning* between checkpoints was quite common before radio navigation. Its principal aim was to account for wind velocity, but an attempt could also be made to correct position, based on the estimate for $\overline{\mathbb{PC}}$ and the *true bearing* shown. In any case the method was not precise; typical error levels were on the order of a minute of time, miles of distance, and a few degrees of heading. Another shortcoming of this approach is its reliance upon visual recognition of certain landmarks. The risk of losing a crucial reference point could be reduced through increased "pilotage," in which portions of flight could follow recognizable railroad tracks, waterways, etc., but exclusive dependence on visual terrain observations is undesirable.

A better approach to navigation is provided by VOR (Very high frequency OmniRange) stations as depicted in Fig. 1-10. At any point $\mathbb{Q}$

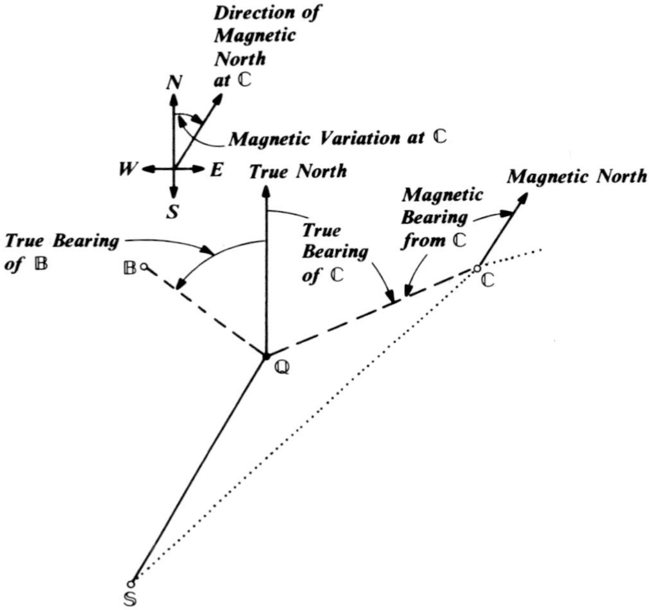

FIG. 1-10. Plan view of course leg with radio navigation. Dotted line shows preflight planned route.

along the path from $\mathbb{S}$, the *magnetic bearing* of the aircraft from a station at point $\mathbb{C}$ can be received. Complete position information can be obtained by either of two methods:

(1) Convert the received magnetic bearing,

$$\text{mag brg from aircraft to } \mathbb{C} = 180^\circ - (\text{mag brg from } \mathbb{C} \text{ to } \mathbb{Q}) \quad (1\text{-}34)$$

and add the local *magnetic variation* shown, to obtain true bearing. When the VOR station contains DME provisions (Distance Measuring Equipment), the polar coordinate representation for a horizontal position vector is complete.

(2) Determine the true bearing from both ℂ and a station at another known location (point 𝔹; not on the line $\overline{\mathbb{QC}}$ or its extension). Draw the two dashed lines and find their intersection at ℚ.

Accuracy of this radio nav approach may be typified by 2 nm in each of two perpendicular directions (for bearing from two stations, each 80 nm away, with sightlines at right angles from ℚ), or by a 2 nm × 0.2 nm "error rectangle" (for VOR/DME data from one station at 80-nm range). These figures do not account for possible advantages of averaging nonsimultaneous fixes; for this reason, the procedures now being discussed are regarded as elementary.

Another simple updating method is based on the ancient concept of "fixed" stars. From (1-22) and (1-23) the star sightline in the rotating earth frame can be written

$$\begin{bmatrix} c_{E1} \\ c_{E2} \\ c_{E3} \end{bmatrix} = \begin{bmatrix} \cos(\mathrm{DEC})\cos(\mathrm{SHA} + \Xi) \\ -\cos(\mathrm{DEC})\sin(\mathrm{SHA} + \Xi) \\ \sin(\mathrm{DEC}) \end{bmatrix} \tag{1-35}$$

In combination with (1-26) it is concluded (Ex. 1-6) that the star elevation angle in Fig. 1-11 is

$$\begin{aligned} Y_{\mathrm{EL}} = \arccos\{&-\cos(\mathrm{DEC})\cos\lambda\cos(\mathrm{SHA} + \Xi + \varphi) \\ &- \sin(\mathrm{DEC})\sin\lambda\} - \pi/2 \end{aligned} \tag{1-36}$$

Figure 1-12 depicts a special case of zero elevation for a hypothetical star and aircraft position ℝ in the equatorial plane. In general, both λ and φ could be determined from elevation measurements of two suitably chosen stars; position error sensitivity would be of order 1 nm/min of arc (sighting and leveling) and 15 nm/min of time.

These are just a few examples of nav measurement types to be listed shortly. At this point, however, another comment is in order regarding the

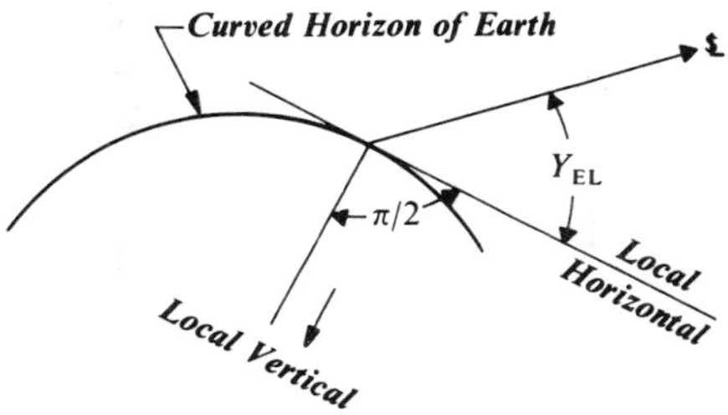

FIG. 1-11. Elevation of star sightline (𝕃).

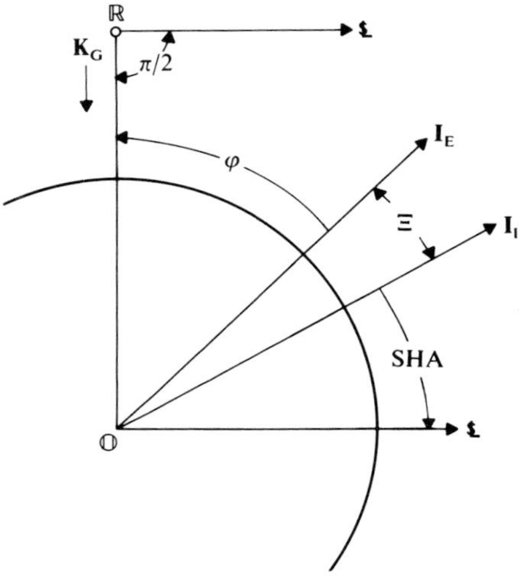

FIG. 1-12. Longitude indication by star elevation measurement—special case ($\mathfrak{L}$ denotes star sightline; $\mathbb{O}$ is at earth center).

orientation of this book. The topic of navigation systems analysis need not involve techniques of mechanizing the star sightline measurements, receiver tuning for acquisition of radio data, or the various displays, charts, projections, and plots used for landmark sightings, star identification, or radio fixes. Adequate conventions and analytical relations will be provided for *functional* characterization of software equations and for assessment of expected performance. Relationships will be expressed in terms of feasible computational algorithms, irrespective of methods used for navigation solutions in the past (Ex. 1-7). Questions regarding recognition, acquisition, mechanization, and software programming will be subordinate to the central issue of functional nav data processing, which will now be outlined.

1.3.2 Advanced Navigation Concepts

When nav systems lack a means of combining multiple (and not necessarily simultaneous) observations, undue sacrifices in capability result. Instrumenting requirements can become unnecessarily stringent, as exemplified by comparing simultaneous versus sequential stellar observations. Also, exclusive reliance on the latest fix forfeits all information remaining from extrapolation of previous data, and makes the system vulnerable to occasional large errors.

As an alternative, suppose that data mixing computations enable updating by full or partial fixes from an arbitrary array of sensors read at any time. Between fixes the position can be extrapolated from "continuous" velocity data, i.e., velocity information obtained at high data rates. Each fix can be weighted according to its accuracy in comparison with the accuracy of the preceding extrapolation and according to sensitivity of the *observable* (bearing, distance, etc.) to each navigation variable (e.g., λ, φ). A broad concept of computer-oriented navigation now emerges, calling for an identification of system inputs, outputs, and data processing. For the first of these items, a general list of information sources can include the following:

(1) *Inertial instruments* In accordance with the conventions established in Chapter 2, an inertial measuring unit (IMU) can be mechanized in either a gimballed platform or a strapdown configuration (Chapters 3 and 4). It is assumed here that any inertial navigation system (INS) contains provisions for digital readouts of incremental velocity, attitude, and heading.

(2) *Local inertial data supplements* Unless a strapdown INS is already provided, additional accelerometers and rate gyros may be body-mounted at various aircraft locations for flight control purposes. Because the aerodynamics (Chapter 4) will produce accompanying motions due to aircraft deformation (i.e., components of motion that are undesirable for control loop mechanization) supplementary inertial instruments will be notched as well as low-pass filtered ahead of the data interface.

(3) *Air data* While sometimes used as nav information, air data also serve functions such as flight control and air-to-ground ballistics computation. As defined herein, air data will be all-inclusive, covering airspeed, angle of attack, barometric pressure, density, etc.

(4) *Navaids* For purposes of this book, a navaid is any measuring device that provides update information to an estimation scheme (Chapter 5), in any of the modes under consideration (Chapter 6). A complete array of devices can include star trackers, magnetic heading indicators, radio navaids (LORAN, Omega, TACAN, VOR/DME), landing aids (ILS), satellite communication, altimeters [all types other than barometric, which are included in the air data of item (3)], imaging systems [synthetic aperture radar (SAR); forward-looking radar or infrared (FLR or FLIR, respectively); television (TV) sensors], doppler radar, etc. These are described at length in Chapter 7.

(5) *Radar tracking data* Radar information (whether used for air-to-air tracking or landmark updating) may include range, range rate, line-of-sight (LOS), and LOS rate data (but excludes SAR data, since SAR falls under the category of imaging systems).

(6) *External references* This class includes information to be inserted

by computer dial-in provisions (such as known position at takeoff), separately observed orientation (e.g., aircraft carrier reference data), or any other source of information not generated on board the aircraft itself.

Nav system *outputs* of primary interest here are:

(1) *Position* Current aircraft location will be parametrized according to the current operating mode. For earth nav, position is generally defined by geodetic latitude, longitude, and altitude above mean sea level. Point nav will typically define position by a Cartesian vector from a target-centered locally level coordinate frame. For air-to-air tracking, the relative position vector between tracked object and tracking aircraft will define position according to a chosen Cartesian or spherical convention.

(2) *Velocity* Earth nav typically calls for velocity in geographic coordinates, while point nav may use a locally level frame with an independent azimuth reference (e.g., a landing strip center line). For air-to-air tracking, velocity as well as position is relative.

(3) *Aircraft orientation* Roll, pitch, and heading angles constitute a typical parametrization but, since the north direction is ill-defined at the poles, another azimuth reference is adopted at high latitudes.

Additional desired outputs may include higher derivatives (e.g., acceleration of tracked object in air-to-air mode) or simple functions of the above quantities, such as

$$\text{dive angle} = -(\text{flight path angle}) = \arctan\{V_V/(V_N{}^2 + V_E{}^2)^{1/2}\} \tag{1-37}$$

$$V_H = \text{groundspeed} = (V_N{}^2 + V_E{}^2)^{1/2} \tag{1-38}$$

$$\text{drift angle} = \zeta - \Psi \tag{1-39}$$

where ζ is defined as the *ground track angle*,

$$\zeta = \arctan(V_E/V_N) \tag{1-40}$$

Outputs may also include other information already listed with inputs (airspeed, angle of attack, raw data from IMU accelerometers). Output data usage is summarized in Appendix III.

To provide the position and velocity output information as accurately as possible, the aforementioned inputs are processed by advanced techniques. The first departure from the dead reckoning scheme described earlier involves integration of the continuously monitored velocity information, thus allowing application to general flight paths. Further modernization to allow mixing of data from multiple sources improves flexibility (e.g., acceptance of partial fixes, typified by incomplete reception from radio navaids already itemized), as well as accuracy. This step, however, previously used methods no longer adequate for many applications. A comparison will highlight the benefits of discrete estimation methods in aircraft navigation.

1.3.3 Conventional versus Modern Approaches

A clear demonstration of classical data mixing with fixed transfer functions is provided in Fig. 1-13. Vertical velocity can theoretically be equated to either the time integral of vertical acceleration or the time derivative of

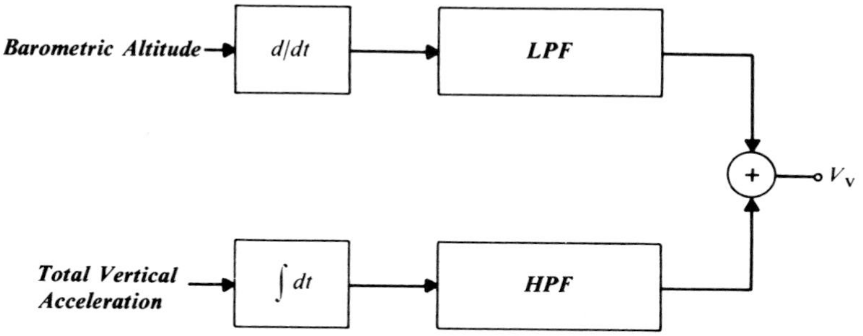

FIG. 1-13. Vertical velocity with conventional (fixed transfer function) filtering. LPF, low-pass filter; HPF, high-pass filter.

altitude above a chosen datum. Since mechanized integration involves drift and true differentiation is impractical with noise present, filters are added as shown. The summing circuit is intended to combine short-term benefits of acceleration data with long-term barometric accuracy. Unfortunately the scheme often performs poorly in the presence of altitude transients. While guarding against differentiation of noise and integration of bias error, the filters also compromise the *desired* differentiating and integrating operations to be performed on the respective signals. Any quantitative selection of poles and zeros will be suitable for only a restrictive class of input waveforms. Traditional approaches have relied upon obtaining a suitable "match" between response functions and input spectra (both digital and analog forms have been implemented; in the limit this leads to physically realizable portions of transfer functions determined from both signal and noise spectra). Clearly, however, spectra evolve in time and cannot be considered as essentially fixed (and therefore cannot be "matched" by any fixed transfer functions) until a steady state condition is reached. Theory thus suggests that, in *any* system with fixed poles and zeros, degradations in performance can be expected temporarily after mode switchings, maneuvers, or other transients—and these expectations have been amply verified in conventional doppler-damped inertial systems, range filters used in air-to-ground radar, etc. Moreover, when input data have nonstationary error properties or time-varying sensitivities (e.g., successive radar fixes under changing geometry), the suitability of fixed transfer functions is further compromised. When data rates vary widely (e.g., intermittent blackouts or loss of lock-on), it may even become arguable as to which sensing mode is currently in operation—and fixed transfer function filters are designed in accordance with specific sensor operating modes and data rates.

The improved vertical channel scheme shown in Fig. 1-14 is conceptually simpler than the classical approach. Total vertical acceleration is integrated

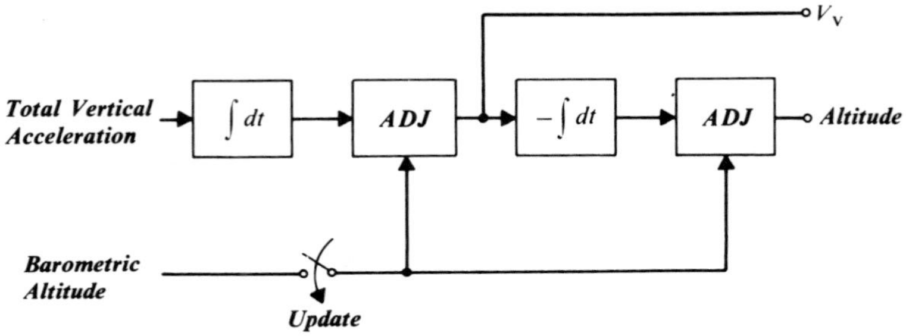

FIG. 1-14. Vertical channel updating approach (ADJ, adjustment).

continuously, while discrete adjustments in the "constants of integration" (h and V_V) are made upon sudden barometer readings. The amount of adjustment is influenced (Chapter 5) by the relative magnitude of measurement tolerance and the rms estimation error before the barometer reading. Time variations in data rates and adjustment sensitivities, which would have taxed an analog design, present no problem in a digital system. Note that repeated barometric observations separated by known time intervals indicate corrections for velocity as well as position. Principles of this approach are easily extended to multidimensional systems involving more variables. All observations can be discrete or discret*ized*, associated with the time center of a prefiltering interval (i.e., a chosen sensor output integration period, during which the observation error may change but the signal can safely be averaged). Occasional fixes of any type can then be used to adjust constants of integration, in a running digital solution of the system's kinematical equations.

The concept of filtering now acquires a general meaning, providing multiple outputs from multiple inputs, in continuous and/or sampled form, with complete multimode flexibility essential for backup provisions and adaptability to changing operating conditions. The task at all times, regardless of conditions, is to provide the aforementioned outputs from whatever input combination is available. Fulfillment of this assignment is influenced by the fact that navigation quantities in general are not measured continuously (e.g., fixes can be characterized by discrete data INSERT operations) or directly (e.g., no "latitude sensor" is present but various nav observations bear a *functional relationship* to position). The set of all applicable functional relationships collectively constitutes a *model* whereby time histories of available sensor data can be combined in a filtering operation; this procedure satisfies the nav function requirements while accommodating all existing information sources in their actual incoming form. By weighting each incoming observation immediately, on the basis of its information content, minimum variance estimates are obtained.

These benefits in performance are matched by system flexibility. Estimation algorithms can be programmed into a unified formalism accommodating every possible sensor combination, for whatever widely variable data rates, accuracies, and sensitivities that may correspond to each sensor, providing multiple outputs at rates specified individually for each output function. Complete backup is not only readily provided but integrated into an overall automatic mode switching arrangement; in marked contrast to previous conventional designs (requiring different filters for different sensing modes) the estimation approach affords an unprecedented degree of integration, requiring only straightforward software rerouting logic for mode switching. The characteristic flexibility also facilitates extension to include any growth capability, in-flight calibration provisions, etc., which may be deemed desirable at any time. Also inherent in the data weighting computations is a probabilistic measure of updated navigation accuracy (specifically, the error covariance matrix, used to compute current values for filter weights). This produces benefits both as an error analysis tool in design evaluation and in actual flight where various decisions (which may affect flight mission objectives) and procedures (such as automatic failure recognition or data editing) depend upon current navigation accuracy. Flight testing can also be conducted at reasonable cost, with the use of airborne recorders. By subsequent ground processing, prospective changes in computational procedures can be tested merely by recycling the recorded data through an appropriately modified computational simulation. With a properly executed overall program the result is a realistic evaluation of system performance under dynamic flight conditions, which include, as necessary, switched modes and/or variable geometry for navigation measurements, with *changing* data rates, sensitivities, and accuracy of observation.

Continuing influence of earlier navigation methods, unfortunately, can limit the capability of operational systems. For example, a doppler radar typically "reads out" groundspeed and drift angle, defined by (1-38) and (1-39), instead of the directly measured frequency shifts. Elimination of this undesired transformation would reduce doppler radar cost, size, weight, and power consumption, while simultaneously enhancing reliability and accuracy (since these two equations are nonlinear, and the estimation scheme to be discussed is linearized). Still, the technique to be used will of course make maximum use of all information, at whatever accuracy and in whatever form it is made available. For clarity it is instructive to adopt the concept of a *sensor/computer interface*, with all navigation data characterized as including various errors with certain properties (irrespective of the reasons why these errors are present). Given this interface, an integrated navigation scheme provides independently operable modes, with coordinated interaction between all functions of the avionics system (e.g., sensing, data transmission, power, display, control actuation, computation).

Actual operation of the estimation scheme has thus far been considered only for the vertical channel (Fig. 1-14). Attention is now focused on a simplified description of one horizontal channel with updates provided by occasional fixes.

1.4 RESTRICTIVE DISCUSSION OF AN INERTIAL SYSTEM

A description of terrestrial inertial navigation in three dimensions might suggest monitoring of total nongravitational acceleration, vector addition of computed gravity, and integrations performed to determine velocity and position. Full kinematical considerations and the mechanics of inertial instruments and platforms will be tentatively deferred; this section will consider a planar path under restrictive conditions whereby a level accelerometer with its sensitive axis along $\mathbf{J}_G$ would sense $\dot{V}_E$. For eastbound equatorial flight at a fixed low altitude,

$$V_N = V_V \equiv 0 \text{ fps}, \qquad \lambda \equiv 0, \qquad \mathscr{R} \doteq R_E \tag{1-41}$$

a drift-free airborne platform could be maintained horizontal by a leveling command of V_E/R_E (Fig. 1-15). With inexact initialization, however, observed east velocity $\hat{V}_E$ would not be correct despite perfect inertial instruments. An instantaneous uncertainty ($\tilde{V}_E \triangleq V_E - \hat{V}_E$) would cause an imperfect leveling command $\hat{V}_E/R_E$, overdriving the platform at a rate

$$\dot{\psi}_N = -\tilde{V}_E/R_E \tag{1-42}$$

where ψ_N is a small tilt of the platform about the north axis in Fig. 1-15. At the same time, velocity error is affected by the tilt; a platform-mounted accelerometer (Chapter 3) with its sensitive axis along $\mathbf{J}_P$ instead of $\mathbf{J}_G$ would sense a small component of the 1-g aerodynamic lift. Since upward lift forms an obtuse angle with $\mathbf{J}_P$ in Fig. 1-15, the unwanted acceleration component has an apparent retarding effect:

$$\dot{\hat{V}}_E = \dot{V}_E - g\psi_N, \qquad \dot{\tilde{V}}_E = g\psi_N \tag{1-43}$$

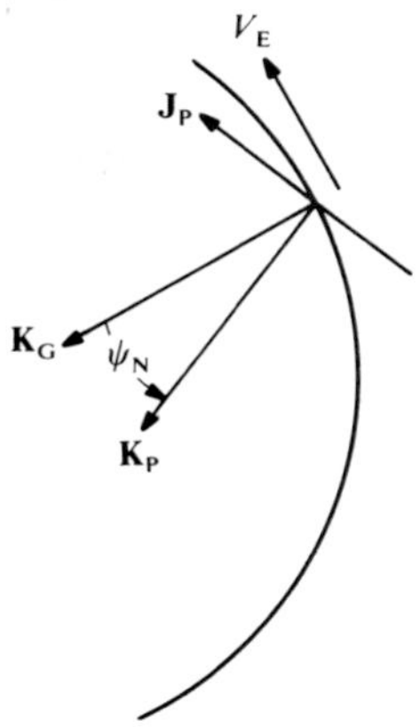

FIG. 1-15. Platform tilt (north axis $\mathbf{I}_G$ is the upward normal to the plane of the figure; rate of rotation of $\mathbf{K}_G$ is V_E/R_E).

The last two equations can be combined by differentiation and substitution to yield

$$\ddot{\tilde{V}}_E + \frac{g}{R_E}\tilde{V}_E = \ddot{\psi}_N + \frac{g}{R_E}\psi_N = 0 \tag{1-44}$$

These equations represent sinusoids at the *Schuler* frequency $(g/R_E)^{1/2}$ rad/sec (Ex. 1-8). The aforementioned retardation is a saving feature for terrestrial inertial navigation; note the absence of errors that continue to grow with time. Even with biases n_1 and n_2 added to the right of (1-42) and (1-43), velocity and tilt errors would not grow continuously (Exs. 1-9, 1-10). In particular, the velocity error would be

$$\tilde{V}_E = \tilde{V}_{E(0)} \cos\left(\frac{g}{R_E}\right)^{1/2} t + \left[(gR_E)^{1/2}\psi_{N(0)} + n_2\left(\frac{R_E}{g}\right)^{1/2}\right] \sin\left(\frac{g}{R_E}\right)^{1/2} t$$
$$+ R_E n_1 \left[1 - \cos\left(\frac{g}{R_E}\right)^{1/2} t\right] \tag{1-45}$$

There *would* be a linearly increasing longitude error, however, since in this example

$$\dot{\tilde{\varphi}} \triangleq \dot{\varphi} - \dot{\hat{\varphi}} = V_E/R_E - \hat{V}_E/R_E = \tilde{V}_E/R_E \tag{1-46}$$

which contains a constant drift rate n_1. To counteract this growing position error, suppose that occasional stellar fixes were taken in conformance to Fig. 1-12. An updating procedure might then be devised, in analogy with the vertical channel scheme of Fig. 1-14. Suppose that additional observations with different accuracies were available from VOR stations at various nonequatorial positions. More generally, there may be unrestricted geometric relations applicable to multiple arbitrarily scheduled measurements having unknown bias and/or random errors, in the presence of accompanying errors in other channels (e.g., ψ_E, ψ_A; drifts in every inertial instrument), with unconstrained flight paths instead of the restricted conditions of (1-41). Furthermore, absence of INS information may call for substituting doppler radar or air data plus wind as the source of velocity time history to be integrated between fixes. Despite added complexity of these general conditions, the rationale from the foregoing examples remains applicable: Known kinematical equations are solved between observations that, properly weighted, provide reinitializations counteracting the growth of position errors obtained by imperfect derivative data integration. This principle is used for all modes of horizontal navigation as well as the vertical channel, and is applicable to any avionics configuration, however extravagant or austere. Guidelines for presenting the necessary material will now be discussed.

1.5 VIEWPOINT FOR SUBSEQUENT CHAPTERS

At this point the reader is aware of the topics that will receive emphasis here and is familiar with some applicable concepts (e.g., patterns of motion; data mixing and updating; inertial, radio, and celestial navigation). Additional considerations are now introduced that further define the scope of this book. First, to assist in mode categorization, five time scales of interest for nav systems will be defined as follows:

(1) frame time for typical SAR imaging (e.g., on the order of a second; Section 7.1.3);
(2) point nav operation intervals [e.g., a few minutes; less than a tenth of the *Schuler period* $2\pi(R_E/g)^{1/2}$];
(3) a "semiopen loop" interval for inertial navigation, during which time-varying attitude reference errors exert appreciable influence on translational nav performance, but effects of imperfections in computed correction terms are neglected (e.g., time spans comparable to the Schuler period, or at least an appreciable fraction thereof);
(4) time durations of a few Schuler periods; and
(5) total durations that are very large compared with the Schuler period (of interest in unaided inertial navigation for transoceanic flight).

In this book, (2)–(4) will receive major emphasis and, since the main topic here is data mixing, (5) will not be considered at all. For this reason, precise geodetic studies (which involve detailed gravitational models, influencing only very long-term dynamic behavior) are ruled out of consideration; the aforementioned ellipsoid is adopted as a *geoid* model. Departures from this model can be added to navigation equations or accepted as acceleration errors. To perform the data mixing, applicable equations are readily linearized (i.e., expressed in terms of first-order Taylor series expansions) for what is often referred to as the "extended Kalman filter" formulation; nonlinear filtering is not necessary for this application. Added operational provisions are described, however, to prevent undue degradation from incomplete fulfillment of underlying theoretical ground rules. While this helps to ensure practicality of mechanizations, this text does not dwell upon actual implementation details. Nor does the discussion stress extension to space applications (though it may be noted that combined aeronautical/astronautical navigation is called for by space shuttle operations); since space navigation by estimation methods has already been studied quite extensively, only an occasional comment in that direction appears in the succeeding chapters. Little emphasis is given to numerical results from computer simulations (which depend upon the detailed conditions under which they were obtained). Rather, considerable effort is made to provide insight by emphasizing *analytical* results. The primary purpose then is to provide a rationale for describing nav performance as a function of instrument and system parameters.

Instead of beginning each chapter with mathematical expressions and terminology that may seem unfamiliar, a gradual approach is adopted whereby a simplified situation is investigated in detail wherever possible at the outset. This provides the understanding necessary for preparation as the more general case evolves. In demonstrating various points, the simplest mathematical tools that suffice have generally been chosen. Detailed proofs have been avoided; rigorous set theory is not used. Where necessary, proofs are demonstrated briefly; more often, to stimulate thought by the reader, they are outlined verbally or left as an exercise at the end of a chapter. Occasionally, proofs are referenced or, if well known, it is understood that many standard reference texts will contain a verification. In many cases where partial reiteration of related expressions occurs in separate derivations, restated equations are put into a somewhat different form; this stimulates further thought and insight.

Although considerable effort is made to provide necessary background material, some familiarity is assumed with fundamental concepts in certain subject areas. A basic understanding of rigid body dynamics and kinematics will be invaluable. Knowledge of Laplace and Fourier transform theory, as well as elementary probability theory, is needed for certain sections. Also, while expertise in matrix analysis is not necessary, there should be no hesitation in using it for summations and simultaneous equations. Appendix II summarizes the matrix relationships necessary for this book.

Since the overall scope of subject matter pertinent to aircraft navigation is quite broad, any attempt to provide complete references and historical background would constitute a survey project of appreciable magnitude. While any book on this subject would perhaps be incomplete without some mention of early contributors (e.g., Ref. 1-4), bibliographical practices adopted here generally follow a pragmatic approach, with only selected references cited to meet an immediate need; thus no special recognition is implied and no slight is intended toward pertinent contributions that are not referenced. Papers are often given preference over books unless a great deal of applicable material appears under one cover. Presentation of this material is not intended to imply that the theory originated with the author of this book; aside from an explicit "data window" formulation for linear estimation, and analysis of certain long-term error propagations in strapdown systems (particularly the commutativity effect and several additional interactive rectifications in Chapter 4), no claim to originality is intended.

EXERCISES

1-1 (a) Principles of mechanics governing the earth's rotation (which for practical purposes is independent of its orbital motion) dictate a substantially constant angular rate. Sketch the earth revolving around the sun, allowing a slight ellipticity in the orbit. Due to this and other astronomical factors, the earth/sun sightline rotation rate (denoted here as ω) varies within the yearly period. Write ω_s as the resultant of ω and a rotational rate of the earth *relative* to the earth/sun line. If these two constituents have appreciable but nearly opposite variations, how is the sidereal rate characterized?

(b) At 0^h UT on July 1, 1968, the mean hour angle of the vernal equinox was $18^h36^m26.166^s$; 2375 days later, the reference value in Section 1.2.1 is obtained. Show that this change conforms very closely to (1-21). How does that equation avoid computational difficulties in forming the product $\omega_s(t - t_0)$?

(c) Do the foregoing conclusions seem to be significantly affected by slight astronomical irregularities, which perturb the ecliptic, equinoxes, etc.?

(d) Discuss artificiality of the mean solar second as a standard time unit. Name a more natural time unit.

1-2 Show that the matrices in (1-22) and (1-28) obey the *orthogonality* relations, $\mathbf{M}^T\mathbf{M} = \mathbf{I}$, $\mathbf{M}\mathbf{M}^T = \mathbf{I}$. Matrices of this type are said to be *orthogonal* (Appendix II).

1-3 Express (1-28) as a product $\mathbf{M}_1\mathbf{M}_2\mathbf{M}_3$ of three orthogonal matrices, representing elevation and azimuth transformations (about y and z axes for latitude and longitude, respectively), in combination with an "axis shift" operation of the form

$$\mathbf{M}_1 = \begin{bmatrix} 0 & 0 & 1 \\ 0 & 1 & 0 \\ -1 & 0 & 0 \end{bmatrix}$$

Interpret this in relation to Fig. 1-4, including consideration of special cases (zero latitude, zero longitude).

1-4 In an earth-centered coordinate frame, the position vector $\mathbf{R}_0$ of a surface point instantaneously below the aircraft can be expressed as $-\mathbf{K}_G R_E$. With the latitude and longitude of this point denoted λ_E and φ_E, respectively, use (1-22) and (1-28) to show that

$$\begin{aligned}\mathbf{R}_0 = R_E[&\cos\lambda_E \cos\varphi_E(\mathbf{I}_I \cos\Xi + \mathbf{J}_I \sin\Xi) \\ &+ \cos\lambda_E \sin\varphi_E(-\mathbf{I}_I \sin\Xi + \mathbf{J}_I \cos\Xi) + \sin\lambda_E\,\mathbf{K}_I]\end{aligned}$$

Differentiate this expression to determine velocity [*HINT*: Since the coordinates of this point are fixed on the earth (i.e., $\dot{R}_E = \dot{\lambda}_E = \dot{\varphi}_E = 0$), its velocity is attributable entirely to variations in Ξ.] Through resubstitution of relations involving axes, express the velocity as

$$-\omega_s R_E \cos\lambda_E[\sin(\Xi + \varphi_E)\mathbf{I}_I - \cos(\Xi + \varphi_E)\mathbf{J}_I]$$

and show that this is purely eastward.

1-5 An easy way to determine the geographic velocity of (1-17) is to express aircraft position in earth coordinates,

$$-\mathbf{K}_G(R_E + h) = (R_E + h)(\mathbf{I}_E \cos\lambda \cos\varphi + \mathbf{J}_E \cos\lambda \sin\varphi + \mathbf{K}_E \sin\lambda)$$

and to differentiate the right of this expression, while ignoring variations in $(\mathbf{I}_E, \mathbf{J}_E, \mathbf{K}_E)$. Perform this differentiation and reexpress the result in geographic coordinates, to obtain $\mathbf{I}_G V_N + \mathbf{J}_G V_E + \mathbf{K}_G V_V$.

1-6 (a) Verify (1-36) by the approach suggested in the text, through formation of a dot product of vectors expressed in the earth frame. Repeat with the vectors expressed in the geographic frame.

(b) Assume that 1 nm on the curved earth surface of Figs. 1-11 and 1-12 subtends an angle of one minute of arc (1 $\widehat{\text{min}}$). Illustrate that a bubble level accurate to 20 $\widehat{\text{sec}}$ would yield errors of order 2000 ft in position determination. What would be the effect of 20 $\widehat{\text{sec}}$ sighting error?

(c) From part (b) and (1-20), show that timing error would produce position uncertainties of order 15 nm/min.

(d) Imagine an aircraft with initially estimated position ℝ of Fig. 1-12 (in the equatorial plane) at a known instant of time. What could be concluded from an elevation measurement 0.1° larger than anticipated, for the star sightline shown? How would confidence in this conclusion be affected by measurement tolerance? What could be gained by observing the elevation angle of another star (Polaris) with a sightline nearly along $\mathbf{K}_E$? What is implied by drawing a star sightline in a fixed direction, regardless of whether it originates from point ℝ or 𝕆?

1-7 (a) Consider (1-36) expressed for each of two elevation measurements (Y_1, Y_2) representing simultaneous observations for elevation angles of two known stars at known sidereal time. Briefly discuss the solution of these two equations in two unknowns λ and φ. Would charts be helpful?

(b) By expanding (1-36) into a first-order Taylor series show that, for small variations $\delta\lambda$ and $\delta\varphi$,

$$(\cos Y_{EL})\,\delta Y_{EL} \doteq \delta\lambda[\sin(\mathrm{DEC})\cos\lambda - \cos(\mathrm{DEC})\cos(\mathrm{SHA} + \Xi + \varphi)\sin\lambda] - \delta\varphi[\cos(\mathrm{DEC})\cos\lambda\sin(\mathrm{SHA} + \Xi + \varphi)]$$

Assuming an initial approximation of position to be available, consider using this relation for determining λ and φ from two elevation measurements. Are charts needed? Are there any sightline geometries for which a solution cannot be obtained? Apply the principles of simultaneous linear equations to establish preferred geometries.

1-8 Solve (1-44) to obtain expressions of the form

$$\psi_N = \psi_{N(0)}\cos(\ ) - (gR_E)^{-1/2}\tilde{V}_{E(0)}\sin(\ )$$

Does the speed have to be constant for these solutions to be valid? Suppose that, instead of a retarding phenomenon, the gravity effect in (1-43) were somehow physically reversed while (1-42) remained unchanged. What would be the prospects for unaided long-range terrestrial inertial navigation?

1-9 Add fixed bias errors n_1 and n_2 to the right of (1-42) and (1-43), respectively. Derive (1-45) and a similar relation for the tilt angle. Differentiate the solutions and illustrate conformance to the differential equations. Illustrate conformance to the initial conditions as well. Are there any other solutions that conform to the initial conditions and also to the differential equations?

1-10 (a) Let $R_E = 3444$ nm and $g = 32.2$ fps^2. How much initial tilt is required to produce 2.6-fps velocity error 21 min later?

(b) Sketch a 1-hr time history of velocity error $\tilde{V}_E$ resulting from an initial platform tilt of $\frac{1}{4}°$. Does this seem to represent acceptable navigation performance?

REFERENCES

1-1. Kalman, R. E., "A New Approach to Linear Filtering and Prediction Problems," *J. Basic Eng.* **82** (1), March 1960, pp. 35–45.
1-2. *American Ephemeris and Nautical Almanac*, U.S. Government Printing Office, Washington, D.C., 1974, p. 19.
1-3. Bowditch, N., *American Practical Navigator*, (U.S. Hydrographic Office), U.S. Government Printing Office, Washington D.C., 1962, pp. 965–972.
1-4. Richman, J., and Friedland, B., "Design of Optimum Mixer-Filter for Aircraft Navigation Systems," presented at 19th Ann. NAECON Conf., Dayton, Ohio, May 15–17, 1967.

Chapter 2

Coordinate Transformations and Kinematics

2.0

Early in the first chapter, several modes of translation were defined and applicable kinematics were described. This chapter will present analytical relationships and kinematics applicable to rotation. The development will concentrate on coordinate frames already introduced, emphasizing the angular *orientation* of axes rather than the location of their intersection point.

Importance of coordinate rotation in navigation analysis is easily established. Since vector operations (addition, formation of a cross product, etc.) require components expressed in the same frame, a means of transformation is necessary when vectors are given in different frames (Ex. 2-1). Furthermore, transformation between inertial and rotating frames is the basis for much kinematical theory. The beginning of Section 1.2 stated a need for determining acceleration in the geographic frame. A basis for this will be initiated through a formulation that allows properties of (1-15) to be exploited. Let $\mathbf{U}_\mathrm{I}$ denote vehicle velocity with respect to the center of the earth,

$$\mathbf{U}_\mathrm{I} = \begin{bmatrix} U_{\mathrm{I}1} \\ U_{\mathrm{I}2} \\ U_{\mathrm{I}3} \end{bmatrix} = \mathbf{I}_\mathrm{I} U_{\mathrm{I}1} + \mathbf{J}_\mathrm{I} U_{\mathrm{I}2} + \mathbf{K}_\mathrm{I} U_{\mathrm{I}3} \tag{2-1}$$

as expressed in inertial coordinates. With U_{G1} defined as the projection of this velocity on $\mathbf{I}_G$,

$$U_{G1} = (\mathbf{I}_G \cdot \mathbf{I}_I)U_{I1} + (\mathbf{I}_G \cdot \mathbf{J}_I)U_{I2} + (\mathbf{I}_G \cdot \mathbf{K}_I)U_{I3} \tag{2-2}$$

and, with all three axes taken under consideration (Ex. 2-2),

$$\mathbf{U}_G = \mathbf{T}_{G/I}\,\mathbf{U}_I, \qquad \mathbf{T}_{G/I} = \begin{bmatrix} \mathbf{I}_G \cdot \mathbf{I}_I & \mathbf{I}_G \cdot \mathbf{J}_I & \mathbf{I}_G \cdot \mathbf{K}_I \\ \mathbf{J}_G \cdot \mathbf{I}_I & \mathbf{J}_G \cdot \mathbf{J}_I & \mathbf{J}_G \cdot \mathbf{K}_I \\ \mathbf{K}_G \cdot \mathbf{I}_I & \mathbf{K}_G \cdot \mathbf{J}_I & \mathbf{K}_G \cdot \mathbf{K}_I \end{bmatrix} \tag{2-3}$$

If this procedure had been carried out with inertial and geographic axis designations reversed (Ex. 2-3), the result would have been

$$\mathbf{U}_I = \mathbf{T}_{I/G}\,\mathbf{U}_G \tag{2-4}$$

where $\mathbf{T}_{I/G}$ is the transpose of $\mathbf{T}_{G/I}$ in (2-3). According to this subscript convention, $\mathbf{T}_{G/I}$ and $\mathbf{T}_{I/G}$ denote transformations from inertial to geographic, and from geographic to inertial coordinates, respectively. One method of computing each transformation has already been supplied through (1-22) and (1-28). Since the time derivative of $\mathbf{U}_I$ is acceleration, application of product differentiation [see (II-18) and (II-19)]* yields total acceleration in inertial coordinates,

$$d\mathbf{U}_I/dt = \dot{\mathbf{U}}_I = \mathbf{T}_{I/G}\,\dot{\mathbf{U}}_G + \dot{\mathbf{T}}_{I/G}\,\mathbf{U}_G \tag{2-5}$$

Expressed in geographic coordinates, this total acceleration, denoted $d\mathbf{U}_G/dt$, is given by

$$d\mathbf{U}_G/dt = \mathbf{T}_{G/I}\,\dot{\mathbf{U}}_I \tag{2-6}$$

which is equivalent to

$$d\mathbf{U}_G/dt = \dot{\mathbf{U}}_G + \mathbf{T}_{G/I}\,\dot{\mathbf{T}}_{I/G}\,\mathbf{U}_G \tag{2-7}$$

These popular conventions for time differentiation operators as applied to vectors are used throughout this book. Equation (2-6) serves as a definition for the total time derivative of *any* vector expressed in any coordinate frame. The operator d/dt denotes intrinsic variation experienced in a non-rotating frame; components are in effect resolved along axes of any desired coordinate frame *after* differentiation. The operator $(\dot{\ })$, in contrast, implies a time derivative of each separate element [recall (II-19)] with product differentiation applicable as in (2-5). Only in an inertial frame will the operators d/dt and $(\dot{\ })$ produce equivalent vector derivatives.

* Refer to Appendix II for any necessary background theory in matrix analysis. Note that $\mathbf{U}_G$ cannot represent the geographic velocity vector $\mathbf{V}$ of (1-17), since the "ground east" convention does not conform to the definition preceding (2-1).

Completion of a kinematical formalism calls for further development of transformation matrices and their derivatives as typified in (2-7). This chapter presents that development from a viewpoint of aeronautical convention. Euler angle, direction cosine, and four-parameter formulations are analyzed for geometric properties, mathematical behavior, kinematics, and *inter*-relationships. Reflective transformations are then briefly investigated, followed by appendices containing classical relations in rotation analysis. This material was assembled using information from Ref. 2-1 with considerable modification, rearrangement, omission, and additions from other sources. Various idealizations (e.g., fixed equinoxes for the inertial frame) are made for brevity wherever permissible.

2.1 FUNDAMENTALS OF ROTATIONAL TRANSFORMATIONS

Definitions for the five coordinate frames introduced in Chapter 1 are now reiterated. Wherever only two reference directions are stated below, the third is supplied by (1-18).

Inertial coordinates are defined such that $\mathbf{K}_I$ points along the north geodetic pole and $\mathbf{I}_I$ points outward from earth center to the vernal equinox.

Earth coordinates are defined with $\mathbf{K}_E \equiv \mathbf{K}_I$ and $\mathbf{I}_E$ points outward from earth center along the equator to the Greenwich meridian.

Geographic coordinates consist of $\mathbf{I}_G$, $\mathbf{J}_G$, and $\mathbf{K}_G$ pointing along actual local north, east, and downward directions, respectively.

Platform coordinates are defined by axes $\mathbf{I}_P$, $\mathbf{J}_P$, and $\mathbf{K}_P$, which constitute *apparent* local north, east, and downward directions.

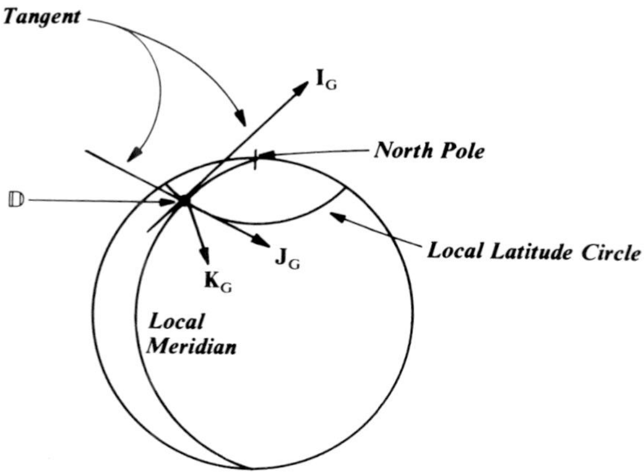

FIG. 2-1. Geographic coordinates ($\mathbb{D}$ is the point of tangency directly below the current aircraft position).

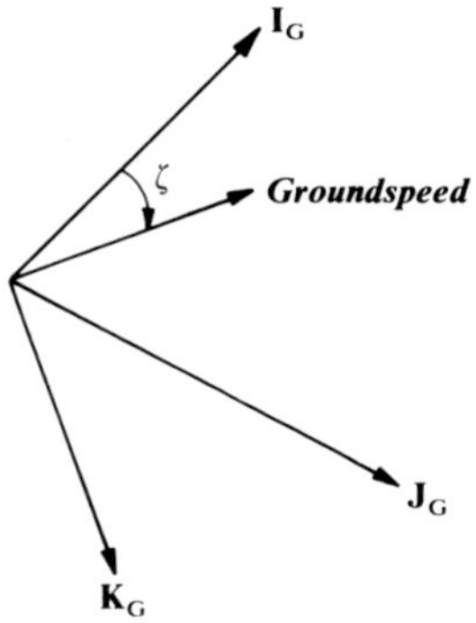

FIG. 2-2. Ground track angle (groundspeed is the horizontal projection of the aircraft velocity vector).

Aircraft coordinates $\mathbf{I}_A$, $\mathbf{J}_A$, and $\mathbf{K}_A$ point forward along the fuselage axis, laterally outward along the right wing, and toward the floor of the aircraft, respectively.

Other triads (e.g., coordinate frames with a particular azimuth offset from the geographic axes, such as ground track and *wander azimuth* references) will be added in later chapters; however, a formulation based on the geographic reference will be presented first. Coordinate axes have already been illustrated in Figs. 1-4–1-8; further demonstrations of their relationships appear in Figs. 2-1–2-4.

It is convenient to define the transformation between platform and aircraft axes in terms of a notation to be adopted here as standard: Let $[\theta]_u$ denote a positive rotation of θ rad about the "u" axis ("u" being

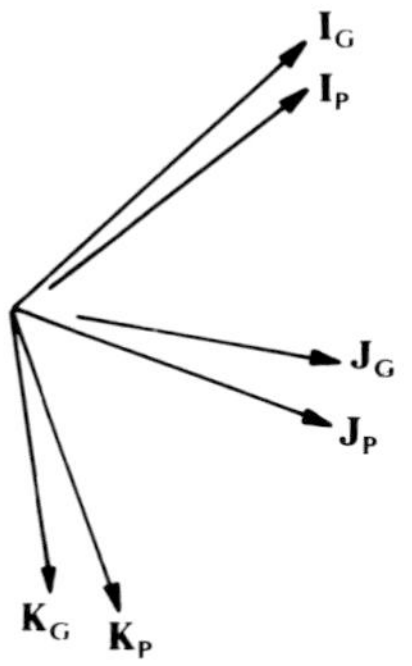

FIG. 2-3. Relation of platform to geographic coordinates ($\mathbf{K}_G \cdot \mathbf{J}_P = \psi_N$, $\mathbf{I}_G \cdot \mathbf{K}_P = \psi_E$, $\mathbf{J}_G \cdot \mathbf{I}_P = \psi_A$).

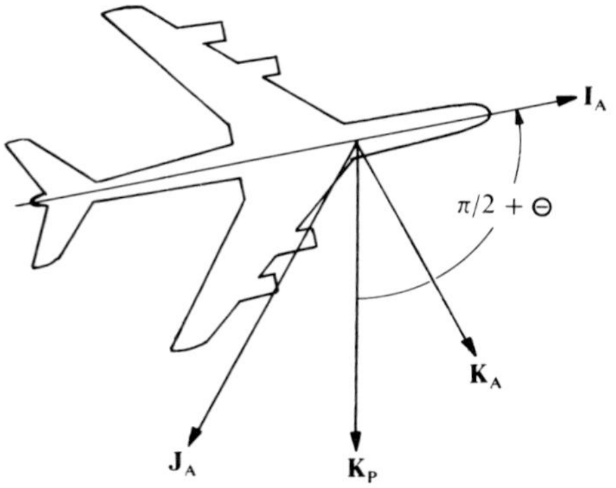

FIG. 2-4. Aircraft coordinates ($\Phi = \arcsin\{\mathbf{K}_P \cdot \mathbf{J}_A \sec \Theta\}$, $\Psi = \arcsin\{\mathbf{J}_P \cdot \mathbf{I}_A \sec \Theta\}$).

replaceable by x, y, or z), so that

$$[\theta_1]_x \triangleq \begin{bmatrix} 1 & 0 & 0 \\ 0 & \cos\theta_1 & \sin\theta_1 \\ 0 & -\sin\theta_1 & \cos\theta_1 \end{bmatrix}, \qquad [\theta_2]_y \triangleq \begin{bmatrix} \cos\theta_2 & 0 & -\sin\theta_2 \\ 0 & 1 & 0 \\ \sin\theta_2 & 0 & \cos\theta_2 \end{bmatrix},$$

$$[\theta_3]_z \triangleq \begin{bmatrix} \cos\theta_3 & \sin\theta_3 & 0 \\ -\sin\theta_3 & \cos\theta_3 & 0 \\ 0 & 0 & 1 \end{bmatrix} \tag{2-8}$$

This convention is used with the roll, pitch, and heading angles Φ, Θ, and Ψ that would be measured from a three-gimbal configuration arranged in that order, with the inner element aligned to the mechanized geographic reference (recall Figs. 1-6–1-8 and see Fig. 2-5, shown with cascaded gimbals at nominal orientation). The transformation from platform to aircraft coordinates is then a 3×3 orthogonal matrix product containing nine direction cosine elements,

$$\mathbf{T}_{A/P} = [\Phi]_x[\Theta]_y[\Psi]_z \tag{2-9}$$

and the transformation from aircraft to platform coordinates is (Ex. 2-4)

$$\mathbf{T}_{P/A} = \mathbf{T}_{A/P}^{-1} = [\Psi]_z^T[\Theta]_y^T[\Phi]_x^T = [-\Psi]_z[-\Theta]_y[-\Phi]_x \tag{2-10}$$

By a development paralleling Ex. 2-2,

$$\begin{bmatrix} \mathbf{I}_A \\ \mathbf{J}_A \\ \mathbf{K}_A \end{bmatrix} = \begin{bmatrix} \mathbf{I}_A \cdot \mathbf{I}_P & \mathbf{I}_A \cdot \mathbf{J}_P & \mathbf{I}_A \cdot \mathbf{K}_P \\ \mathbf{J}_A \cdot \mathbf{I}_P & \mathbf{J}_A \cdot \mathbf{J}_P & \mathbf{J}_A \cdot \mathbf{K}_P \\ \mathbf{K}_A \cdot \mathbf{I}_P & \mathbf{K}_A \cdot \mathbf{J}_P & \mathbf{K}_A \cdot \mathbf{K}_P \end{bmatrix} \begin{bmatrix} \mathbf{I}_P \\ \mathbf{J}_P \\ \mathbf{K}_P \end{bmatrix} \tag{2-11}$$

and, for any vector defined in aircraft coordinates (e.g., a known displacement $\mathbf{d}_A$ between two fixed points on the aircraft), the components are

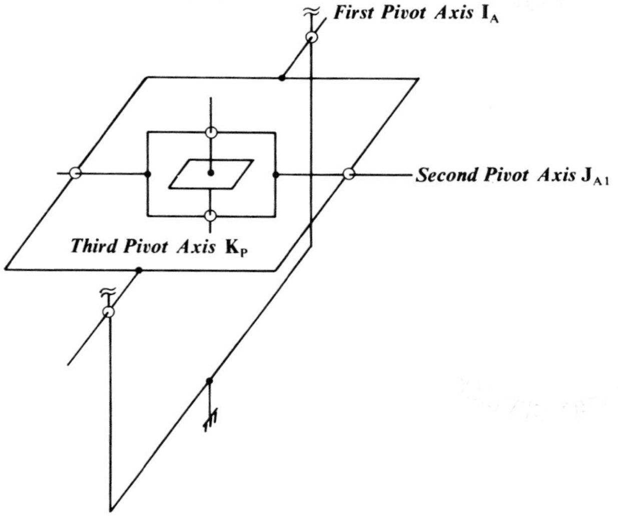

FIG. 2-5. Three cascaded gimbals.

resolved along platform axes here by

$$\mathbf{d}_P = \mathbf{T}_{P/A}\,\mathbf{d}_A \tag{2-12}$$

which of course has the same magnitude as $\mathbf{d}_A$ but in general a different distribution of components. More broadly, $\mathbf{d}_A$ can be reexpressed in any coordinate frame desired, by means of transformation products. For example, with $\mathbf{T}_{G/P}$, $\mathbf{T}_{E/G}$, and $\mathbf{T}_{I/E}$ used to denote the orthogonal transformations from platform to geographic, geographic to earth, and earth to inertial coordinates, respectively, the previously defined displacement vector expressed in inertial coordinates is

$$\mathbf{d}_I = \mathbf{T}_{I/A}\,\mathbf{d}_A, \qquad \mathbf{T}_{I/A} = \mathbf{T}_{I/E}\,\mathbf{T}_{E/G}\,\mathbf{T}_{G/P}\,\mathbf{T}_{P/A} \tag{2-13}$$

or the elements could have been resolved into any other frame at an intermediate stage of cascaded matrix products. The first two matrix factors in (2-13), in agreement with Section 1.2, are

$$\mathbf{T}_{I/E} = [\Xi]_z^T, \qquad \mathbf{T}_{E/G} = [\varphi]_z^T[(\pi/2) + \lambda]_y \tag{2-14}$$

For the remaining factor it is noted that, at any permissible departure between geographic and platform axes, small angle approximations are accurate. Since the sign conventions in (1-30) associate positive angles with positive rotations from geographic to platform axes (Ex. 2-5),

$$\mathbf{T}_{G/P} \doteq \begin{bmatrix} 1 & -\psi_A & \psi_E \\ \psi_A & 1 & -\psi_N \\ -\psi_E & \psi_N & 1 \end{bmatrix} \tag{2-15}$$

This transformation is the sum of an identity and a vector product ($\boldsymbol{\psi}\times\ $) operator. With these definitions complete, attention is directed toward further relations of interest in navigation.

2.2 PROPERTIES OF ROTATIONAL TRANSFORMATIONS

The preceding section illustrated basic characteristics regarding xyz Euler angle and direction cosine parametrizations. While there are several alternative formulations for representing rotations (see, e.g., Ref. 2-2), not all are equally applicable to current navigational procedures. Consequently the subsequent development will not include all possible parametrizations, nor all properties of those formulations chosen for consideration. Selected transformation operators are analyzed here for geometric interpretation, presence or absence of singularities or discontinuities, uniqueness and orthogonality properties as applicable, kinematic behavior in relation to vehicle angular rates, and relations between different parameter sets.

2.2.1 Euler Angle Sequences

The matrix $\mathbf{T}_{P/A}$ in (2-10) exemplifies a direction cosine transformation expressed in terms of the xyz Euler angle sequence $(-\Phi, -\Theta, -\Psi)$, respectively [the order of rotations in (2-10) is read from right to left for the transformation *from* aircraft *to* platform coordinates]. An expanded version of (2-10) can be written (Ex. 2-6)

$$\mathbf{T}_{P/A} = \begin{bmatrix} \cos\Psi\cos\Theta & -\sin\Psi\cos\Phi + \cos\Psi\sin\Theta\sin\Phi & \sin\Psi\sin\Phi + \cos\Psi\sin\Theta\cos\Phi \\ \sin\Psi\cos\Theta & \cos\Psi\cos\Phi + \sin\Psi\sin\Theta\sin\Phi & -\cos\Psi\sin\Phi + \sin\Psi\sin\Theta\cos\Phi \\ -\sin\Theta & \cos\Theta\sin\Phi & \cos\Theta\cos\Phi \end{bmatrix} \tag{2-16}$$

It is of interest to consider an alternative Euler angle convention; e.g., suppose it had been desirable to express the same transformation in terms of a zxz sequence of Euler angles $(-\theta_1', -\theta_2', -\theta_3')$:

$$[-\theta_3']_z[-\theta_2']_x[-\theta_1']_z$$

$$= \begin{bmatrix} \cos\theta_1'\cos\theta_3' - \cos\theta_2'\sin\theta_3'\sin\theta_1' & -\sin\theta_1'\cos\theta_3' - \cos\theta_2'\sin\theta_3'\cos\theta_1' & \sin\theta_3'\sin\theta_2' \\ \cos\theta_1'\sin\theta_3' + \cos\theta_2'\cos\theta_3'\sin\theta_1' & -\sin\theta_1'\sin\theta_3' + \cos\theta_2'\cos\theta_3'\cos\theta_1' & -\cos\theta_3'\sin\theta_2' \\ \sin\theta_2'\sin\theta_1' & \sin\theta_2'\cos\theta_1' & \cos\theta_2' \end{bmatrix} \tag{2-17}$$

A basic difference between these conventions is the point at which an irregularity occurs. When $\theta_2{}' = 0$ or π in (2-17) it is easy to see that (1) the class of transformations that can be expressed is restricted to $\mathscr{z}$ axis rotations, and (2) for any transformation in this restricted class, $\theta_1{}'$ and $\theta_3{}'$ are not unique, i.e., the same transformation matrix can be obtained from different combinations of these two angles; only the net sum influences the overall transformation. An analogous situation holds in (2-10) and (2-16) when $\Theta = \pm\pi/2$; the phenomenon is often referred to as "gimbal lock," in reference to the singularity present in three-gimbal platform mechanizations. This point will arise again in an ensuing description of rotational kinematics.

For all nonsingular cases it should be noted that the same matrix $\mathbf{T}_{P/A}$ could be obtained from several different Euler angle conventions. Two sets of angles (θ_j, $\theta_j{}'$; $1 \le j \le 3$) could be found, for example, subject to the conditions

$$\begin{aligned} -\sin\theta_2 &= \sin\theta_2{}' \sin\theta_1{}' \\ \cos\theta_2 \sin\theta_3 &= \cos\theta_1{}' \sin\theta_3{}' + \cos\theta_2{}' \cos\theta_3{}' \sin\theta_1{}' \\ \cos\theta_2 \sin\theta_1 &= \sin\theta_2{}' \cos\theta_1{}', \qquad \theta_2{}' \ne 0, \quad \theta_2 \ne \pi/2 \end{aligned} \tag{2-18}$$

and the overall transformation could be the same for either of the two previously defined Euler angle sequences. Physically it is easy to visualize how different combinations of rotations can have the same initial and final orientations. This should not, however, be regarded as an ambiguity; apart from singularities, an Euler angle triad is unique *once the convention has been specified*. The roll–pitch–heading convention is given preference here because of its common usage in aircraft navigation ($\mathscr{z x z}$ sequences are commonly used to describe satellite orbits except in cases of near-singularity, i.e., low inclination).

2.2.2 Kinematics of Rotations

DYNAMIC BEHAVIOR OF EULER ANGLES

In airborne control applications the situation occurs repeatedly in which angular motion sensors are employed in the generation of rotational commands about displaced axes (i.e., rotations are defined about axes non-coincident with sensor axis orientation). This requires a kinematical re-expression of angular rates, an example of which follows:

A gyro triad with input axes along the aircraft roll, pitch, and yaw axes would be sensitive to the aircraft *inertial* rate vector $\boldsymbol{\omega}_A$, i.e., the *absolute* angular rate of the aircraft with respect to the *inertial* ($\mathbf{I}_I$, $\mathbf{J}_I$, $\mathbf{K}_I$) coordinate frame, with components expressed along instantaneous *aircraft* ($\mathbf{I}_A$, $\mathbf{J}_A$, $\mathbf{K}_A$) axes. Clearly, the components of $\boldsymbol{\omega}_A$ in general are *not* equal to the time derivatives of the Euler angles Φ, Θ, Ψ, primarily because $\mathbf{J}_{A1}$ and $\mathbf{K}_{A2}$ of Figs. 1-7 and 1-8 can deviate from $\mathbf{J}_A$ and $\mathbf{K}_A$, respectively. A kinematical relation is obtained by associating each Euler angle with its axial direction. Let $\boldsymbol{\omega}_P$ be defined as the absolute angular rate of the platform in the $(\mathbf{I}_P, \mathbf{J}_P, \mathbf{K}_P)$ coordinate frame, so that the angular rate of the aircraft *relative to the platform reference*, expressed in aircraft coordinates, is (Ex. 2-6)

$$\zeta = \begin{bmatrix} \mathscr{p} \\ \mathscr{q} \\ \mathscr{r} \end{bmatrix} = \boldsymbol{\omega}_A - \mathbf{T}_{A/P}\,\boldsymbol{\omega}_P \tag{2-19}$$

where the script notation distinguishes $\mathscr{p}$, $\mathscr{q}$, $\mathscr{r}$ from the variables p, q, r normally identified with the absolute rate $\boldsymbol{\omega}_A$. The relation between ζ and the gimbal angle derivatives, with the aforementioned axial directions, is

$$\begin{aligned} \zeta = {} & \dot{\Phi}\mathbf{I}_A + \dot{\Theta}(\mathbf{J}_A \cos\Phi - \mathbf{K}_A \sin\Phi) \\ & + \dot{\Psi}[-\mathbf{I}_A \sin\Theta + \cos\Theta(\mathbf{J}_A \sin\Phi + \mathbf{K}_A \cos\Phi)] \end{aligned} \tag{2-20}$$

which corresponds to the two equivalent relations

$$\begin{bmatrix} \mathscr{p} \\ \mathscr{q} \\ \mathscr{r} \end{bmatrix} = \begin{bmatrix} \dot{\Phi} - \dot{\Psi}\sin\Theta \\ \dot{\Theta}\cos\Phi + \dot{\Psi}\sin\Phi\cos\Theta \\ \dot{\Psi}\cos\Phi\cos\Theta - \dot{\Theta}\sin\Phi \end{bmatrix} \tag{2-21}$$

$$\begin{bmatrix} \dot{\Phi} \\ \dot{\Theta} \\ \dot{\Psi} \end{bmatrix} = \begin{bmatrix} \mathscr{p} + \tan\Theta(\mathscr{q}\sin\Phi + \mathscr{r}\cos\Phi) \\ \mathscr{q}\cos\Phi - \mathscr{r}\sin\Phi \\ \sec\Theta(\mathscr{q}\sin\Phi + \mathscr{r}\cos\Phi) \end{bmatrix} \tag{2-22}$$

Once again note that a singularity exists at $\Theta = \pm\pi/2$, i.e., a 90° pitch in Fig. 2-5 would align the inner and outer pivot axes. Mathematically, an Euler angle triad does not have global properties; physically, a three-gimbal platform does not have all-attitude capability without a change in convention to circumvent singularity.

Immediately a question arises in regard to alternative rotational parametrizations that do *not* suffer from undesirable properties such as singularities. Although there are three-parameter sets other than Euler angle triads, none are completely free of all singularities or discontinuities (see Ref. 2-2). For sets of more than three parameters, one obvious possibility is a sequence of four angles according to a specified convention (e.g., $\mathscr{xyxz}$) as mentioned in Section 3.2, in connection with a four-gimbal platform mechanization. Another possibility that comes readily to mind is the set of nine direction cosines, which will now be scrutinized.

DIRECTION COSINE MATRIX KINEMATICS

In addition to a clear geometric interpretation, direction cosines afford beneficial mathematical properties such as orthogonality and freedom from singularities or discontinuities. Furthermore, their relation to other rotational parameter sets is easily defined (as already illustrated in connection with the Euler angles) and their time derivatives are defined in terms of differential equations far simpler than the highly nonlinear (2-22). The desired kinematics can easily be derived by noting (as in Ref. 2-1) that, although rotational operators in general do not exhibit vector properties, an *infinitesimal* rotation is expressible as the product of a vector (e.g., $\boldsymbol{\omega}_A$) times a scalar (e.g., differential time dt); thus it is permissible to define a vector $\mathbf{d\theta}$ of infinitesimal magnitude having a direction along the instantaneous axis of rotation:

$$\mathbf{d\theta} \triangleq \boldsymbol{\omega}_A \, dt \tag{2-23}$$

At this point it is instructive to consider three coordinate frames: (1) the inertial set ($\mathbf{I}_I$, $\mathbf{J}_I$, $\mathbf{K}_I$), (2) aircraft axes at time t, (3) aircraft axes at time $t + dt$. The transformation from the first to the second frame is denoted $\mathbf{T}_{A/I}(t)$. The transformation from the second to the third is obtained by combining three infinitesimal rotations $[d\theta_1]_x$, $[d\theta_2]_y$, and $[d\theta_3]_z$, in the notation of (2-8); note that these three commute [i.e., $\cos(d\theta) = 1$ and $\sin(d\theta) = d\theta$ exactly for infinitesimal angles, and products of infinitesimals are negligible], so that the resulting combined transformation is independent of the rotational sequence:

$$\mathbf{T}_{t+dt/t} = [d\theta_1]_x [d\theta_2]_y [d\theta_3]_z = \begin{bmatrix} 1 & d\theta_3 & -d\theta_2 \\ -d\theta_3 & 1 & d\theta_1 \\ d\theta_2 & -d\theta_1 & 1 \end{bmatrix} \tag{2-24}$$

By combination of the last two expressions,

$$\mathbf{T}_{t+dt/t} = \mathbf{I}_{3\times 3} + \mathbf{\Omega}_A \, dt, \qquad \mathbf{\Omega}_A \triangleq \begin{bmatrix} 0 & \omega_{A3} & -\omega_{A2} \\ -\omega_{A3} & 0 & \omega_{A1} \\ \omega_{A2} & -\omega_{A1} & 0 \end{bmatrix} \tag{2-25}$$

The transformation from the first to the third frame is simply the product $\mathbf{T}_{t+dt/t} \mathbf{T}_{A/I}(t)$, which can now be written

$$\mathbf{T}_{A/I}(t + dt) = \mathbf{T}_{A/I}(t) + \mathbf{\Omega}_A \mathbf{T}_{A/I}(t) \, dt \tag{2-26}$$

and, by definition of the time derivative of a matrix,

$$\dot{\mathbf{T}}_{A/I} = \mathbf{\Omega}_A \mathbf{T}_{A/I} \tag{2-27}$$

By a similar derivation,

$$\dot{\mathbf{T}}_{I/A} = -\mathbf{T}_{I/A} \mathbf{\Omega}_A \tag{2-28}$$

To extend (2-28) to the case of a noninertial reference [e.g., the slowly rotating ($\mathbf{I}_G$, $\mathbf{J}_G$, $\mathbf{K}_G$) frame] the same relation is applied to the geographic frame:

$$\dot{\mathbf{T}}_{I/G} = -\mathbf{T}_{I/G}\,\mathbf{\Omega}_G \tag{2-29}$$

where, in analogy with $\mathbf{\Omega}_A$ in (2-25), $\mathbf{\Omega}_G$ is skew-symmetric with elements of the absolute angular rate $\boldsymbol{\omega}_G$ of the ($\mathbf{I}_G$, $\mathbf{J}_G$, $\mathbf{K}_G$) frame, expressed in geographic coordinates. By differentiation of the product

$$\mathbf{T}_{G/A} = \mathbf{T}_{G/I}\,\mathbf{T}_{I/A} \tag{2-30}$$

it follows that

$$\dot{\mathbf{T}}_{G/A} = -\mathbf{T}_{G/A}\,\mathbf{\Omega}_A + \mathbf{\Omega}_G\,\mathbf{T}_{G/A} \tag{2-31}$$

To underline the importance of this analysis, an immediate application will be made to (2-7), and the basis for a standard nav formulation will be established.

EFFECTS OF COORDINATE ROTATIONS ON TRANSLATIONAL KINEMATICS

Consider the relation (Ex. 2-7),

$$\mathbf{\Omega}_G = -(\boldsymbol{\omega}_G \times \quad) \tag{2-32}$$

in combination with (2-7) and (2-29):

$$d\mathbf{U}_G/dt = \dot{\mathbf{U}}_G + \boldsymbol{\omega}_G \times \mathbf{U}_G \tag{2-33}$$

This expression and all preceding equations used to derive it are standard analysis tools, to be understood thoroughly (Ex. 2-7 is aimed at further detailed exposition).

As applied to position dynamics, let $\mathbf{R}$ represent the vector in (1-25). With components resolved along inertial axes, this becomes

$$\mathbf{R}_I = \mathbf{T}_{I/G}\,\mathbf{R} \tag{2-34}$$

By product differentiation, using (2-29),

$$d\mathbf{R}_I/dt = \mathbf{T}_{I/G}(\dot{\mathbf{R}} - \mathbf{\Omega}_G\,\mathbf{R}) \tag{2-35}$$

and, by another differentiation,

$$d^2\mathbf{R}_I/dt^2 = \mathbf{T}_{I/G}(\ddot{\mathbf{R}} - \dot{\mathbf{\Omega}}_G\,\mathbf{R} - 2\mathbf{\Omega}_G\,\dot{\mathbf{R}} + \mathbf{\Omega}_G\,\mathbf{\Omega}_G\,\mathbf{R}) \tag{2-36}$$

When $\mathbf{\Omega}_G$ is replaced by the vector product operator $(-\boldsymbol{\varpi} \times \quad)$ and both sides of the equation are premultiplied by $\mathbf{T}_{G/I}$, the result is a general form free of subscripts:

$$d^2\mathbf{R}/dt^2 = \ddot{\mathbf{R}} + \dot{\boldsymbol{\varpi}} \times \mathbf{R} + 2\boldsymbol{\varpi} \times \dot{\mathbf{R}} + \boldsymbol{\varpi} \times (\boldsymbol{\varpi} \times \mathbf{R}) \tag{2-37}$$

where the notation on the left is defined as absolute acceleration expressed in whatever coordinate frame corresponds to **R**, while $\boldsymbol{\varpi}$ denotes the profile rate of that frame with respect to the inertial (nonrotating) axes ($\mathbf{I}_I$, $\mathbf{J}_I$, $\mathbf{K}_I$). Common nomenclature for the last three terms on the right are tangential, Coriolis, and centripetal acceleration, respectively.

Although (2-37) is a general relation, its usage in this book will concentrate on the ($\mathbf{I}_G$, $\mathbf{J}_G$, $\mathbf{K}_G$) reference application. Despite the background derivation involving (1-11) as well as (1-25), **R** will be defined as a position vector in geographic coordinates throughout, i.e., these kinematics will be used with the nonspherical earth. The next chapter develops that theme and uses (2-37) to derive geographic velocity, by adjusting absolute accelerations observed in a rotating frame (recall the beginning of Section 1.2).

This matrix differentiation approach to kinematical analysis is brief, direct, and free of "mysterious rules." Clarity is enhanced by avoiding lengthy accumulations for derivatives of individual component projections and of the coordinate axis vectors themselves.

A useful alternate formulation for rotation operators calls for investigation of another transformation property; following is a development involving principal directions.

2.2.3 Principal Axis Relationships

A fundamental property of orthogonal matrix transformations is the invariance of the vector magnitude. This property is consistent with the mathematical definition of matrix orthogonality in Appendix II and, physically, a vector magnitude is obviously independent of the coordinate frame in which it is expressed. Consider now the possibility that a vector **E** is also unchanged in *direction* after transformation by an orthogonal matrix **T**, i.e.,

$$\mathbf{T}\mathbf{E} = \mathbf{E} \tag{2-38}$$

so that the vector components are the same as expressed in either of the two frames involved in the transformation. This is a special (unit root) case of the eigenvalue equation (II-20). Simple examples include the matrices in (2-8), operating on vectors pointed along x, y, and z axes, respectively. For *any* transformation from one right-hand orthogonal frame to another, the root (+1) always satisfies the eigenvalue equation, and **E** is its *only* corresponding eigenvector in nontrivial cases (Ex. 2-8). This *principal eigenvector*, easily computed from (II-27), defines a direction for total rotation incurred by the transformation **T**; regardless of the rotational sequence whereby **T** was generated, any combination of multiple Euler angle transformations can be reduced to a single rotation about **E**. In general, of course, this axis will have nonzero components in more than one coordinate direction and/or can be time-varying; thus no simple convention is available that will consistently replace **T** by a single-axis transformation of the type in (2-8). It is always possible, however, to perform a similarity transformation* whereby **T** is related to a rotation about one coordinate axis; significantly, the *trace* (i.e., sum of diagonal elements) of a matrix is invariant under a similarity transformation. For any angle θ in a single-axis rotation it is obvious from (2-8) that the trace is equivalent to $1 + 2\cos\theta$. Thus an orthogonal matrix **T** corresponds to rotation about **E**, with an angle

$$\theta = \arccos[\tfrac{1}{2}(T_{11} + T_{22} + T_{33} - 1)] \text{ rad} \tag{2-39}$$

The quadrant for this angle and the order of vectors in (II-27) must be compatibly chosen. This principal axis development will now be applied to another useful parametrization.

2.3 FOUR-PARAMETER ROTATION OPERATORS

Every direction cosine matrix corresponds to a unique set of four parameters exemplified by

$$\mathbf{T}_{A/I} = \begin{bmatrix} b_4^2 + b_1^2 - b_2^2 - b_3^2 & 2(b_1 b_2 + b_4 b_3) & 2(b_1 b_3 - b_4 b_2) \\ 2(b_1 b_2 - b_4 b_3) & b_4^2 - b_1^2 + b_2^2 - b_3^2 & 2(b_2 b_3 + b_4 b_1) \\ 2(b_1 b_3 + b_4 b_2) & 2(b_2 b_3 - b_4 b_1) & b_4^2 - b_1^2 - b_2^2 + b_3^2 \end{bmatrix} \tag{2-40}$$

where these *b*-parameters conform to the vector differential equation

$$\dot{\boldsymbol{b}} = \boldsymbol{\Psi b}, \qquad \boldsymbol{\Psi} \triangleq \frac{1}{2}\begin{bmatrix} 0 & \omega_{A3} & -\omega_{A2} & \omega_{A1} \\ -\omega_{A3} & 0 & \omega_{A1} & \omega_{A2} \\ \omega_{A2} & -\omega_{A1} & 0 & \omega_{A3} \\ -\omega_{A1} & -\omega_{A2} & -\omega_{A3} & 0 \end{bmatrix} \tag{2-41}$$

The simplicity of (2-41) in comparison with (2-27) or (2-28) suggests a preference for a four-parameter formulation in certain applications. This is significant where the main computational effort is spent in updating the attitude itself, by a running integration for the angular orientation time history of a vehicle subjected to rapid irregular angular rates. Although this advantage is offset in other situations requiring frequent *transformation*

* A similarity transformation simply consists of pre- and postmultiplication by a matrix and its inverse (the latter being equivalent to its transpose for an orthogonal matrix), respectively. Examples appear in Appendix II, and further details regarding this subject appear in Ref. 2-1, pp. 105, 123–124.

of vector components (Ex. 2-9), there are enough four-parameter applications to warrant further development below.

Consider the 4×1 vector with a partitioned form (see b and $\boldsymbol{b}$ in Appendix I),

$$\boldsymbol{b} = \begin{bmatrix} b_1 \\ b_2 \\ b_3 \\ b_4 \end{bmatrix} = \begin{bmatrix} \mathbf{b} \\ \text{---} \\ \ell \end{bmatrix} = \begin{bmatrix} b_1 \\ b_2 \\ b_3 \\ \ell \end{bmatrix} = \begin{bmatrix} \mathbf{E}\sin(\theta/2) \\ \text{-------------} \\ \cos(\theta/2) \end{bmatrix} \tag{2-42}$$

where $\mathbf{E}$ and θ conform to definitions in Section 2.2.3. This is an all-attitude parametrization, free of singularities and discontinuities. It also has the orthogonality property

$$b_1{}^2 + b_2{}^2 + b_3{}^2 + b_4{}^2 = 1 \tag{2-43}$$

to enhance its ready geometric interpretation. Reversal of either the angle or the axis of rotation would transpose the matrix in (2-40); reversal of both would leave the matrix unchanged. Conformance to basic transformations under special conditions, such as $[\theta]_x$ for $E_1 = 1$, $E_2 = E_3 = 0$, is easily verified. Despite such logical associations, alternate four-parameter formulations are often presented with permutations of indices and other variations in convention. The definitions used here are recommended for applications pertinent to this book.

It remains to establish the validity of (2-40) and (2-41) under general conditions. The ensuing development can be traced to an earlier century, and has been largely replaced by the relations already presented for actual *usage* of four-parameter sets. As a validation approach, however, consider an operator $\boldsymbol{B}$ having a scalar "component" $\cos(\theta/2)$ and a vector "component" $\mathbf{E}\sin(\theta/2)$, combined in accordance with the definition

$$\boldsymbol{B} = \cos(\theta/2) + \mathbf{E}\sin(\theta/2) = b_4 + \mathbf{i}b_1 + \mathbf{j}b_2 + \mathbf{k}b_3 \tag{2-44}$$

where $(\mathbf{i}, \mathbf{j}, \mathbf{k})$ represent the basis vectors for either of the two coordinate frames involved in the transformation. This operator, known as a *quaternion* (Ref. 2-3), has an inverse

$$\boldsymbol{B}^{-1} = \cos(\theta/2) - \mathbf{E}\sin(\theta/2) \tag{2-45}$$

and the product of two quaternions ($\boldsymbol{B}_1 = \ell_1 + \mathbf{b}_1, \boldsymbol{B}_2 = \ell_2 + \mathbf{b}_2$) is defined as

$$\boldsymbol{B}_1\boldsymbol{B}_2 = \ell_1\ell_2 - \mathbf{b}_1 \cdot \mathbf{b}_2 + \ell_1\mathbf{b}_2 + \ell_2\mathbf{b}_1 + \mathbf{b}_1 \times \mathbf{b}_2 \tag{2-46}$$

Any three-dimensional vector $\mathbf{v}$ can be regarded as a quaternion $\boldsymbol{V}$ with a zero scalar component. To illustrate the application of quaternions to rotations, $\mathbf{c}_\mathrm{I}$ and $\mathbf{c}_\mathrm{A}$, defining a star sightline expressed in inertial and aircraft coordinates, are written as quaternions $\boldsymbol{C}_\mathrm{I}$ and $\boldsymbol{C}_\mathrm{A}$, respectively.

Whereas rotational transformation through matrices can be represented by one premultiplicative operator, rotation via quaternions takes the form

$$\boldsymbol{C}_{\mathrm{I}} = \boldsymbol{B}\boldsymbol{C}_{\mathrm{A}}\boldsymbol{B}^{-1} \tag{2-47}$$

so that, by combination with the three preceding equations (Ex. 2-10),

$$\mathbf{c}_{\mathrm{I}} = (\cos\theta)\mathbf{c}_{\mathrm{A}} + (1-\cos\theta)(\mathbf{E}\cdot\mathbf{c}_{\mathrm{A}})\mathbf{E} + (\sin\theta)\mathbf{E}\times\mathbf{c}_{\mathrm{A}} \tag{2-48}$$

To derive (2-41), differentiation of (2-47) yields

$$\dot{\boldsymbol{C}}_{\mathrm{I}} = \boldsymbol{O} = \dot{\boldsymbol{B}}\boldsymbol{C}_{\mathrm{A}}\boldsymbol{B}^{-1} + \boldsymbol{B}\dot{\boldsymbol{C}}_{\mathrm{A}}\boldsymbol{B}^{-1} + \boldsymbol{B}\boldsymbol{C}_{\mathrm{A}}\dot{\boldsymbol{B}}^{-1} \tag{2-49}$$

By combination with the identity (Ex. 2-11)

$$(\boldsymbol{B}^{-1}\dot{\boldsymbol{B}})^{-1} = \dot{\boldsymbol{B}}^{-1}\boldsymbol{B} \tag{2-50}$$

solution of (2-49) for $\dot{\boldsymbol{C}}_{\mathrm{A}}$ yields

$$\dot{\boldsymbol{C}}_{\mathrm{A}} = -\{[\boldsymbol{B}^{-1}\dot{\boldsymbol{B}})\boldsymbol{C}_{\mathrm{A}} - \boldsymbol{C}_{\mathrm{A}}^{-1}(\boldsymbol{B}^{-1}\dot{\boldsymbol{B}})^{-1}\} \tag{2-51}$$

Since (Ex. 2-7)

$$\dot{\mathbf{c}}_{\mathrm{A}} = -\boldsymbol{\omega}_{\mathrm{A}}\times\mathbf{c}_{\mathrm{A}} \tag{2-52}$$

it follows that

$$\dot{\boldsymbol{C}}_{\mathrm{A}} = -(\boldsymbol{U}\boldsymbol{C}_{\mathrm{A}} - \boldsymbol{C}_{\mathrm{A}}^{-1}\boldsymbol{U}^{-1}) \tag{2-53}$$

where $\boldsymbol{U}$ denotes a quaternion corresponding to $\frac{1}{2}\boldsymbol{\omega}_{\mathrm{A}}$:

$$\boldsymbol{U} \triangleq 0 + \tfrac{1}{2}\boldsymbol{\omega}_{\mathrm{A}}, \qquad \boldsymbol{U}^{-1} = -\boldsymbol{U} \tag{2-54}$$

From (2-51), (2-53), and (2-54),

$$\dot{\boldsymbol{B}} = \boldsymbol{B}\boldsymbol{U} \tag{2-55}$$

For a term-by-term expression of the quaternion element derivatives, (2-46) can be written in either of two partitioned (matrix × vector) forms,

$$\left[\begin{array}{c|c} \ell_1 & -\mathbf{b}_1^{\mathrm{T}} \\ \hline \mathbf{b}_1 & \ell_1\mathbf{I}_{3\times3} + (\mathbf{b}_1\times\ \ ) \end{array}\right]\left[\begin{array}{c} \ell_2 \\ \hline \mathbf{b}_2 \end{array}\right] = \left[\begin{array}{c|c} \ell_2 & -\mathbf{b}_2^{\mathrm{T}} \\ \hline \mathbf{b}_2 & \ell_2\mathbf{I}_{3\times3} - (\mathbf{b}_2\times\ \ ) \end{array}\right]\left[\begin{array}{c} \ell_1 \\ \hline \mathbf{b}_1 \end{array}\right] \tag{2-56}$$

and, as applied to (2-55), the right side of this expression produces (2-41).

By expressing (2-48) in the form

$$\mathbf{c}_{\mathrm{I}} = \{(2\ell^2-1)\mathbf{I} + 2\mathbf{b}\mathbf{b}^{\mathrm{T}} + 2\ell(\mathbf{b}\times\ \ )\}\mathbf{c}_{\mathrm{A}} \tag{2-57}$$

the matrix in (2-40) readily follows. Thus, both of the relations at the start of this section are established.

2.4 REFLECTIVE TRANSFORMATIONS

An orthogonal 3 × 3 matrix with a negative determinant (−1) could represent an inversion from a right- to a left-handed coordinate frame. Such an "improper" transformation cannot correspond to any rigid body rotation operator (Ref. 2-1, p. 122) but, when combined with another inversion, the result is a proper rotation, physically obtainable with pairs of reflectors. Consider a vector **D** incident upon a mirror surface with unit normal N. With $\boldsymbol{M}$ denoting any unit vector in the plane of the mirror, **D** can be resolved along N, $\boldsymbol{M}$, and $N \times \boldsymbol{M}$:

$$\mathbf{D} = (\mathbf{D} \cdot N)N + (\mathbf{D} \cdot \boldsymbol{M})\boldsymbol{M} + (\mathbf{D} \cdot N \times \boldsymbol{M})N \times \boldsymbol{M} \tag{2-58}$$

The mirror principle states that the reflection must be

$$\boldsymbol{D} = -(\mathbf{D} \cdot N)N + (\mathbf{D} \cdot \boldsymbol{M})\boldsymbol{M} + (\mathbf{D} \cdot N \times \boldsymbol{M})N \times \boldsymbol{M} \tag{2-59}$$

or

$$\boldsymbol{D} = \mathbf{D} - 2(\mathbf{D} \cdot N)N \tag{2-60}$$

From (II-9) and the fact that **D** is arbitrary, the reflective transformation is (Ex. 2-12)

$$\boldsymbol{\Gamma} = \mathbf{I} - 2NN^{\mathrm{T}} = \begin{bmatrix} 1 - 2N_1^2 & -2N_1N_2 & -2N_1N_3 \\ -2N_1N_2 & 1 - 2N_2^2 & -2N_2N_3 \\ -2N_1N_3 & -2N_2N_3 & 1 - 2N_3^2 \end{bmatrix} \tag{2-61}$$

This matrix is both orthogonal (with negative determinant) and symmetric; thus no directional subscripts are required to denote the direction of the transformation. However, in combination with another reflective operation, such as

$$\boldsymbol{\Gamma}_1 = \begin{bmatrix} -1 & 0 & 0 \\ 0 & 1 & 0 \\ 0 & 0 & 1 \end{bmatrix} \tag{2-62}$$

an orthogonal matrix with positive determinant (+1) results:

$$\mathscr{T} = \boldsymbol{\Gamma}\boldsymbol{\Gamma}_1 \tag{2-63}$$

These relations are clearly demonstrated by the configuration in Fig. 2-6. Suppose that a point source at the center of the movable plate sends a spherical wave to the fixed plate, with a polarization that causes it to be reflected. The reflected wave is also refocused as a plane wave, with every ray in the direction of $(-\mathbf{I}_A)$, regardless of the movable plate orientation. The plane wave is again reflected, with an overall polarization change needed

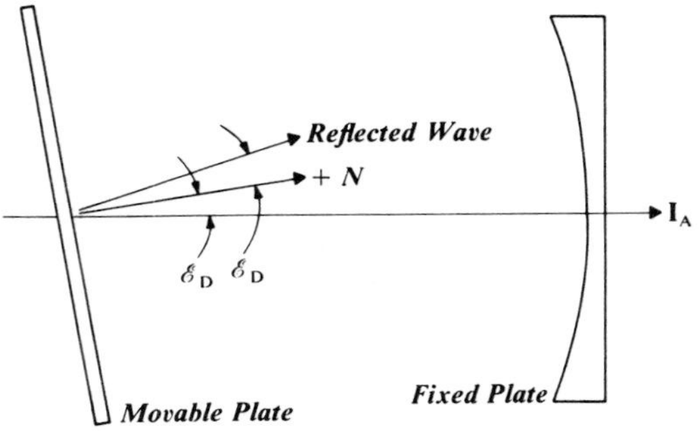

FIG. 2-6. Illustrative reflector geometry ($\mathbf{I}_A$, forward normal from fixed plate, along aircraft roll axis; $+N$, forward normal from movable plate).

for transmission through the fixed plate. The complete transformation obtained by movable plate rotation is easily constructed through five closely related coordinate frames. These are defined with the $+y$ axis along $\mathbf{J}_A$ (outward normal to the plane of Fig. 2-6), and $+x$ axes along the following directions:

reference vector emanating from the point source $(+\mathbf{I}_A)$,
reflection from the fixed plate $(-\mathbf{I}_A)$,
component of the backward reflected wave that is normal to the movable plate $(-N)$,
forward normal from movable plate $(+N)$,
travel of plane wave forward from the movable plate.

The transformation from each of these frames to the next is expressible through one of the basic matrix forms already introduced. Each step of the overall transformation, obtained by successive factors as in (2-13), can be recognized from the cascaded product

$$[2\mathscr{E}_D]_y = [\mathscr{E}_D]_y \mathbf{\Gamma}_1 [\mathscr{E}_D]_y^T \mathbf{\Gamma}_1 \tag{2-64}$$

which represents the transformation from aircraft to beam-referenced coordinates. The double angle is the combined result of equal angles of incidence and reflection for the movable plate. Alternatively, (2-63) can represent the inverse transformation by the rearrangement,

$$[-2\mathscr{E}_D]_y = [\mathscr{E}_D]_y^T \mathbf{\Gamma}_1 [\mathscr{E}_D]_y \mathbf{\Gamma}_1 \tag{2-65}$$

Addition of an azimuth gimbal to this assembly and rotation through an angle $\mathscr{A}_D$ produces

$$\mathbf{\Gamma} = [\mathscr{A}_D]_x^T [\mathscr{E}_D]_y^T \mathbf{\Gamma}_1 [\mathscr{E}_D]_y [\mathscr{A}_D]_x \tag{2-66}$$

so that, by combination with (2-62) and (2-63), a second reflection yields the transformation from beam-referenced axes to aircraft coordinates:

$$\mathscr{T} = \begin{bmatrix} 2\cos^2 \mathscr{A}_{\mathrm{D}} \cos^2 \mathscr{E}_{\mathrm{D}} - 1 & -\cos^2 \mathscr{E}_{\mathrm{D}} \sin 2\mathscr{A}_{\mathrm{D}} & \cos \mathscr{A}_{\mathrm{D}} \sin 2\mathscr{E}_{\mathrm{D}} \\ \cos^2 \mathscr{E}_{\mathrm{D}} \sin 2\mathscr{A}_{\mathrm{D}} & 1 - 2\sin^2 \mathscr{A}_{\mathrm{D}} \cos^2 \mathscr{E}_{\mathrm{D}} & \sin \mathscr{A}_{\mathrm{D}} \sin 2\mathscr{E}_{\mathrm{D}} \\ -\cos \mathscr{A}_{\mathrm{D}} \sin 2\mathscr{E}_{\mathrm{D}} & \sin \mathscr{A}_{\mathrm{D}} \sin 2\mathscr{E}_{\mathrm{D}} & \cos 2\mathscr{E}_{\mathrm{D}} \end{bmatrix} \tag{2-67}$$

The principal eigenvector of this orthogonal matrix has a vanishing x component, traceable to the conventions chosen. Obviously this example represents one of many possible conventions and physical configurations.

2.5 APPLICATION TO SUBSEQUENT CHAPTERS

Conventions already presented are adopted as standard throughout this book, and the kinematical equations developed here are used repeatedly for numerous dynamic analyses to follow. The four-parameter formulation is applied to a general locally level navigation reference frame in Chapter 3, and also mentioned in connection with a strapdown attitude parametrization [this latter usage is not fully presented, however, for reasons related to the discussion following (2-41)]. The reflective transformation analysis is applicable to some navigation and tracking modes (e.g., Section 6.6).

APPENDIX 2.A: ILLUSTRATIVE SOLUTIONS TO DIFFERENTIAL EQUATIONS OF ROTATION

Dynamics and kinematics of angular motion cannot generally be determined without resorting to computational algorithms. Certain restrictive conditions, however, allow closed-form solutions providing valuable insight into time-varying rotational behavior. Two classical cases will now be presented.

2.A.1 Rotation about a Fixed Axis

When a vehicle is subjected to a time-varying absolute angular rate $\boldsymbol{\omega}_{\mathrm{A}}(t)$ about a fixed axis, the matrix $\boldsymbol{\Omega}_{\mathrm{A}}$ of (2-28) commutes with its time integral. The solution to that equation, in analogy with (II-54), is then (Ex. 2-13)

$$\mathbf{T}_{\mathrm{I/A}}(t) = \mathbf{T}_{\mathrm{I/A}}(t_0) \exp\left\{ -\int_{t_0}^{t} \boldsymbol{\Omega}_{\mathrm{A}}(\mathrm{r})\, d\mathrm{r} \right\} \tag{2-68}$$

where the matrix exponential in this case takes the form

$$\exp\left\{-\int_{t_0}^{t} \mathbf{\Omega}_A(\mathfrak{r})\, d\mathfrak{r}\right\} = \mathbf{I} - \int_{t_0}^{t} \mathbf{\Omega}_A(\mathfrak{r})\, d\mathfrak{r} + \frac{1}{2!}\left[\int_{t_0}^{t} \mathbf{\Omega}_A(\mathfrak{r})\, d\mathfrak{r}\right]^2 - \frac{1}{3!}\left[\int_{t_0}^{t} \mathbf{\Omega}_A(\mathfrak{r})\, d\mathfrak{r}\right]^3 + \cdots \tag{2-69}$$

Since the axis of rotation is fixed, $\mathbf{\Omega}_A(t)$ must be expressible as the product of a fixed matrix $\mathbf{X}$ with a scalar function of time; for example, let the scalar function be a cosinusoid,

$$\mathbf{\Omega}_A(t) = \mathbf{X}\omega_0 \cos \omega_X(t - t_0), \qquad \mathbf{X} = (-\mathbf{E} \times \ ) \tag{2-70}$$

so that $\mathbf{X}$ is dimensionless and the root sum square of its positive off-diagonal elements is unity; then

$$\int_{t_0}^{t} \mathbf{\Omega}_A(\mathfrak{r})\, d\mathfrak{r} = \mathbf{X}\mathfrak{x}, \qquad \mathfrak{x} \triangleq (\omega_0/\omega_X) \sin \omega_X(t - t_0) \tag{2-71}$$

and this can be substituted into (2-68) and (2-69):

$$\mathbf{T}_{I/A}(t) = \mathbf{T}_{I/A}(t_0)\left(\mathbf{I} - \mathfrak{x}\mathbf{X} + \frac{\mathfrak{x}^2}{2!}\mathbf{X}^2 - \frac{\mathfrak{x}^3}{3!}\mathbf{X}^3 + \cdots\right) \tag{2-72}$$

By noting that $\mathbf{X}^3 = -\mathbf{X}$ and performing the resulting compression of the series,

$$\mathbf{T}_{I/A}(t) = \mathbf{T}_{I/A}(t_0)[\mathbf{I} - (\sin \mathfrak{x})\mathbf{X} + (1 - \cos \mathfrak{x})\mathbf{X}^2] \tag{2-73}$$

which, as readily verified through differentiation, conforms to (2-28). The solution remains valid in the limiting case of zero oscillation frequency, where $\mathfrak{x} = \omega_0(t - t_0)$.

To apply the four-parameter formulation here, note that

$$\boldsymbol{\omega}_A(t) = \mathbf{E}\omega_0 \cos \omega_X(t - t_0) = \mathbf{E}\dot{\mathfrak{x}} \tag{2-74}$$

from which it follows that

$$\boldsymbol{\Psi} = \frac{\dot{\mathfrak{x}}}{2}\left[\begin{array}{c|c} \mathbf{X} & \mathbf{E} \\ \hline -\mathbf{E}^T & 0 \end{array}\right] \tag{2-75}$$

As a product of a scalar with a constant matrix, this must obviously commute with its time integral. Thus (II-54) is applicable to the 4×1 vector $\boldsymbol{b}$ with initial components (0, 0, 0, 1) as dictated by an initial angle of $\mathfrak{x}_0 = 0$ from (2-71). After integrating (2-75) by inspection and noting that

$$\left[\int_{t_0}^{t} \boldsymbol{\Psi}\, d\mathfrak{r}\right]^2 = -(\mathfrak{x}/2)^2\mathbf{I} \tag{2-76}$$

substitution into (II-54) yields

$$\boldsymbol{b} = \left\{ \mathbf{I}\cos(\mathfrak{x}/2) + \left[\begin{array}{c:c} \mathbf{X} & \mathbf{E} \\ \hdashline -\mathbf{E}^{\mathrm{T}} & 0 \end{array}\right] \sin(\mathfrak{x}/2) \right\} \begin{bmatrix} 0 \\ 0 \\ 0 \\ 1 \end{bmatrix} \tag{2-77}$$

which immediately reduces to

$$\boldsymbol{b} = \left[\begin{array}{c} \mathbf{E}\sin(\mathfrak{x}/2) \\ \hdashline \cos(\mathfrak{x}/2) \end{array}\right] \tag{2-78}$$

in conformance with (2-41) and (2-42) [Ex. 2-14].

Rotation about an arbitrary but fixed axis is thus governed by the dynamics of the scalar $\mathfrak{x}$. A somewhat more involved rotational pattern will now be examined.

2.A.2 Classical Intrinsic Precession

According to Newtonian mechanics, the time derivative of linear momentum is equal to applied force in an inertial coordinate frame. Similarly, for rotation with respect to inertial axes, the time derivative of angular momentum is equal to applied torque. In general this leads to nonlinear dynamic equations, since angular momentum (e.g., for the aircraft) is defined as

$$\mathbf{H}_{\mathrm{A}} \triangleq \mathscr{I}_{\mathrm{A}} \boldsymbol{\omega}_{\mathrm{A}} \tag{2-79}$$

where $\mathscr{I}_{\mathrm{A}}$ represents the aircraft inertia matrix, which is obviously defined most naturally in *body* axes. In fact, it is common practice to choose principal axes of inertia for the triad $(\mathbf{I}_{\mathrm{A}}, \mathbf{J}_{\mathrm{A}}, \mathbf{K}_{\mathrm{A}})$ so that $\mathscr{I}_{\mathrm{A}}$ is diagonal as well as time-invariant:

$$\mathbf{H}_{\mathrm{A}} = \begin{bmatrix} \mathscr{I}_{\mathrm{A1}} \omega_{\mathrm{A1}} \\ \mathscr{I}_{\mathrm{A2}} \omega_{\mathrm{A2}} \\ \mathscr{I}_{\mathrm{A3}} \omega_{\mathrm{A3}} \end{bmatrix} \tag{2-80}$$

But, since the torque vector (denoted $\mathbf{L}_{\mathrm{I}}$ as expressed in inertial coordinates) must conform to

$$\int \mathbf{L}_{\mathrm{I}}\, dt = \mathbf{T}_{\mathrm{I/A}} \mathbf{H}_{\mathrm{A}} \tag{2-81}$$

then, by product differentiation using (2-28),

$$\mathbf{L}_{\mathrm{I}} = \mathbf{T}_{\mathrm{I/A}}(\dot{\mathbf{H}}_{\mathrm{A}} - \boldsymbol{\Omega}_{\mathrm{A}} \mathbf{H}_{\mathrm{A}}) \tag{2-82}$$

Through premultiplication of both sides by $\mathbf{T}_{A/I}$ the torque vector as defined in principal axes is

$$\mathbf{L} = \begin{bmatrix} \mathcal{I}_{A1}\dot{\omega}_{A1} \\ \mathcal{I}_{A2}\dot{\omega}_{A2} \\ \mathcal{I}_{A3}\dot{\omega}_{A3} \end{bmatrix} + \begin{bmatrix} (\mathcal{I}_{A3} - \mathcal{I}_{A2})\omega_{A2}\omega_{A3} \\ (\mathcal{I}_{A1} - \mathcal{I}_{A3})\omega_{A3}\omega_{A1} \\ (\mathcal{I}_{A2} - \mathcal{I}_{A1})\omega_{A1}\omega_{A2} \end{bmatrix} \tag{2-83}$$

This simple derivation offers an explanation for what is sometimes regarded as a curious phenomenon. The last term on the right is perpendicular to $\boldsymbol{\omega}_A$. Thus the dynamics of a rotating body will include (Ex. 2-15) a "gyroscopic" reaction normal to the direction of its absolute angular rate. The source of this reaction is inherent in the physical law, (2-81).

Although aircraft notation was used in the above derivation, the subscript A could be dropped and the principles applied to other rigid bodies (including gyro rotors, as described in Section 4.2). Of more immediate interest, however, the phenomenon associated with the last term of (2-83) will now be scrutinized further, for an arbitrary torque-free rigid body with dynamic symmetry about its x axis ($\mathfrak{I}_2 = \mathfrak{I}_3 = I$), in which case the equations of motion reduce to

$$\dot{\omega}_1 = 0, \qquad \dot{\omega}_2 = -\left(\frac{\mathfrak{I}_1 - I}{I}\right)\omega_1\omega_3, \qquad \dot{\omega}_3 = \left(\frac{\mathfrak{I}_1 - I}{I}\right)\omega_1\omega_2 \tag{2-84}$$

The solution to these equations can be expressed in terms of a spin rate ω_α, a retrograde* precession rate ω_β, the angle γ between the spin axis $\mathbf{i}$ and the angular momentum vector $\mathbf{H}$, and a spin phase angle $\omega_\alpha t_\alpha$. These constants are defined by the initial angular rate vector $\boldsymbol{\omega}_0$ as follows:

$$\omega_\alpha = -\xi\omega_{01} \tag{2-85}$$

$$\omega_\beta = \mathfrak{I}_1\omega_\alpha \sec\gamma/(\mathfrak{I}_1 - I) = -(\mathfrak{I}_1/I)\omega_{01} \sec\gamma \tag{2-86}$$

$$\gamma = \arctan\{I\omega_{yz}/(\mathfrak{I}_1\omega_{01})\}, \qquad 0 \le \gamma \le \pi \tag{2-87}$$

$$\omega_\alpha t_\alpha = \arctan(\omega_{02}/\omega_{03}) \tag{2-88}$$

where

$$\xi \triangleq (\mathfrak{I}_1 - I)/I \tag{2-89}$$

$$\omega_{yz} \triangleq +(\omega_{02}^2 + \omega_{03}^2)^{1/2} \tag{2-90}$$

and the inverse tangent of (2-88) is defined to lie in the quadrant dictated by signs of the numerator and the denominator. In terms of these constants,

* The case $\mathfrak{I}_1 > I$, corresponding to spin stabilization about the major principal axis, is typical of many space applications.

the solution to (2-84) may be written

$$\omega_1 = \omega_{01} = -\omega_\alpha/\xi \tag{2-91}$$

$$\omega_2 = -\alpha_s \omega_\alpha(\xi + 1)(\tan \gamma)/\xi \tag{2-92}$$

$$\omega_3 = -\alpha_c \omega_\alpha(\xi + 1)(\tan \gamma)/\xi \tag{2-93}$$

in which α_s and α_c are defined as the sine and cosine, respectively, of the composite angle $\omega_\alpha(t + t_\alpha)$. Also note that

$$\omega_2 = \omega_{02} \cos(\xi\omega_1 t) - \omega_{03} \sin(\xi\omega_1 t) \tag{2-94}$$

$$\omega_3 = \omega_{02} \sin(\xi\omega_1 t) + \omega_{03} \cos(\xi\omega_1 t) \tag{2-95}$$

This formulation corresponds to the xyx Euler angle sequence

$$\mathbf{T} = [\omega_\beta t]_x[-\gamma]_y[-\omega_\alpha(t + t_\alpha)]_x \tag{2-96}$$

and it can easily be verified that the derivative of this matrix is equal to the product

$$\dot{\mathbf{T}} = \begin{bmatrix} \gamma_c & \gamma_s \alpha_s & \gamma_s \alpha_c \\ -\gamma_s \beta_s & \alpha_c \beta_c + \gamma_c \alpha_s \beta_s & -\alpha_s \beta_c + \gamma_c \alpha_c \beta_s \\ -\gamma_s \beta_c & -\alpha_c \beta_s + \gamma_c \alpha_s \beta_c & \alpha_s \beta_s + \gamma_c \alpha_c \beta_c \end{bmatrix} \begin{bmatrix} 0 & \gamma_s \alpha_c \omega_\beta & -\gamma_s \alpha_s \omega_\beta \\ -\gamma_s \alpha_c \omega_\beta & 0 & -\omega_\alpha + \gamma_c \omega_\beta \\ \gamma_s \alpha_s \omega_\beta & \omega_\alpha - \gamma_c \omega_\beta & 0 \end{bmatrix} \tag{2-97}$$

where the subscripts s and c again denote the sine and cosine, respectively, of the angles contained in (2-96). The correspondence between the off-diagonal terms of the above skew-symmetric matrix and the angular rates of (2-91)–(2-93) is easily established from the defining relationships given earlier. It follows that the premultiplying factor on the right of (2-97) is a closed-form solution for vehicle attitude, complete to within a premultiplicative constant matrix. Obviously, to satisfy the initial conditions, the complete transformation from vehicle to inertial coordinates must be

$$\mathbf{T}_{I/V} = \mathbf{T}_R[\omega_\alpha t_\alpha]_x[\gamma]_y[\omega_\beta t]_x[-\gamma]_y[-\omega_\alpha(t + t_\alpha)]_x \tag{2-98}$$

where $\mathbf{T}_R$ is the value of $\mathbf{T}_{I/V}$ at the reference time.

The following comments will facilitate a rigorous interpretation of the preceding analysis:

(1) The angular momentum vector in vehicle coordinates,

$$\mathbf{H} = \mathbf{i}\mathfrak{I}_1\omega_1 + I(\mathbf{j}\omega_2 + \mathbf{k}\omega_3) \tag{2-99}$$

has a magnitude $\mathfrak{I}_1\omega_{01} \sec \gamma$. The conventions adopted here ensure that, since ω_{01} and $\sec \gamma$ always have the same sign, this expression cannot be negative. It is easily verified that $\mathbf{TH}$ has only an x component.

(2) The phase angle $\omega_\alpha t_\alpha$ is equal to the angle between **H** and the initial **k** axis. By including this in the transformation, it is ensured that the intermediate y axis is indeed perpendicular to the plane of **H** and the initial vehicle x axis. This paves the way for the middle transformation in (2-96).

(3) By definition, the algebraic signs of ω_{01} and $\cos \gamma$ must agree. From (2-85) it follows that when ξ is positive, ω_α is positive for obtuse γ and negative for acute γ. From (2-86) it can be seen that ω_β always turns out negative with the conventions adopted here. Since the spin and precession rates are of opposite sense in (2-96), it follows that positive values of ξ produce retrograde precession.

Additional development of this topic by elementary methods appears in Ref. 2-1, pp. 143–156, 159–162. The phenomenon described here is often referred to as "intrinsic" precession, to distinguish it from precession of the angular momentum vector itself by applied torques in the more general case. The foregoing analytical results are quite well known, but are not presented elsewhere with all these details collected together.

APPENDIX 2.B: ORTHOGONALIZATION

When a four-parameter set is integrated by a computational algorithm that does not guarantee conformance to (2-43), a reasonable orthogonalizing procedure is division by the root sum square (Ex. 2-16). For the analogous problem with a 3 × 3 real positive definite matrix $\mathbf{C}$, the following derivation was obtained from the author of Ref. 2-2: Define $\mathbf{M}$ as the orthogonal matrix that minimizes the sum of squares of difference matrix elements $(\mathbf{M} - \mathbf{C})$. In Ex. 2-16 this is expressed as a maximization for the trace,

$$\mathrm{trace}(\mathbf{M}^{\mathrm{T}}\mathbf{C}) = \mathrm{trace}(\mathbf{N}\mathbf{M}^{\mathrm{T}}\mathbf{L}\mathbf{N}^{\mathrm{T}}\mathbf{D}) \tag{2-100}$$

where $\mathbf{L}$ and $\mathbf{N}$ are orthogonal and $\mathbf{D}$ is diagonal. Since the matrix product that premultiplies $\mathbf{D}$ is orthogonal, the trace is maximized when the coefficients of the diagonal matrix elements are all equal to unity. Thus $\mathbf{M} = \mathbf{L}$ and, from the association

$$\mathbf{C} \equiv \mathbf{C}(\mathbf{C}^{\mathrm{T}}\mathbf{C})^{-1/2}(\mathbf{C}^{\mathrm{T}}\mathbf{C})^{1/2} \tag{2-101}$$

it follows that the unique $\mathbf{L}$-matrix must be

$$\mathbf{M} = \mathbf{C}(\mathbf{C}^{\mathrm{T}}\mathbf{C})^{-1/2} \tag{2-102}$$

EXERCISES

2-1 (a) Let $\boldsymbol{R}_S$ denote the position vector [with respect to an origin at (0, 0, 0)] of a specified point located on an airborne IMU, expressed in a reference $(\mathbf{I}_0, \mathbf{J}_0, \mathbf{K}_0)$ coordinate frame. Suppose that a radar antenna is separated from the IMU by a displacement vector $\mathbf{d}_{N/R}$, expressed in the aircraft frame $(\mathbf{I}_A, \mathbf{J}_A, \mathbf{K}_A)$. With $\mathbf{T}_{0/A}$ representing a transformation operator, whereby a vector defined in aircraft coordinates is resolved along the reference axes, find the position vector $\boldsymbol{R}_R$ of the radar antenna, as expressed in the reference frame.

(b) Let $\mathbf{F}$ represent the total nongravitational acceleration experienced by an IMU, expressed in aircraft coordinates, while $\boldsymbol{\psi}_F$ characterizes imperfect knowledge of true vehicle orientation, as resolved along *geographic* reference axes. If a meaningful interactive force can be formed by the cross product of $\mathbf{F}$ with an attitude uncertainty vector, and $\mathbf{T}_{A/G}$ denotes the transformation from geographic to aircraft coordinates, find the components of this interactive force in the $(\mathbf{I}_A, \mathbf{J}_A, \mathbf{K}_A)$ frame.

2-2 Any vector in physical space can be resolved along any chosen triad of orthogonal axes. In particular, $\mathbf{I}_G = (\mathbf{I}_G \cdot \mathbf{I}_I)\mathbf{I}_I + (\mathbf{I}_G \cdot \mathbf{J}_I)\mathbf{J}_I + (\mathbf{I}_G \cdot \mathbf{K}_I)\mathbf{K}_I$. Repeat this coordinate resolution for $\mathbf{J}_G$ and $\mathbf{K}_G$. Collect the equations into a form reminiscent of (1-22) and (1-28). Note that the matrix, which can be denoted $\mathbf{T}_{G/I}$, serves as a transformation for vector components expressed in different frames [as in (2-3) and (2-4)] and for the axes themselves.

2-3 Derive (2-4) as suggested in the text, including a matrix $\mathbf{T}_{I/G}$ analogous to $\mathbf{T}_{G/I}$ in (2-3). Show that these are orthogonal [Appendix II, (II-23)]. Verify and interpret the relation

$$\mathbf{T}_{G/I} \cdot \mathbf{T}_{I/G} = \mathbf{I}_{3\times 3}$$

and use this with (2-5) to obtain (2-7) from (2-6).

2-4 (a) Expand (1-31)–(1-33) individually into scalars and provide a geometric interpretation. Write each in a form analogous to (2-11), and correlate algebraic signs for dot products involving acute and obtuse angles with the off-diagonal elements of (2-8). Include consideration of the angle directions in Figs. 1-6–1-8, versus the text of Section 1.2.3 (why the difference?).

(b) Write (2-8) for roll, pitch, and heading angles of $-45°$, $+5°$, and $210°$, respectively.

(c) Obtain (2-9) from (1-31)–(1-33), and verify all relations implied by (2-10).

2-5 (a) Show that

$$\mathbf{T}_{G/P} = \mathbf{I}_{3\times 3} + \mathscr{Y}$$

where $\mathscr{Y}$ is *skew-symmetric*, i.e.,

$$\mathscr{Y}_{ji} = -\mathscr{Y}_{ij}$$

Can any general statement be made in regard to the diagonal elements of a skew-symmetric matrix? Show that any vector product operator is equivalent to a premultiplicative skew-symmetric matrix.

(b) For acute angles in roll, pitch, and heading, why do the off-diagonal elements in (2-8) have signs opposite to those in corresponding positions of (2-15)?

(c) Are the small angle approximations implied by (2-15) reasonable for Ex. 1-10?

(d) Express the peripheral velocity from Ex. 1-4 in terms of a vector product involving

$$\boldsymbol{\omega}_s \triangleq (\mathbf{I}_G \cos\lambda - \mathbf{K}_G \sin\lambda)\omega_s$$

2-6 (a) Verify (2-16) and show that the matrix is orthogonal.

(b) For the case $\psi = 0$, use selected elements from (2-16) to find a unit vector in the direction of

$$(\mathbf{I}_A \cdot \mathbf{I}_G)\mathbf{I}_G + (\mathbf{I}_A \cdot \mathbf{J}_G)\mathbf{J}_G$$

Does the horizontal projection of $\mathbf{I}_A$ make an angle of Ψ with $\mathbf{I}_G$, as in Fig. 1-9?

(c) Draw an analogy between (2-19) and Ex. 2-1a. Note that at $\Phi = \Theta = 0$, $\mathbf{T}_{A/P}$ reduces to a heading transformation, while $\mathbf{J}_{A1}$ and $\mathbf{K}_{A2}$ are parallel to $\mathbf{J}_A$ and $\mathbf{K}_A$, respectively. With aircraft rotation rates suddenly applied under those conditions, would the initial $\boldsymbol{\omega}_A$ vector instantaneously be equal to $\mathbf{I}_A\dot{\Phi} + \mathbf{J}_A\dot{\Theta} + \mathbf{K}_A\dot{\Psi}$?

(d) Show correspondence between (2-20) and the relation

$$\boldsymbol{\zeta} = \dot{\Phi}\mathbf{I}_A + \dot{\Theta}\mathbf{J}_{A1} + \dot{\Psi}\mathbf{K}_{A2}$$

Now reconsider the last question in part (c).

(e) Show that (2-21) and (2-22) are equivalent.

2-7 (a) Recalling the skew-symmetric matrix form in Ex. 2-5a, show that $\boldsymbol{\Omega}_A$ in (2-25) is equivalent to the operator $(-\boldsymbol{\omega}_A \times\ )$. Apply this reasoning to (2-29) also.

(b) Combine the skew-symmetric character of $\boldsymbol{\Omega}_A$ with (II-14) and (II-23), using the principle that the differentiation and transpose operations commute, to obtain (2-28) from (2-27).

(c) Let $\mathbf{c}_I$ and $\mathbf{c}_A$ represent a known star sightline vector expressed in inertial coordinates and the corresponding vector expressed in aircraft coordinates, respectively. By differentiating the relation

$$\mathbf{c}_I = \mathbf{T}_{I/A}\mathbf{c}_A$$

while using (2-28), show that

$$\dot{\mathbf{c}}_A = -\boldsymbol{\omega}_A \times \mathbf{c}_A$$

How does this relation apply to the columns of $\mathbf{T}_{A/I}$? These columns represent unit vectors in which directions? As expressed in what frame? Write (2-27) as a 9×1 vector differential equation in partition form.

(d) Apply the procedure in part (c) to the vectors $\mathbf{d}_I$ and $\mathbf{d}_A$ in (2-13). Show that

$$d\mathbf{d}_A/dt = +\boldsymbol{\omega}_A \times \mathbf{d}_A$$

Establish a correspondence between this example and the columns of a particular matrix in Section 2.2.2.

(e) Show that this vector $\mathbf{d}_A$ conforms to

$$d^2\mathbf{d}_A/dt^2 = \dot{\boldsymbol{\omega}}_A \times \mathbf{d}_A + \boldsymbol{\omega}_A \times (\boldsymbol{\omega}_A \times \mathbf{d}_A)$$

(f) By matrix product differentiation with respect to time, obtain the relation

$$\dot{\mathbf{U}}_G = \mathbf{T}_{G/I}\dot{\mathbf{U}}_I + \dot{\mathbf{T}}_{G/I}\mathbf{U}_I$$

and verify its agreement with (2-33). Discuss the role of (2-6) in this exercise.

(g) Time differentiation, as applied to the *right* of (1-17), would be reminiscent of which operator? Which operator differentiates $\mathbf{U}_G$ as a 3×1 matrix?

2-8 (a) Write $\mathbf{Z}$ of (II-24) for the three dimensional case as

$$\mathbf{Z} = [\mathbf{T}_1 - \mathbf{1}_1 \,\vdots\, \mathbf{T}_2 - \mathbf{1}_2 \,\vdots\, \mathbf{T}_3 - \mathbf{1}_3]$$

where $\mathbf{1}_j$ denotes the jth column of $\mathbf{I}$. Using orthogonality conditions (e.g., $\mathbf{T}_3 \times \mathbf{T}_1 = \mathbf{T}_2$), write $|\mathbf{Z}|$ as a scalar triple product to show that it vanishes. What does this prove?

(b) Is $\mathbf{I}_{3 \times 3}$ orthogonal? Could it represent a rotational transformation? Find all its eigenvalues. Make a statement regarding eigenvector direction.

(c) Show that (II-27) is a special case of (II-28).

(d) Apply (II-24) and (II-27) to the matrices in (2-8) for nonzero angles and interpret results. Discuss any restrictions for selection of columns in the transposed $\mathbf{Z}$-matrix.

(e) Repeat (d) for a 45° roll followed by a 0.01° pitch. Are there any column pair selections for which eigenvector computation is impossible? Are there any column pair selections for which eigenvector computation is difficult?

(f) Discuss uniqueness of $\mathbf{E}$ in (d) and (e).

2-9 (a) Suppose that an aircraft, subjected to irregular angular oscillations, is equipped with instruments to monitor $\boldsymbol{\omega}_A$ completely, exactly, and continuously. Assume that attitude information is *used* only for low-speed operations but, to prevent loss of the orientation reference, integration of kinematical equations must be performed at high speed. If high-speed computations are very costly, compare (2-27) and (2-41).

(b) What further computations are required to perform coordinate transformations of vectors once the direction cosines are obtained by integration of (2-27)? What further computations are required for this purpose, after integration of (2-41)? Reconsider part (a) for applications requiring frequent coordinate transformations.

(c) Consider a set of linear differential equations with a skew-symmetric matrix of coefficients that may be time-varying. Can Euler angle kinematics be described in this form? Direction cosine? Any other?

(d) For a given set of initial conditions and angular rate time histories, is the solution of (2-41) unique?

(e) Substitute (2-42) into (2-40) for the two special cases

$$\mathbf{E} = \begin{bmatrix} 0 \\ 1 \\ 0 \end{bmatrix}, \quad \theta < \pi/2, \qquad \mathbf{E} = \begin{bmatrix} 0 \\ 0 \\ 1 \end{bmatrix}, \quad \pi < \theta < 3\pi/2$$

and discuss algebraic signs of all matrix elements.

(f) Use (2-43) to prove orthogonality of the matrix on the right of (2-40).

2-10 (a) What does (2-46) indicate in regard to commutativity? Apply this relation to the quaternions in the two equations preceding it. Does commutativity hold in this special case?

(b) Let $\mathbf{c}_A$ have projections of 0.3, 0.4, and 0.866 length units along $\mathbf{I}_A$, $\mathbf{J}_A$, and $\mathbf{K}_A$, respectively. Use (2-48) to find components of this vector after rotation through 30° about an axis inclined at 45° in the x–y plane. Compare this with the result of transformation via $\mathbf{T}_{I/A}$ as computed from (2-40).

(c) Does part (b) require knowledge of quaternion operator theory? Does (2-48) contain any restrictions regarding vectors to be transformed, rotational axis directions, or angles of rotation?

2-11 (a) Use (2-44)–(2-46) to verify (2-50).

(b) Use (2-46) to derive (2-53) from (2-52).

(c) Why does (2-55) follow from (2-51) and (2-53)?

(d) Use the right of (2-56) to derive (2-41).

(e) Use trigonometric identities and (II-9) to obtain (2-57) from (2-48).

(f) Combine (2-55) with the left of (2-56) to obtain a coefficient matrix of quaternion elements, premultiplying a 4×1 matrix with components $(0, \frac{1}{2}\omega_{A1}, \frac{1}{2}\omega_{A2}, \frac{1}{2}\omega_{A3})$.

2-12 (a) Use $(N_1{}^2 + N_2{}^2 + N_3{}^2 = 1)$ to show that $\mathbf{\Gamma}$ in (2-61) is orthogonal.

(b) Interpret (2-62) as a special case of (2-61), and (2-64) as a modification of (2-67).

(c) What rearrangement of matrix factors in (2-64), other than that in (2-65), can invert the result?

(d) Expand (2-66) to verify (2-67).

(e) Show that the x axis component of the principal eigenvector for (2-67) is zero.

2-13 (a) Transpose (2-68) and find a dynamic expression applicable to each column of $\mathbf{T}_{A/I}$. Compare with (II-54).

(b) Sum the series in (2-72) to obtain (2-73).

(c) Differentiate (2-73) and show conformance to (2-28).

2-14 (a) Obtain (2-77) by summing an exponential series, using (2-76).

(b) Differentiate (2-77) and use (2-75) to verify conformance to (2-41). Demonstrate validity of this solution under more general four-parameter initial conditions.

2-15 (a) Use a dot product to show that the gyroscopic term in (2-83) is perpendicular to $\boldsymbol{\omega}_A$.

(b) Show that the angular rates in (2-91)–(2-93) conform to (2-84).

(c) Verify (2-97).

(d) Does the premultiplication in (2-98) destroy conformance to (2-28)? Could a postmultiplicative adjustment have been used instead?

(e) Verify the items enumerated at the end of Appendix 2.A. If alternate conventions had been used (e.g., restriction of γ to acute angles), would the matrix solution be affected?

2-16 (a) Let $\boldsymbol{a}$ be a 4×1 vector with $|\boldsymbol{a}| \neq 1$. Show that, for any specified value of $|\boldsymbol{b}|$, the choice of $\boldsymbol{b}$ that maximizes $\boldsymbol{b}^{\mathrm{T}}\boldsymbol{a}$ will minimize $|\boldsymbol{b} - \boldsymbol{a}|$. Given $\boldsymbol{a}$, what choice of $\boldsymbol{b}$ satisfies this condition and also (2-43)?

(b) Express the trace of $(\mathbf{M} - \mathbf{C})^{\mathrm{T}}(\mathbf{M} - \mathbf{C})$ as a sum of squares of difference matrix elements. When $\mathbf{M}$ is orthogonal how is this sum affected by maximizing the trace of $\mathbf{M}^{\mathrm{T}}\mathbf{C}$?

(c) There is a unique pair of positive definite matrices $\mathbf{L}$ (orthogonal) and $\mathbf{Y}$ (symmetric) for which $\mathbf{C} = \mathbf{LY}$; also $\mathbf{Y}$ can be factored into $\mathbf{N}^{\mathrm{T}}\mathbf{DN}$, where $\mathbf{N}$ is orthogonal and $\mathbf{D}$ is diagonal. Use this and a trace invariance property to obtain (2-100).

(d) Demonstrate orthogonality of (2-102). Show that the post-multiplying factor is equal to

$$(\mathbf{C}^{\mathrm{T}}\mathbf{C})^{-1/2} = \mathbf{N}^{\mathrm{T}}\mathbf{D}^{-1}\mathbf{N}$$

which allows all inversion and square root operations to be performed on elements of the diagonal matrix $\mathbf{D}^2$.

REFERENCES

2-1. Goldstein, H., *Classical Mechanics.* Reading, Massachusetts: Addison-Wesley, 1950.
2-2. Stuelpnagel, J. C., "On the Parametrization of the Three-Dimensional Rotation Group," *SIAM Rev.* **6** (4), Oct., 1964, pp. 422–430.
2-3. Robinson, A. C., "On the Use of Quaternions in Simulation of Rigid Body Motion," WADC Tech. Rep. 58-17, Dayton, Ohio, Dec., 1958.

Chapter 3

Principles of Inertial Navigation

3.0

For purposes of systems analysis, inertial navigation theory is explainable in terms of selected fundamentals. Following is a collection of the basic mathematics, kinematics, mechanization concepts, navigation equations, and error analysis required for a functional description. This presentation involves little physics and astronomy, and no relativistic (non-Newtonian) mechanics or philosophical overtones. Many other important subjects (e.g., passive stabilization by large gyroscopic torques, redundant inertial sensor arrays, test procedures, fabrication techniques, guidance and control) are outside the scope of this book. For those topics as well as further analytical developments the reader is referred to Refs. 3-1–3-4.

An *inertial measuring unit* (IMU) is defined here as an assembly of instruments capable of providing full three-dimensional monitoring for *absolute* rotation and nongravitational acceleration [i.e., angular velocity with respect to a nonrotating ($\mathbf{I}_I$, $\mathbf{J}_I$, $\mathbf{K}_I$) frame, and translational acceleration relative to a point that is stationary or moving with constant velocity in space]. For terrestrial applications gravitational effects must be added, and time integrals of acceleration (velocity and position) are expressed in a convenient reference such as the geographic convention. The *inertial navigation system* (INS), in any case, includes the IMU plus all provisions for stabilizing and processing the sensor outputs to derive position and velocity in the chosen reference. Inertial instruments may be mounted on a servo-driven gimballed platform, or along axes attached to a vehicle (so that their outputs are stabilized computationally instead of mechanically; this is referred to as the *gimballess* or *strapdown* approach). These two alternatives, applicable to various instrument types and choices of reference frames, are presented from a viewpoint fitting within a general integrated navigation system.

3.1 BASIC TRANSLATIONAL DYNAMICS

The output of an accelerometer will first be given with ideal conditions assumed. Consider an assembly of negligible dimensions in which one end of a spring is connected to a proof mass, while the other is attached to a case. Under error-free conditions (conformance to Hooke's law, zero null, etc.), observed strain of the spring is proportional to the force imparted to it by the mass. When the spring is unconstrained along one axis only, the assembly acts as a *single-degree-of-freedom* (SDF) accelerometer, and the unconstrained direction is called its *sensitive axis* or *input axis* (IA). Imagine the accelerometer in vacuum free fall along a line parallel to this IA, under the gravitational influence of a large homogeneous sphere that is completely stationary and isolated (i.e., the only appreciable mass in the universe). With the case and its contents falling together, the accelerometer output would then be zero. On the other hand, with the case held stationary, the accelerometer acts as a spring scale weighing the proof mass; an output of magnitude G (denoting gravitational acceleration) would then be obtained from a scale calibrated in acceleration units.

More generally, if the altitude h of the accelerometer above the sphere were allowed to have an arbitrary time history, the instantaneous output would be

$$A = d^2(-h)/dt^2 - G \tag{3-1}$$

with a positive convention for downward acceleration. Let it now be assumed that two more of these accelerometers are placed in the same location with the first and the sensitive axes are pointed along ($\mathbf{I}_\text{I}$, $\mathbf{J}_\text{I}$, $\mathbf{K}_\text{I}$). With vectors expressed in this frame denoted by the subscript I, the output of the triad at position denoted by $\mathbf{R}_\text{I}$ is

$$\mathbf{A}_\text{I} = d^2\mathbf{R}_\text{I}/dt^2 - \mathbf{G}_\text{I} \tag{3-2}$$

Before this reasoning is applied to navigation over the earth (which is *not* an isolated stationary homogeneous sphere), note the following:

(1) Accelerometers measure *specific force*, defined as total nongravitational acceleration to which they are subjected, and are insensitive to all* gravitational accelerations.

(2) Principles stated here are unaffected by the types of accelerometers considered [e.g., substitution of other devices for the spring–mass assembly and usage of *two-degree-of-freedom* (2DF) instruments]; the term "accelerometer triad" hereafter will denote any configuration capable of measuring specific force along three mutually perpendicular axes.

(3) The accelerations in the above example are *absolute*, i.e., defined with respect to the sphere. However, if a product (constant velocity vector × time) were added to $\mathbf{R}_I$, (3-2) would remain valid. Instead of a stationary sphere, a linearly translating sphere with constant speed could have been assumed; if the constants of integration applicable to (3-2) were modified accordingly, a fixed velocity vector could be superimposed upon the whole system. On the basis of observed accelerations, which would still be absolute, this situation would be indistinguishable from the case of fixed sphere position. Obviously an arbitrary constant displacement from the chosen origin could also be introduced. It follows that position and velocity are *not* absolute; they are defined *relative* to a chosen reference (Exs. 3-1, 3-2).

As explained in the next chapter, it is permissible to use the earth center as a position reference for terrestrial inertial navigation. Immediately this suggests a possible approach for implementation: Mount an accelerometer triad on a platform having provisions for maintaining orientation along ($\mathbf{I}_I$, $\mathbf{J}_I$, $\mathbf{K}_I$), or along any other frame with known fixed directions. Equip the system with provisions for (1) computing gravitation in the platform frame, (2) integrating this gravitation vector, and (3) adding this integrated result to the time integral of specific force, to obtain the velocity $\mathbf{U}_I$ introduced in Section 2.0:

$$d\mathbf{R}_I/dt = \mathbf{U}_I = \int \mathbf{A}_I \, dt + \int \mathbf{G}_I \, dt \qquad (3\text{-}3)$$

These integrations are separated here because the first is often mechanized ahead of the instrument output; an *integrating accelerometer* provides velocity increments for the first term on the right of (3-3). In any event, for terrestrial navigation it is convenient to define a *hover velocity*

* For example, if the instruments were in a trajectory completely governed by superposition of two separate sources of gravitation, the outputs would still be zero. Alternate interpretations exist, but "equivalence principles" and related theories are not needed in this book.

$\boldsymbol{\omega}_E \times \mathbf{R}_I$ due to sidereal rate, where (Ex. 3-3)

$$\boldsymbol{\omega}_E = \begin{bmatrix} 0 \\ 0 \\ \omega_s \end{bmatrix} \tag{3-4}$$

and to express the geographic velocity of (1-17) as

$$\mathbf{V} = \mathbf{T}_{G/I}(\mathbf{U}_I - \boldsymbol{\omega}_E \times \mathbf{R}_I) \tag{3-5}$$

in preparation for geographic position determination.

The mechanization described above characterizes a gimballed platform with a nonrotating orientation reference. As one alternative to that approach, the accelerometer triad could be rigidly attached to the $(\mathbf{I}_A, \mathbf{J}_A, \mathbf{K}_A)$ frame and the computations for (3-3) could be mechanized in strapdown inertial form (Ex. 3-4),

$$\mathbf{U}_I = \int (\mathbf{T}_{I/A}\,\mathbf{A}_A)\,dt + \int \mathbf{G}_I\,dt \tag{3-6}$$

In principle this expression is straightforward, but the presence of high angular rates affects both performance and mechanization requirements. Accurate integration of (2-28) as well as (3-6) calls for fast computation. Also, various inertial instrument degradations, virtually absent from the relatively benign dynamic environment of a gimballed platform, are aggravated by rotations experienced. Furthermore, the maximum anticipated angular rate exerts direct influence on required operating ranges of the instruments. Before full development of these considerations, the means of maintaining an orientation reference must be explained. The next section describes both mechanical (gimballed platform) and computational (strapdown) stabilization with gyros, including modifications for INS mechanization in the geographic reference frame.

3.2 IMU STABILIZATION

Velocity increments from integrating accelerometers, *when summed in a reference frame*, provide a meaningful association of position and velocity with time integrals of acceleration. Following is a description of alternate configurations in which the *gyroscope* ("gyro") is the instrument used to maintain the computational or physical orientation reference. The nonrotating $(\mathbf{I}_I, \mathbf{J}_I, \mathbf{K}_I)$ reference is considered first, followed by extension to a geographic reference frame.

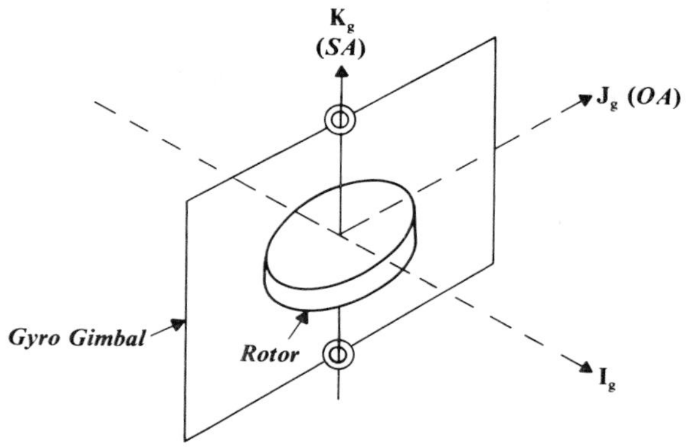

FIG. 3-1. SDF gyro in undisturbed state.

Figure 3-1 illustrates a spinning wheel mounted in an SDF *gyro gimbal**; although shown unenclosed, this assembly can be visualized within a case. An SDF gyro gimbal can rotate relative to the case about one axis only; this gyro gimbal axis, normal to the rotor *spin axis* (SA) in Fig. 3-1, is called the *output axis* (OA). Physically an angular rate perpendicular to both OA and SA will cause a gyroscopic moment to be developed about the OA. Suppose that the rotor in Fig. 3-1, typically spinning at several thousand rad/sec but otherwise undisturbed, is suddenly subjected to an angular rate $\dot{\phi}$ about the input axis (IA). The resulting gyroscopic moment is

$$L_{(\text{developed})} = H\dot{\phi} \tag{3-7}$$

where H is the *angular momentum* of the undisturbed rotor.† Since the SA in Fig. 3-1 is attached to the gyro gimbal, the developed torque is mechanically transmitted to this gimbal from the rotor.

Gyro gimbal response to this gyroscopic reaction is influenced primarily by the type of restraining action for OA rotation. Of immediate concern is the behavior of the relative gimbal-to-case OA displacement angle θ, for a restraining torque proportional to $\dot{\theta}$, with rotation of the case itself and all other undesirable effects again tentatively ignored (see Chapter 4). For a proportionality constant B, net OA torque is

$$L_y = H\dot{\phi} - B\dot{\theta} \tag{3-8}$$

* The identifying nomenclature "gyro gimbal" and "platform gimbal" will be used for the remainder of this chapter. This prevents ambiguities that could arise, for example, in considering a triad of accelerometers and gyros mounted together on the platform of Fig. 2-5. Again, although SDF instruments are used first for illustration, application of 2DF gyros is included subsequently.

† For a more detailed derivation, see Section 4.2.

A first approximation for characterizing the SDF *rate integrating gyro* is obtained by neglecting this net torque, in which case

$$H\dot{\phi} - B\dot{\theta} \doteq 0 \tag{3-9}$$

or, upon integration,

$$\theta/\phi \doteq H/B \tag{3-10}$$

Although the gyro is inherently sensitive to absolute angular rate, its output θ is essentially proportional to the *angle* of rotation for this type of gimbal restraining mechanism (a spring restraint produces a "rate gyro," of no interest for precision INS applications). The output angle θ is sensed by a *pickoff*, converting the mechanical rotation into a signal input for either (1) a platform servo configuration, or (2) a strapdown gyro *rebalancing* command, used to produce an opposing displacement $(-\theta)$ and also fed to a high-speed computational algorithm.

If standard terrestrial INS mechanizations employed the $(\mathbf{I}_I, \mathbf{J}_I, \mathbf{K}_I)$ frame, this discussion of the orientation reference would now be near termination. The algorithm just mentioned would be identified as (2-28), with the computed elements substituted into (3-6), while the rebalancing command would be the input signal for the gyro *torquer* in Fig. 3-2. For gimballed

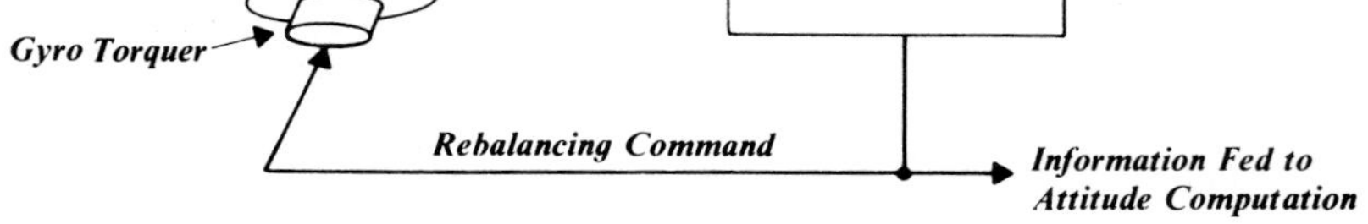

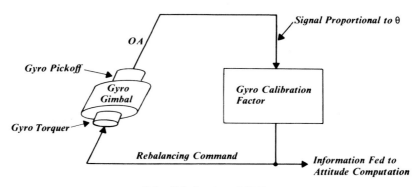

FIG. 3-2. Rebalancing of SDF gyro.

platform configurations, a simplified explanation could refer to Fig. 3-3. The gyro pickoff signal, proportional to θ (and therefore, essentially proportional to ϕ), is fed back to counteract platform rotation; full provisions for three-dimensional stabilization would then lead to determination of $\mathbf{U}_I$ by (3-3). Insertion of $\mathbf{U}_I$ into (3-5) and subsequent position computations could be identical for the gimballed and gimballess mechanizations (Exs. 3-5, 3-6).

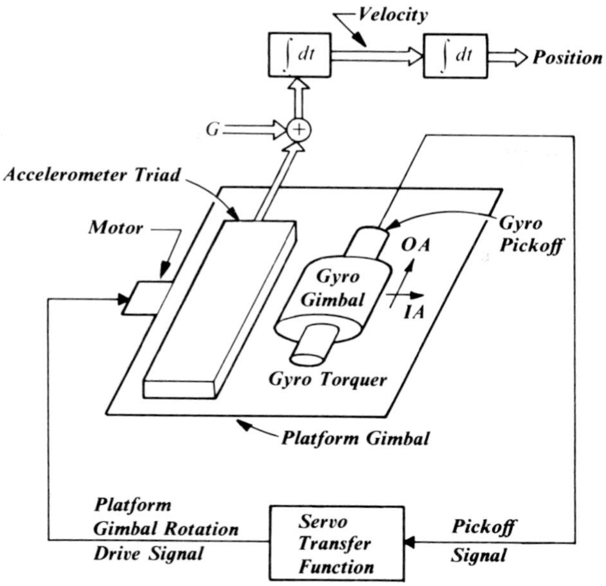

FIG. 3-3. Simplified inertially stabilized platform with rotation about one axis.

GEOGRAPHIC REFERENCE

Inconveniences of an inertial reference for terrestrial applications include the need for a gravitational representation in the inertial frame as well as subsequent transformations of velocity and/or position. A popular alternative is to compute position and velocity in the geographic frame, from specific force *as obtained in that frame*. This necessitates modification of both computation and mechanization, due to rotation of the geographic axes. An introduction to the required adjustments will now be given, leading to a more complete description in the ensuing sections.

Figure 3-4 depicts a geographic frame at initial time t_0 and again after translation along a meridian; $\mathbf{J}_G$ at both times is the outward normal to the plane shown. Clearly the frame rotates at a rate $(-\dot{\lambda})$ about $\mathbf{J}_G$. With reference axis angular rates denoted by ϖ,

$$\varpi_{\text{east}} = -\dot{\lambda} \tag{3-11}$$

Now suppose that the gyro IA and the platform gimbal axis in Fig. 3-3 are initially pointed east and a *torquing command*, with a value of

$$L = H\varpi_{\text{east}} \tag{3-12}$$

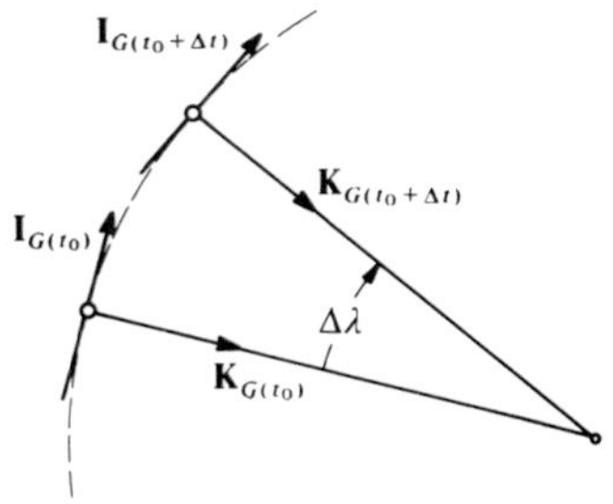

FIG. 3-4. Rotation of $\mathbf{I}_G$ and $\mathbf{K}_G$ during northbound flight.

is added, applied to the gyro torquer so that (3-9) is replaced by

$$B\dot{\theta} = H\dot{\phi} - L = H(\dot{\phi} - \varpi_{\text{east}}) \tag{3-13}$$

In this case the platform gimbal will receive a drive signal unless the instruments are rotating with the local vertical; i.e., given an initial orientation along $\mathbf{J}_G$, the control scheme just described would maintain it. By a similar development for total rotation about $\mathbf{K}_E$ (Ex. 3-7),

$$\mathbf{K}_E\,\varpi_{\text{pole}} = (\mathbf{I}_G \cos\lambda - \mathbf{K}_G \sin\lambda)(\omega_s + \dot{\varphi}) \tag{3-14}$$

and, by combination with (3-11), the angular rate of the geographic frame is

$$\boldsymbol{\varpi} = \begin{bmatrix} \varpi_1 \\ \varpi_2 \\ \varpi_3 \end{bmatrix} = \boldsymbol{\omega}_G = \begin{bmatrix} (\omega_s + \dot{\varphi}) \cos\lambda \\ -\dot{\lambda} \\ -\varpi_1 \tan\lambda \end{bmatrix} \tag{3-15}$$

Immediately this suggests generalization of (3-12) to define a torquing command vector,

$$\mathbf{L} = H\boldsymbol{\varpi} \tag{3-16}$$

so that, given an initial orientation along ($\mathbf{I}_G$, $\mathbf{J}_G$, $\mathbf{K}_G$), the platform in Fig. 2-5 is "slaved" to all three axes. The configuration is then said to be *locally level* and *north-slaved*, while inertial instruments are designated by their IA directions (e.g., "north gyro," "east gyro," and "azimuth gyro" for SDF instruments). Navigation computations for this configuration bypass (3-3) and (3-5), in favor of a combined expression for $\mathbf{V}$ to be presented subsequently. Also, the gimbal drive system needed for Fig. 3-5 is more complex than the scheme shown in Fig. 3-3. For the arrangement of Fig. 2-5, the azimuth gyro (IA downward) would control the inner platform gimbal, while an azimuth *resolver* would transform the other two gyro signals through the heading angle ψ (Fig. 3-6). These resolved signals would be the appropriate commands for the roll and pitch gimbals if the aircraft were level (Ex. 3-8). More generally, leveling servo commands could be

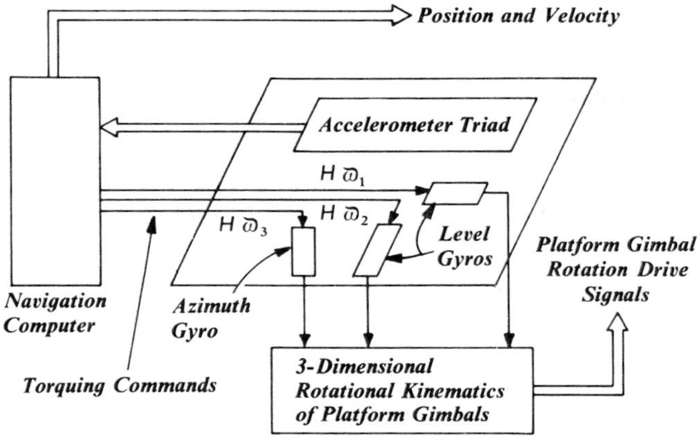

FIG. 3-5. Gimballed platform slaved to a rotating reference.

modified to account for gimbal axis pitching; theoretically this would maintain a more nearly uniform proportionality between excitation (aircraft rotation) and response (platform rotation) in all three axes. These possibilities are not pursued further here, because of the singularity encountered in a three-gimbal configuration. Chapter 5 of Ref. 3-1 describes both the three-gimbal system and a four-gimbal configuration commonly used to prevent singularity by holding the three inner gimbal axes nearly orthogonal.

A simple basis for initially erecting a platform can now be described: On a stable support ($\mathbf{V} \equiv \mathbf{0}$) the horizontal specific force is zero; thus the platform can be rotated about two perpendicular axes until two accelerometer outputs are nulled. The platform is then said to be *leveled*, and prepared for a *gyrocompassing* operation, which will align its axes along the geographic reference. A cursory description of that operation will suffice

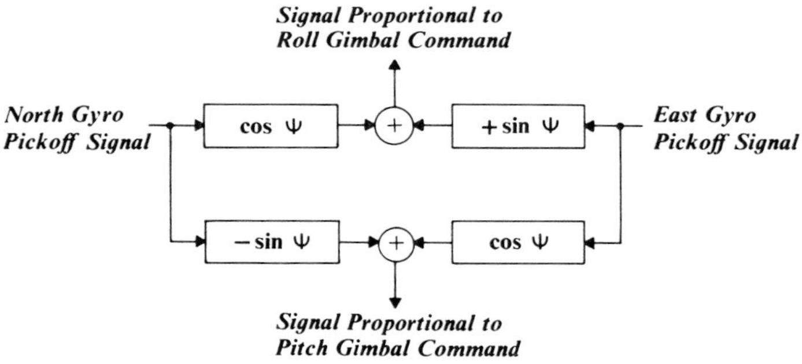

FIG. 3-6. Resolution of level gyro signals for platform gimbal commands.

here: a level platform is rotated in azimuth until one level gyro senses no component of earth rotation. With the other level gyro (IA now pointed north) and the azimuth gyro then torqued at ($H\omega_s \cos \lambda$) and ($-H\omega_s \sin \lambda$), respectively, it follows from (3-15) and (3-16) that initial alignment conditions are satisfied (Ex. 3-9).

All that remains in this section is extension of the basic strapdown approach to a geographic reference frame. This can be accomplished by the same scheme in Fig. 3-2, with (2-31) used to compute orientation with respect to the reference driven by angular rates of (3-15). Initial alignment also parallels the procedure outlined above, with two modifications:

(1) Physical rotations are replaced by computational adjustments of $\mathbf{T}_{G/A}$.

(2) Instead of nulling directly measured instrument outputs, these adjustments null the *computationally resolved* level specific forces and angular rate about the east axis.

3.3 MECHANIZATION CONSIDERATIONS

At high latitudes the north direction changes more rapidly until, in the limit, the azimuth torquing rate in (3-15) becomes infinite. Consequently another locally level reference formulation, used for near-polar flight, is presented at the end of this chapter. For navigation at low and moderate latitudes, however, concepts underlying the previous two sections include a broad scope of application, as discussed below.

Several inertial instrument types are consistent with analytical relations given here. An outer gyro gimbal could be added in Fig. 3-1, for example, to allow rotation with respect to the case about *two* axes normal to the SA. Such 2DF gyros are commonly used in inertial systems; while affecting the detailed design of torquing and stabilizing controls, they do not alter the basic equations presented here. The formulation accommodates 2DF accelerometers as well. Both analog and digital readouts for gyros and accelerometers are also included; note that the latter can facilitate velocity computation as follows: A *digital integrating accelerometer* can produce a calibrated pulse every time it experiences an increment of υ fps integrated specific force. That integration is thus reduced to pulse counting, at rates up to A_{max}/υ updates per second, where A_{max} denotes maximum specific force experienced by an accelerometer. Similarly, a *pulse-rebalanced rate integrating gyro* can use a fixed small increment (ϕ rad) as its basis of calibration; torquing rates for Fig. 3-2 would then have to provide this resolution at the maximum angular rate of the vehicle (e.g., a roll rate of 5 rad/sec and ~ 10 $\widehat{\text{sec}}$ resolution implies torque pulsing rates up to 10^5 Hz).

The above digital gyro example was purposely slanted toward a strapdown application. Despite the intended rotational isolation feature of platform gimbals, friction imposes requirements of high angular rate measurement for short durations. These requirements are often conveniently met by analog stabilization designs. Note also the possibility of analog platform servos with digital gyro torquing.

Table 3-1 highlights some major characteristics of gimballed platform and strapdown mechanizations. Only a general comparison is intended, exclusive of (1) considerations common to both systems (e.g., shock mounting) and (2) accuracy requirements. Shock mounts provide partial isolation of instruments from the dynamic environment, but can introduce instrument limit cycling or degrade observations by relative motion between INS and other equipment (e.g., antennas). For the second item above, an obvious example is gyro torquer calibration. A 0.1% error in the command of (3-12) at 600 kt, by reason of (3-11) and (1-12), corresponds to drift

TABLE 3-1

GENERAL COMPARISON OF GIMBALLED PLATFORM AND STRAPDOWN SYSTEMS

	Gimballed platform mechanization	Strapdown mechanization
Mounting	Inertial instruments (accelerometers and gyros) are mounted together on a platform, essentially isolated from vehicle rotations	Instruments are mounted together, along a vehicle-based orientation reference
Rotations experienced	Instruments are subjected to small or short-lived angular rates of platform	Instruments are subjected to vehicle rotation
Gyro torquing requirements	Gyros are equipped with small torquers	Either larger OA rotational displacements must be allowed ("wide-angle" gyros) or large torquers are needed for rebalancing
Function performed by gyros	Gyro pickoffs drive gimbal servos, to counteract undesirable rotations mechanically transmitted (e.g., via gimbal friction) to platform	Gyro rebalancing signals are fed to fast computer for angular orientation
Basis of velocity determination	Accelerometer IA directions are mechanically held along reference axes	Accelerometer outputs are computationally resolved into reference coordinates for summing

rates of order 0.01°/hr. The same calibration error in a strapdown gyro, however, would produce nearly 0.1° INS misorientation after a right angle turn.

Although a broader comparison of gimballed platform versus strapdown system accuracy requires analysis yet to be presented (Chapter 4), certain considerations regarding mechanization and attitude will be brought to attention before development of the navigation equations. These equations expressed in rotating reference frames are applicable to both gimballed and strapdown mechanizations, with the following conventions adopted:

For a gimballed platform $\mathbf{J}_P$ lies along one level accelerometer IA, while $\mathbf{K}_P$ is normal to the plane containing both $\mathbf{J}_P$ and the other level accelerometer IA. For a strapdown system having instrument triads nominally oriented with IA directions along roll (#1), pitch (#2), and yaw (#3) axes, $\mathbf{J}_A$ coincides with accelerometer IA #2, while $\mathbf{K}_A$ is normal to the plane of $\mathbf{J}_A$ and accelerometer IA #1. These definitions ensure orthogonality of all coordinate frames while maintaining a true physical representation. Inertial instrument misalignments with respect to *each other* can be taken into account without imposing unnecessary constraints (such as conformance of the aforementioned nominal axes to an aerodynamic reference; see Ex. 3-10). With either convention the IMU misorientation $\boldsymbol{\psi}$ represents the deviation between a true reference frame and its apparent directions, e.g.,

$$\mathbf{I} - (\boldsymbol{\psi} \times \quad) = \hat{\mathbf{T}}_{G/A}\, \mathbf{T}_{G/A}^{T} \tag{3-17}$$

where the circumflex denotes apparent quantities. The apparent directions of $(\mathbf{I}_G, \mathbf{J}_G, \mathbf{K}_G)$ in a gimballed platform correspond to $(\mathbf{I}_P, \mathbf{J}_P, \mathbf{K}_P)$, respectively; thus (3-17) is consistent with (2-15) (see Ex. 3-11). For a strapdown "analytic platform," $\hat{\mathbf{T}}_{G/A}$ represents the *mechanized* solution to (2-31), obtained from an airborne attitude computer and used to transform accelerometer outputs prior to summing.

In strapdown applications $\boldsymbol{\psi}$ contains the full effect of attitude uncertainty in the $(\mathbf{I}_A, \mathbf{J}_A, \mathbf{K}_A)$ axes, whereas platform gimbal pickoff imperfections (e.g., resolution) dominate errors in vehicle orientation as indicated by a gimballed platform. In any case, it is $\boldsymbol{\psi}$ that governs the dynamics of IMU position and velocity uncertainty. Gimballed platform stabilization accuracy is typically superior to that of the gimbal pickoffs; in many applications the orientation and position of the IMU are of paramount importance, while orientation of the vehicle itself is of secondary importance for navigation. This might have already been inferred from the definitions of (Φ, Θ, Ψ) on the basis of an apparent, rather than a true, geographic reference (2-9).

Preparation is now complete for a description of geographic navigation, applicable to both gimballed and gimballess mechanizations.

3.4 NAVIGATION IN GEOGRAPHIC COORDINATES

Equations relating geographic velocity and accelerations differ from the corresponding expressions in the ($\mathbf{I}_I$, $\mathbf{J}_I$, $\mathbf{K}_I$) frame. The integrated vector sum of gravitation and specific force, as obtained in a rotating frame, cannot be equated to velocity along the reference axes (e.g., recall the discussion at the beginning of Section 1.2). Also, in many systems, gravitation for the oblate earth is not formulated explicitly. The necessary expressions are developed here, by starting with considerations of spheroidal geometry and kinematics. Navigation equations are then followed by an analysis of performance with biased inertial instruments.

3.4.1 Motion over the Ellipsoid

Figure 3-7 shows, with exaggerated eccentricity and altitude, an elliptical meridian plane in which λ is the geodetic latitude used for mapping (and henceforth for all pertinent mathematical expressions in this book).

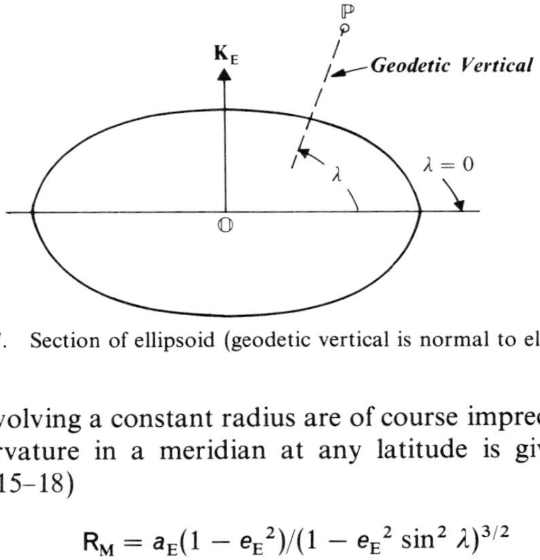

FIG. 3-7. Section of ellipsoid (geodetic vertical is normal to ellipsoid).

Equations involving a constant radius are of course imprecise; however the radius of curvature in a meridian at any latitude is given by (see, e.g., Ref. 3-4, pp. 15–18)

$$\mathsf{R}_M = a_E(1 - e_E^2)/(1 - e_E^2 \sin^2 \lambda)^{3/2} \tag{3-18}$$

where a_E and e_E denote equatorial radius and eccentricity:

$$a_E \doteq (3444)(6080) \text{ ft}, \qquad e_E^2 \doteq 6.7(10)^{-3} \tag{3-19}$$

so that translation along a meridian can be described by

$$\dot{\lambda} = V_N/(\mathsf{R}_M + h) \tag{3-20}$$

Similarly (Ex. 3-12),

$$\dot{\varphi} = V_E \sec \lambda / (\mathsf{R}_P + h) \tag{3-21}$$

where

$$\mathsf{R}_P = a_E / (1 - e_E{}^2 \sin^2 \lambda)^{1/2} \tag{3-22}$$

For a sphere of radius R_E the translational kinematics would of course reduce to (1-12) and (1-13):

$$\dot{\lambda} \doteq V_N / \mathscr{R}, \qquad \dot{\varphi} \doteq V_E \sec \lambda / \mathscr{R} \tag{3-23}$$

Combination of (3-20) and (3-21) with (3-15) yields

$$\boldsymbol{\varpi} = \begin{bmatrix} \omega_s \cos \lambda + V_E / (\mathsf{R}_P + h) \\ -V_N / (\mathsf{R}_M + h) \\ -\omega_s \sin \lambda - V_E \tan \lambda / (\mathsf{R}_P + h) \end{bmatrix} \tag{3-24}$$

and the matrix in (2-29) and (2-31) is

$$\boldsymbol{\Omega}_G = (-\boldsymbol{\varpi} \times \ ) \tag{3-25}$$

With this relation used in a strapdown attitude computer, or with (3-16) and (3-24) used to torque the gyros on a gimballed platform, specific force components are obtained in geographic coordinates. To relate these components to geographic velocity, (3-2) is transformed into the $(\mathbf{I}_G, \mathbf{J}_G, \mathbf{K}_G)$ frame and combined with (2-37):

$$\mathbf{A} + \mathbf{G} = \ddot{\mathbf{R}} + \dot{\boldsymbol{\varpi}} \times \mathbf{R} + 2\boldsymbol{\varpi} \times \dot{\mathbf{R}} + \boldsymbol{\varpi} \times (\boldsymbol{\varpi} \times \mathbf{R}) \tag{3-26}$$

It is convenient to write the last term as

$$\boldsymbol{\varpi} \times (\boldsymbol{\varpi} \times \mathbf{R}) \doteq \boldsymbol{\omega}_s \times (\boldsymbol{\omega}_s \times \mathbf{R}) + \begin{bmatrix} 2\omega_s V_E \sin \lambda + \dfrac{V_E{}^2}{\mathscr{R}} \tan \lambda \\ -\omega_s V_N \sin \lambda - \dfrac{V_N V_E}{\mathscr{R}} \tan \lambda \\ 2\omega_s V_E \cos \lambda + \dfrac{V_N{}^2 + V_E{}^2}{\mathscr{R}} \end{bmatrix} \tag{3-27}$$

where the approximation (3-23) has been used (Ex. 3-13), and the first term on the right is the centripetal acceleration due to sidereal rotation:

$$\boldsymbol{\omega}_s \times (\boldsymbol{\omega}_s \times \mathbf{R}) \doteq \begin{bmatrix} \mathscr{R}\omega_s{}^2 \cos \lambda \sin \lambda \\ 0 \\ \mathscr{R}\omega_s{}^2 \cos^2 \lambda \end{bmatrix}, \qquad \boldsymbol{\omega}_s = \begin{bmatrix} \cos \lambda \\ 0 \\ -\sin \lambda \end{bmatrix} \omega_s \tag{3-28}$$

This vector combines with the gravitation **G** to produce *gravity* along $\mathbf{K}_G$ (see Appendix 3.A),

$$\mathbf{g} = \begin{bmatrix} 0 \\ 0 \\ g \end{bmatrix} = \mathbf{G} - \boldsymbol{\omega}_s \times (\boldsymbol{\omega}_s \times \mathbf{R}) \tag{3-29}$$

By combination of the last four equations with the derivatives

$$\dot{\mathbf{R}} = \begin{bmatrix} 0 \\ 0 \\ V_V \end{bmatrix}, \qquad \ddot{\mathbf{R}} = \begin{bmatrix} 0 \\ 0 \\ \dot{V}_V \end{bmatrix} \tag{3-30}$$

and, again using (3-23),

$$\dot{\boldsymbol{\varpi}} \doteq \begin{bmatrix} (-\omega_s \sin \lambda)\left(\dfrac{V_N}{\mathscr{R}}\right) + \dfrac{\dot{V}_E}{\mathscr{R}} + \dfrac{V_E V_V}{\mathscr{R}^2} \\ -\dfrac{\dot{V}_N}{\mathscr{R}} - \dfrac{V_N V_V}{\mathscr{R}^2} \\ \dot{\varpi}_3 \end{bmatrix} \tag{3-31}$$

the standard nav equations are obtained (Exs. 3.14–3.16),

$$\begin{bmatrix} \dot{V}_N \\ \dot{V}_E \\ \dot{V}_V \end{bmatrix} = \begin{bmatrix} A_N \\ A_E \\ A_V \end{bmatrix} + \begin{bmatrix} 0 \\ 0 \\ g \end{bmatrix} + \begin{bmatrix} -\left(2\omega_s \sin \lambda + \dfrac{V_E}{\mathscr{R}} \tan \lambda\right) V_E + \dfrac{V_N V_V}{\mathscr{R}} \\ \left(2\omega_s \sin \lambda + \dfrac{V_E}{\mathscr{R}} \tan \lambda\right) V_N + 2\omega_s V_V \cos \lambda + \dfrac{V_E V_V}{\mathscr{R}} \\ -2\omega_s V_E \cos \lambda - \dfrac{V_N{}^2 + V_E{}^2}{\mathscr{R}} \end{bmatrix} \tag{3-32}$$

Typically these equations are solved by an airborne computer with iteration rates of order 40 Hz (Ex. 3-17). Velocities obtained are then substituted into (3-20) and (3-21), typically at iteration rates near 10 Hz.

3.4.2 Bias Error Analysis

As fed by an IMU with time-invariant gyro and accelerometer biases, an airborne computer integrating (3-32) will produce nav errors as follows: Let the last term of that equation be denoted by **f** so that

$$\dot{\mathbf{V}} = \mathbf{A} + \mathbf{g} + \mathbf{f} \tag{3-33}$$

With accelerometer errors denoted by*

$$\mathbf{n}_a = \mathbf{A} - \hat{\mathbf{A}} \tag{3-34}$$

in combination with directional uncertainty from (3-17), the specific force vector used in the airborne computer is

$$(\mathbf{I} - \boldsymbol{\psi} \times \ )\hat{\mathbf{A}} \doteq \mathbf{A} - \boldsymbol{\psi} \times \mathbf{A} - \mathbf{n}_a \tag{3-35}$$

and the *apparent* derivative of geographic velocity is essentially

$$\dot{\hat{\mathbf{V}}} = \mathbf{A} + \hat{\mathbf{g}} + \hat{\mathbf{f}} - \boldsymbol{\psi} \times \mathbf{A} - \mathbf{n}_a \tag{3-36}$$

Since imperfections in $\hat{\mathbf{f}}$ and $\hat{\mathbf{g}}$ are time-varying errors, deferred to subsequent chapters, subtraction of (3-36) from (3-33) here reduces to

$$\dot{\tilde{\mathbf{V}}} = \dot{\mathbf{V}} - \dot{\hat{\mathbf{V}}} \doteq \boldsymbol{\psi} \times \mathbf{A} + \mathbf{n}_a \tag{3-37}$$

Dynamic behavior of $\boldsymbol{\psi}$ can be closely approximated (Ex. 3-18) by

$$d\boldsymbol{\psi}/dt \doteq \dot{\boldsymbol{\psi}} + \boldsymbol{\varpi} \times \boldsymbol{\psi} \tag{3-38}$$

where the total time derivative includes a *drift* rate $\mathbf{n}_\omega$ (i.e., the amount by which the measured angular rate differs from the true value), and imperfect torquing effects due to $\tilde{\mathbf{V}}$; Appendix 3.B shows that, for near-equatorial flight (Ex. 3-19),

$$\begin{bmatrix} \dot{\psi}_N \\ \dot{\psi}_E \end{bmatrix} \doteq \frac{1}{\mathscr{R}} \begin{bmatrix} -\tilde{V}_E \\ \tilde{V}_N \end{bmatrix} + \begin{bmatrix} n_{\omega 1} \\ n_{\omega 2} \end{bmatrix} + \psi_A \begin{bmatrix} -\varpi_2 \\ \varpi_1 \end{bmatrix}, \qquad \varpi_3 \doteq 0 \tag{3-39}$$

and

$$\psi_A \doteq \psi_{A(0)} + n_{\omega 3}\, t \tag{3-40}$$

By combining the last two equations with (3-37), under restrictive conditions of steady flight at known fixed altitude,

$$\dot{\mathscr{R}} = 0, \qquad A_V = -g, \qquad A_N = A_E = \dot{V}_N = \dot{V}_E = 0 \tag{3-41}$$

* Circumflexes denote apparent (observed or estimated) quantities throughout. The sign convention used in (3-34) is chosen for conformance to subsequent chapters involving estimation.

the following expressions are obtained:

$$\begin{bmatrix} \dot{\tilde{V}}_E \\ \dot{\psi}_N \end{bmatrix} = \begin{bmatrix} 0 & g \\ -\dfrac{1}{\mathscr{R}} & 0 \end{bmatrix} \begin{bmatrix} \tilde{V}_E \\ \psi_N \end{bmatrix} + \begin{bmatrix} n_{a2} \\ n_{\omega 1} - \varpi_2 \psi_{A(0)} \end{bmatrix} + \begin{bmatrix} 0 \\ -\varpi_2 n_{\omega 3} \end{bmatrix} t \tag{3-42}$$

$$\begin{bmatrix} \dot{\tilde{V}}_N \\ \dot{\psi}_E \end{bmatrix} = \begin{bmatrix} 0 & -g \\ \dfrac{1}{\mathscr{R}} & 0 \end{bmatrix} \begin{bmatrix} \tilde{V}_N \\ \psi_E \end{bmatrix} + \begin{bmatrix} n_{a1} \\ n_{\omega 2} + \varpi_1 \psi_{A(0)} \end{bmatrix} + \begin{bmatrix} 0 \\ \varpi_1 n_{\omega 3} \end{bmatrix} t \tag{3-43}$$

The solutions to these equations are easily verified (Ex. 3-20):

$$\begin{bmatrix} \tilde{V}_E \\ \psi_N \end{bmatrix} = \begin{bmatrix} \cos Wt & u \sin Wt \\ -\dfrac{\sin Wt}{u} & \cos Wt \end{bmatrix} \begin{bmatrix} \tilde{V}_{E(0)} \\ \psi_{N(0)} \end{bmatrix} + \begin{bmatrix} \dfrac{\sin Wt}{W} & (1 - \cos Wt)\mathscr{R} \\ \dfrac{\cos Wt - 1}{g} & \dfrac{\sin Wt}{W} \end{bmatrix} \begin{bmatrix} n_{a2} \\ d \end{bmatrix} + \begin{bmatrix} \mathscr{R}\left(\dfrac{\sin Wt}{W} - t\right) \\ \dfrac{\cos Wt - 1}{W^2} \end{bmatrix} \varpi_2 n_{\omega 3} \tag{3-44}$$

and

$$\begin{bmatrix} \tilde{V}_N \\ \psi_E \end{bmatrix} = \begin{bmatrix} \cos Wt & -u \sin Wt \\ \dfrac{\sin Wt}{u} & \cos Wt \end{bmatrix} \begin{bmatrix} \tilde{V}_{N(0)} \\ \psi_{E(0)} \end{bmatrix} + \begin{bmatrix} \dfrac{\sin Wt}{W} & (\cos Wt - 1)\mathscr{R} \\ \dfrac{1 - \cos Wt}{g} & \dfrac{\sin Wt}{W} \end{bmatrix} \begin{bmatrix} n_{a1} \\ d' \end{bmatrix} + \begin{bmatrix} \mathscr{R}\left(\dfrac{\sin Wt}{W} - t\right) \\ \dfrac{1 - \cos Wt}{W^2} \end{bmatrix} \varpi_1 n_{\omega 3} \tag{3-45}$$

where

$$W = \sqrt{\frac{g}{\mathscr{R}}} \doteq \frac{2\pi \text{ rad}}{84(60) \text{ sec}}, \qquad u = \sqrt{g\mathscr{R}} \doteq 26(10^3) \text{ fps}$$
$$d = n_{\omega 1} - \varpi_2 \psi_{A(0)}, \qquad d' = n_{\omega 2} + \varpi_1 \psi_{A(0)} \tag{3-46}$$

With the notation

$$D = \varpi_2 n_{\omega 3}\left(\frac{\sin Wt}{W} - t\right) \tag{3-47}$$

(3-44) can be rewritten as

$$\tilde{V}_E = \tilde{V}_{E(0)} \cos Wt + \left(u\psi_{N(0)} + \frac{n_{a2}}{W}\right) \sin Wt + \mathscr{R}d(1 - \cos Wt) + \mathscr{R}D \tag{3-48}$$

and

$$\psi_N = \psi_{N(0)} \cos Wt + \left(\frac{d}{W} - \frac{\tilde{V}_{E(0)}}{u}\right) \sin Wt - \left(\frac{n_{a2}}{g} + \frac{\varpi_2 n_{\omega 3}}{W^2}\right)(1 - \cos Wt) \tag{3-49}$$

Aside from the effect of azimuth drift rate on velocity error, these expressions consist of appropriately initialized sinusoidal oscillations at $(W/2\pi)$ Hz; this is called the *Schuler* frequency. Of fundamental significance here is the *absence* of continuously increasing tilt or velocity errors as contributed from tilt rates or accelerometer biases. This is directly attributable to the first term on the right of (3-39), previously illustrated in Section 1.4. When a horizontal velocity error is present, it causes the platform to be driven in a direction that produces an opposing error. As further illustration of this characteristic, applicable to interaction between tilt and horizontal velocity errors, consider differentiation of the first relation covered by (3-42), under conditions of no azimuth drift:

$$\ddot{\tilde{V}}_E = g\dot{\psi}_N = -\frac{g}{\mathscr{R}} \tilde{V}_E + gd \tag{3-50}$$

Obviously this produces a sinusoid at the Schuler frequency. Often in discussions on this subject, a parallel is drawn between this mathematical property and the dynamics of a pendulum of length $\mathscr{R}$. Note that, in nonterrestrial applications, the larger values of $\mathscr{R}$ would weaken the effective restraining term $\tilde{V}_E g/\mathscr{R}$ and change the basic behavior of the error [e.g., consider the effect of removing this term from (3-50)].* Actually an analogous situation is observed for fairly short-term ($t < 0.1 \times 2\pi/W$) applications demonstrated below.

* It must not be inferred that nonterrestrial applications merely produce sinusoidal errors at lower characteristic frequencies. The characteristic periods would be too long for many of the simplifications adopted here (e.g., $\mathscr{R}$ known and constant, and omission of various kinematical effects) to remain approximately valid. Note also (Ex. 3-21) that only the *horizontal* velocity errors are characterized by these trigonometric expressions.

DYNAMICS OF POSITION ERROR

The velocity error in (3-48) produces an eastward position error denoted

$$\tilde{x}_{\mathrm{E}} = \tilde{x}_{\mathrm{E}(0)} + \tilde{V}_{\mathrm{E}(0)} \frac{\sin Wt}{W} + \mathscr{R}\left(\psi_{\mathrm{N}(0)} + \frac{n_{\mathrm{a}2}}{g}\right)(1 - \cos Wt)$$
$$+ \mathscr{R}d\left(t - \frac{\sin Wt}{W}\right) + \mathscr{R}\varpi_2 n_{\omega 3}\left(\frac{1 - \cos Wt}{W^2} - \frac{t^2}{2}\right) \tag{3-51}$$

For time durations of less than $2\pi/(10W)$,

$$\sin Wt/W \doteq t - \tfrac{1}{6}(Wt)^2(t), \qquad 1 - \cos Wt \doteq \tfrac{1}{2}(Wt)^2 \tag{3-52}$$

and (3-51) reduces to

$$\tilde{x}_{\mathrm{E}} \doteq \tilde{x}_{\mathrm{E}(0)} + \tilde{V}_{\mathrm{E}(0)} t + \tfrac{1}{2}(g\psi_{\mathrm{N}(0)} + n_{\mathrm{a}2})t^2 + \tfrac{1}{6}\mathscr{R}d(Wt)^2 t \tag{3-53}$$

which by inspection provides an extremely simple interpretation. In sequence, the terms on the right represent initial position error, linearly increasing position error due to initial velocity uncertainty, a quadratic error proportional to initial total acceleration uncertainty, and the result of an effective drift rate $d(Wt)^2/6$. For durations on the order of one-tenth Schuler period this effective drift rate is an order of magnitude less than d. It follows that, if any system needed an INS for only short-term operations, relatively high drift rates (e.g., ten times the allowable drift on a long-term navigator) would be acceptable. At the same time, other INS performance requirements (such as accelerometer accuracy and platform gimbal servo response) would *not* be relaxed by reason of short-term operation. Still, the more lenient demand on gyro capability is significant; the gyro is the most costly item in a precision INS.

Further manipulation of the preceding equations would produce conclusions similar to those already presented. The dynamics involving north/south excursions and tilt about the east axis could be further scrutinized, but (3-45) resembles (3-44) quite closely. A quartic term could be included in (3-52), which would demonstrate that azimuth drift rate effects are significantly attenuated in the short term. Due to extreme simplicity of the solutions obtained here, no additional mathematical manipulation is necessary. Instead, the individual error terms corresponding to (3-51) are sketched in Fig. 3-8, under conditions of the hypothetical parameter values given in Table 3-2. (Note that the slope of the dashed line in Fig. 3-8c is essentially 0.6 nm/hr per $\frac{1}{100}$ deg/hr drift.) The curves are drawn under the assumption that the solution remains applicable for a whole period. After an hour they must be regarded as approximations only, since appreciable changes (e.g., nonzero latitude and general kinematical coupling) occur within the cycle (Exs. 3-22–3-25).

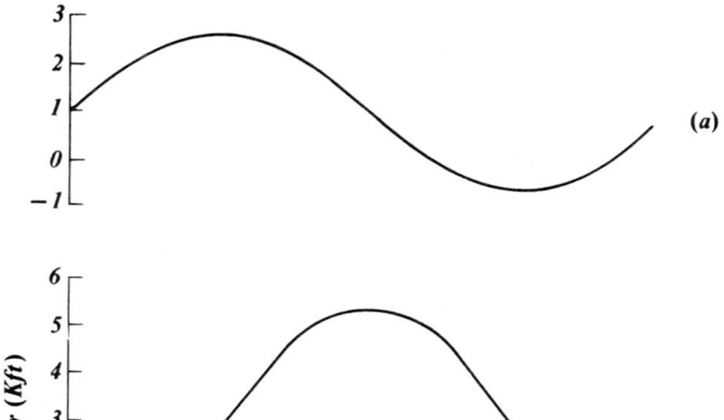

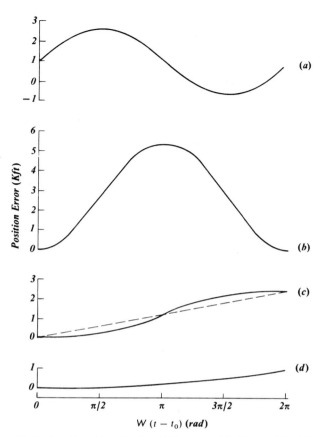

FIG. 3-8. Illustrative time histories for INS errors. (a) Effect of initial position and velocity uncertainty, (b) effect of initial tilt and accelerometer bias, (c) effect of north gyro drift and initial azimuth error, (d) effect of azimuth gyro drift.

TABLE 3-2

PARAMETERS FOR NUMERICAL EXAMPLE

Parameter	Value
V_N	1500 fps
$\tilde{x}_{E(0)}$	1000 ft
$\tilde{V}_{E(0)}$	2 fps
$\psi_{N(0)}$	10^{-4} rad
$\psi_{A(0)}$	10^{-3} rad
n_{a2}	$3(10)^{-5}g = 9.66(10)^{-4}$ fps^2
$n_{\omega 1}$	$-0.01°$/hr $= -4.84(10)^{-8}$ rad/sec
$n_{\omega 3}$	$+0.01°$/hr $= 4.84(10)^{-8}$ rad/sec

It is now appropriate to raise questions regarding performance expected under more general conditions [including full coupling effects corresponding to (3-38), arbitrary flight paths, and the resulting implications upon **A** in (3-37)]. Since many of the possible generalizations lead to time variations in coefficients of the differential equations, it is not surprising that customary approaches to general solutions (Chapter 6) are computer-oriented. Once this dependence on numerical algorithms is accepted, however, a great deal can be achieved in performance evaluation. Moreover, the type of analysis exemplified here can provide some insight applicable to more general flight simulations, either as "special case" tests for computer programs or as a model of approximate behavior to be expected over limited time durations.

3.5 STRAPDOWN COMPUTATIONS

While the preceding equations are applicable to both gimballed and strapdown systems, it remains to be demonstrated here how specific force is resolved along (apparent) geographic axes. As one possibility, analog inertial instrument output waveforms can be sampled at rates commensurate with dynamics experienced, for use in a mechanized numerical integration of (2-31) and of subsequent velocity computations. Alternatively, inertial instrument outputs can be quantized so that these integrations are simplified. To demonstrate that scheme a triad of pulse-rebalanced SDF rate integrating gyros and digital integrating accelerometers will be considered, with nominal mounting directions chosen along the aircraft axes.

Suppose that, at the time t_i starting each computing interval of duration $\mathrm{t} < \upsilon/A_{\max}$, presence or absence of a calibrated pulse (weight υ) is indicated with appropriate polarity from each accelerometer. With an instantaneous apparent transformation $\hat{\mathbf{T}}$ from aircraft to nav reference axes, specific force increments to be accumulated during each interval take the form

$$\int_{t_i}^{t_i+\mathrm{t}} \hat{\mathbf{T}}\hat{\mathbf{A}}_{\mathrm{A}}\, d\mathrm{t} = \begin{bmatrix} K_1\hat{T}_{11} + K_2\hat{T}_{12} + K_3\hat{T}_{13} \\ K_1\hat{T}_{21} + K_2\hat{T}_{22} + K_3\hat{T}_{23} \\ K_1\hat{T}_{31} + K_2\hat{T}_{32} + K_3\hat{T}_{33} \end{bmatrix} \upsilon \tag{3-54}$$

where each K-coefficient is 0 or ± 1, according to the corresponding accelerometer response. With υ calibrated as a negative integral power of two (in fps), each multiplication in a binary mechanization of (3-54) reduces to a binary shift. These fast computations can be implemented by a *digital differential analyzer* (DDA), which is restricted to simple operations (e.g., shift, add; see Ex. 3-26). The DDA output can be processed in a manner similar to velocity summations obtained from a gimballed platform mechanization, whether a nonrotating or a geographic reference is used (Ex. 3-27). Therefore, only the computations involved in maintaining the orientation reference need be covered in the following development.

ATTITUDE COMPUTATION

Under conditions that $\boldsymbol{\Omega}_A$ and $\boldsymbol{\Omega}_G$ commute with their time integrals, the solution to (2-31) is

$$\mathbf{T}_{G/A} = \exp\left\{\int_{t_i}^{t} \boldsymbol{\Omega}_G(\tau)\, d\tau\right\} \mathbf{T}_{G/A}(t_i) \exp\left\{-\int_{t_i}^{t} \boldsymbol{\Omega}_A(\tau)\, d\tau\right\} \tag{3-55}$$

where, in place of reference time t_0, an updated attitude is inserted at the time of each strapdown computer iteration t_i (Ex. 3-28). Despite its commutativity restriction this solution exerts considerable influence on strapdown attitude algorithms because, over short data processing intervals, the underlying conditions are nearly satisfied. As an example, a data processing interval t might be chosen as the ratio of an angular resolution ϕ of 2^{-13} rad to a maximum vehicle angular rate $\omega_{\max}$ of 1 rad/sec. Rotational excursions accumulated during such brief intervals (of order 10^{-4} sec) can be represented by small steps, regarded as integrated effects of short-lived constant angular rates.

In principle the pre- and postmultiplying matrices on the right of (3-55) could be expanded as in (2-69), and attitude could be computed from the series. Actually that is the basis for mechanized DDA computations, but only a few terms are represented in the computed expansion. To illustrate the appr:ach first recall that elements of $\boldsymbol{\Omega}_G$ are much smaller than the vehicle rates that determine t. It follows that the integrated effect of $\boldsymbol{\varpi}$ can produce excursions greater than ϕ only occasionally, so that (3-55) effectively reduces to (2-68) for most attitude updating intervals. Thus the majority of updates are made on the basis of a matrix

$$\boldsymbol{\mu} \triangleq \begin{bmatrix} 0 & \mu_3 & -\mu_2 \\ -\mu_3 & 0 & \mu_1 \\ \mu_2 & -\mu_1 & 0 \end{bmatrix} \tag{3-56}$$

defined separately for each DDA interval such that (1) any off-diagonal element μ_j will be set to zero unless the jth gyro produces a pulse during the current interval, and (2) nonzero elements have algebraic signs dictated by gyro pulse polarity, and magnitudes

$$\phi = 2^{-W} \text{ rad} > \omega_{\max}\,\mathrm{t} \tag{3-57}$$

where W is a positive integer. By substitution into (2-69), truncated after the quadratic term, adjustments to be added to elements of $\hat{\mathbf{T}}_{G/A}$ take the form*

* The reason for the above procedure is that, in the absence of detailed knowledge of precise rotational time histories producing the observed pulses, only an "after-the-fact" adjustment as described here is obtainable. In practice there are certain extensions of this basic approach (e.g., pulse modulation; Ref. 3-5), but equations in this book remain applicable with ϕ defined as *effective* resolution of quantized rotations.

$$\begin{aligned}
\Delta_{i1} &= \hat{T}_{i2}\mu_3 - \hat{T}_{i3}\mu_2 + \tfrac{1}{2}[\mu_1(\hat{T}_{i2}\mu_2 + \hat{T}_{i3}\mu_3) - \hat{T}_{i1}(\mu_2{}^2 + \mu_3{}^2)] \\
\Delta_{i2} &= \hat{T}_{i3}\mu_1 - \hat{T}_{i1}\mu_3 + \tfrac{1}{2}[\mu_2(\hat{T}_{i3}\mu_3 + \hat{T}_{i1}\mu_1) - \hat{T}_{i2}(\mu_3{}^2 + \mu_1{}^2)] \\
\Delta_{i3} &= \hat{T}_{i1}\mu_2 - \hat{T}_{i2}\mu_1 + \tfrac{1}{2}[\mu_3(\hat{T}_{i1}\mu_1 + \hat{T}_{i2}\mu_2) - \hat{T}_{i3}(\mu_1{}^2 + \mu_2{}^2)]
\end{aligned} \tag{3-58}$$

for $i = 1, 2, 3$. As with (3-54), these fast computations require only shift and add operations. From the inequality of (3-57) it follows that some or all gyro updates will be unnecessary for many DDA intervals. During those intervals the accumulated rotation of the reference frame can be examined and, if necessary, matrix adjustments can be made in the form

$$\begin{aligned}
\Delta_{1j} &= \hat{T}_{2j}\nu_3 - \hat{T}_{3j}\nu_2 \\
\Delta_{2j} &= \hat{T}_{3j}\nu_1 - \hat{T}_{1j}\nu_3 \\
\Delta_{3j} &= \hat{T}_{1j}\nu_2 - \hat{T}_{2j}\nu_1
\end{aligned} \tag{3-59}$$

for $j = 1, 2, 3$, where the counterpart to (3-56)

$$\mathbf{v} \triangleq \begin{bmatrix} 0 & \nu_3 & -\nu_2 \\ -\nu_3 & 0 & \nu_1 \\ \nu_2 & -\nu_1 & 0 \end{bmatrix} \tag{3-60}$$

can be computed by slowly integrating the elements of (3-24) and subjecting the result to the same threshold ϕ already defined. Whereas (3-58) is a *second*-order updating algorithm (i.e., with terms of order ϕ^2), only a *first*-order algorithm is needed for adjusting the matrix for reference frame rotation. Only a small percentage of updates will involve (3-59) and, moreover, a first-order algorithm is substantially correct for near-uniform reference angular rates (Exs. 3-29 and 3-30). For the nonuniform vehicle rotations requiring frequent updating via (3-58), however, the use of a second-order algorithm must now be briefly explained.

Omission of cubic and higher order terms from the alternating series of (2-69) produces a *truncation error* of order ϕ^3 for each update by (3-58). This degradation is small, however, in comparison with a *commutation* error produced as follows: Due to quantization, rotational increments up to ϕ rad can be processed in the wrong sequence, as illustrated in Fig. 3-9. For *true* and *apparent* transformations of $[\phi]_y[\phi]_x$ and $[\phi]_x[\phi]_y$, respectively, there is an angular uncertainty (Ex. 3-31) of order ϕ^2,

$$[\phi]_x[\phi]_y \left[[\phi]_y[\phi]_x\right]^{-1} \doteq \mathbf{I}_{3\times 3} + \begin{bmatrix} 0 & -\phi^2 & 0 \\ \phi^2 & 0 & 0 \\ 0 & 0 & 0 \end{bmatrix} \tag{3-61}$$

which illustrates the well-known noncommutativity of finite rotations about nonparallel axes. Accumulation of computational degradations will depend on the angular rates experienced (Chapter 4) but, in any case, (3-61) illustrates the dominant error controlled by ϕ in conformance with (3-57) (Ex. 3-32).

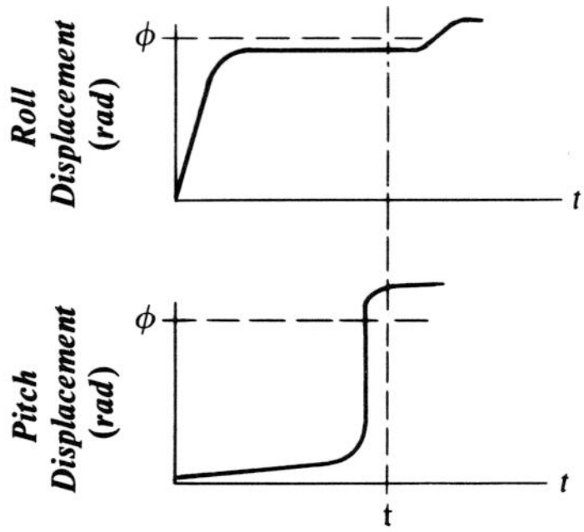

FIG. 3-9. Small rotations with misleading sequence of threshold crossings.

3.6 WANDER AZIMUTH NAVIGATION

Divergence of the azimuth torquing rate in (3-15) and the time derivative of longitude in (3-21), at $\lambda = \pm\pi/2$, can be explained by an Euler angle singularity as follows: Let a locally level *wander azimuth* coordinate frame $(\mathbf{I}_L, \mathbf{J}_L, \mathbf{K}_L)$ be defined such that $\mathbf{K}_L \equiv \mathbf{K}_G$ and the instantaneous eastward deviation of $\mathbf{I}_G$ from $\mathbf{I}_L$ is $\mathfrak{a}$ rad. From (2-14),

$$\mathbf{T}_{L/E} = [\mathfrak{a}]_x^T[\tfrac{1}{2}\pi + \lambda]_y^T[\varphi]_x \tag{3-62}$$

By analogy with (2-17), this angle sequence convention is ill-defined with 0 or π rad in the middle transformation. The matrix $\mathbf{T}_{L/E}$, however, is well defined under all conditions. It is therefore permissible to choose a convenient initialization for this matrix (e.g., with $\mathfrak{a}_0 = 0$ for geographic referencing at an initial position of moderate latitude) and to replace mechanized solutions of (3-20), (3-21) by integration of the differential equation (Ex. 3-33)

$$\dot{\mathbf{T}}_{L/E} = \mathbf{T}_{L/E}(\boldsymbol{\omega}_E \times \ ) - (\boldsymbol{\omega}_L \times \ )\mathbf{T}_{L/E} \tag{3-63}$$

where $\boldsymbol{\omega}_E$ and $\boldsymbol{\omega}_L$ denote absolute angular rates of the earth and wander azimuth frames, respectively, with each rate expressed in its own coordinates. The former has already been given by (3-4), while $\boldsymbol{\omega}_L$ will depend on the particular wander azimuth convention chosen. Some platform mechanizations use continuous azimuth torquing at a slow rate (e.g., about one revolution per hour), to cancel long-term drifts of level gyros. Alternatively, all azimuth torquing can be neglected, using only the leveling commands from (3-24),

$$\mathbf{T}_{L/G}\begin{bmatrix}\varpi_1\\ \varpi_2\\ 0\end{bmatrix}=\begin{bmatrix}\dfrac{V_E \cos \mathsf{a}}{\mathsf{R_P}+h}+\dfrac{V_N \sin \mathsf{a}}{\mathsf{R_M}+h}+\omega_s \cos\lambda \cos \mathsf{a}\\ \dfrac{V_E \sin \mathsf{a}}{\mathsf{R_P}+h}-\dfrac{V_N \cos \mathsf{a}}{\mathsf{R_M}+h}+\omega_s \cos\lambda \sin \mathsf{a}\\ 0\end{bmatrix} \tag{3-64}$$

or, since

$$\mathsf{R_P}=\mathsf{R_M}+r, \qquad r \triangleq \mathsf{R_P}\, e_E{}^2 \cos^2\lambda/(1-e_E{}^2\sin^2\lambda) \tag{3-65}$$

it follows that (Ex. 3-34)

$$\boldsymbol{\omega}_L \doteq \begin{bmatrix}\dfrac{V_{L2}}{\mathsf{R_P}+h}+\omega_s \cos\lambda \cos \mathsf{a}\\ \dfrac{-V_{L1}}{\mathsf{R_M}+h}+\omega_s \cos\lambda \sin \mathsf{a}\\ 0\end{bmatrix}+\frac{r \sin \mathsf{a}}{(a_E+h)^2}\begin{bmatrix}V_{L1}\cos \mathsf{a}+V_{L2}\sin \mathsf{a}\\ V_{L1}\sin \mathsf{a}-V_{L2}\cos \mathsf{a}\\ 0\end{bmatrix} \tag{3-66}$$

where the elements of $\mathbf{V}_L$ are velocity components in the $(\mathbf{I}_L, \mathbf{J}_L, \mathbf{K}_L)$ frame, which for nonpolar positions conform to the relation (Ex. 3-35)

$$\mathbf{V}_L=\begin{bmatrix}V_{L1}\\ V_{L2}\\ V_{L3}\end{bmatrix}=[\mathsf{a}]_z^T\,\mathbf{V} \tag{3-67}$$

It is crucial to note that, although both φ and the *wander angle* a lose their unique identity at the poles, all necessary computations can be mechanized. From (3-65), r vanishes at a pole, so that the last vector on the right of (3-66) is zero. Either (3-18) or (3-22) then provides both $\mathsf{R_M}$ and $\mathsf{R_P}$, which, combined with the wander azimuth system velocity outputs V_{L1} and V_{L2}, provides $\boldsymbol{\omega}_L$. Thus gyro torquing (gimballed platform mechanizations) or attitude computer reference adjustment (strapdown mechanizations), along with integration of (3-63), can be performed under all conditions, while latitude is always obtainable from this latter integration as

$$\lambda=-\text{Arcsin}\, T_{33} \tag{3-68}$$

At nonpolar locations,

$$\alpha = \arccos(T_{13}/\cos\lambda) = \arcsin(T_{23}/\cos\lambda) \tag{3-69}$$

and

$$\varphi = \arccos(-T_{31}/\cos\lambda) = \arcsin(-T_{32}/\cos\lambda) \tag{3-70}$$

so that longitude and heading (Ex. 3-36) can be determined, but these latter computations can be omitted at near-polar latitudes.

Considerations remaining for completion of the wander azimuth description are (1) an alternative attitude reference formulation, (2) the method of computing $\mathbf{V}_L$, and (3) bias error analysis; these are addressed below.

FOUR-PARAMETER ATTITUDE REFERENCE COMPUTATION

Section 2.3 noted a computational reduction obtainable when orthogonal matrix elements are updated at a faster rate than they are used to transform vectors. This is the case for $\mathbf{T}_{L/E}$, although the computational advantage is compromised somewhat by the noninertial character of *both* $(\mathbf{I}_L, \mathbf{J}_L, \mathbf{K}_L)$ and $(\mathbf{I}_E, \mathbf{J}_E, \mathbf{K}_E)$. As one possible four-parameter formulation for these two frames, Ex. 3-37 establishes the 4×1 vector,

$$\boldsymbol{h} \triangleq \left[\begin{array}{c} \mathbf{b}_L \cos(\Xi/2) - \mathbf{1}_3 \ell_L \sin(\Xi/2) - \mathbf{1}_3 \times \mathbf{b}_L \sin(\Xi/2) \\ \hdashline \ell_L \cos(\Xi/2) + b_{L3} \sin(\Xi/2) \end{array}\right] \tag{3-71}$$

where $(\ell_L, \mathbf{b}_L)$ conform to (2-42), obtained by integrating (2-41) with substitution of $\boldsymbol{\omega}_L$ for the angular rate. With components of $\boldsymbol{h}$ used on the right of (2-40), only a selected part of the matrix would be needed for position determination; the computations would include T_{33} for (3-68) and, as desired, the elements in (3-69) and (3-70).

TRANSLATIONAL KINEMATICS

Equations to be integrated for $\mathbf{V}_L$ can be approached through an alternate derivation of (3-32). By differentiation of (3-5) and combination with (3-3),

$$\dot{\mathbf{V}} = \mathbf{T}_{G/I}(\mathbf{A}_I + \mathbf{G}_I - \boldsymbol{\omega}_E \times \mathbf{U}_I) - \boldsymbol{\varpi} \times \mathbf{V} \tag{3-72}$$

which, after another rearrangement using (3-4), (3-5), and (3-29), yields (Ex. 3-38)

$$\dot{\mathbf{V}} = \mathbf{A} + \mathbf{g} - (\boldsymbol{\varpi} + \boldsymbol{\omega}_s) \times \mathbf{V} \tag{3-73}$$

A parallel development in the $(\mathbf{I}_L, \mathbf{J}_L, \mathbf{K}_L)$ frame readily provides the relation (Exs. 3-39 and 3-40)

$$\dot{\mathbf{V}}_L = \mathbf{A}_L + \mathbf{g} - (\boldsymbol{\omega}_L + \boldsymbol{\omega}_{sL}) \times \mathbf{V}_L \tag{3-74}$$

where the L subscripts denote expression of all vectors in wander azimuth coordinates. The absence of singularity in this last equation follows immediately from comparing (3-66) versus (3-24) at high latitudes. Also, if the aforementioned continuous azimuth torquing had been included in (3-66) to counteract steady level gyro drifts, (3-74) would still be valid. With no azimuth torquing, however, (3-66) is substituted into (3-74) with

$$\boldsymbol{\omega}_{\mathrm{L}} + \boldsymbol{\omega}_{\mathrm{sL}} = \begin{bmatrix} \dfrac{V_{\mathrm{L}2}}{\mathrm{R_P} + h} + 2\omega_{\mathrm{s}} \cos\lambda \cos \mathfrak{a} \\[2ex] \dfrac{-V_{\mathrm{L}1}}{\mathrm{R_M} + h} + 2\omega_{\mathrm{s}} \cos\lambda \sin \mathfrak{a} \\[2ex] -\omega_{\mathrm{s}} \sin\lambda \end{bmatrix} + \frac{r \sin \mathfrak{a}}{(a_{\mathrm{E}} + h)^2} \begin{bmatrix} V_{\mathrm{L}1} \cos \mathfrak{a} + V_{\mathrm{L}2} \sin \mathfrak{a} \\ V_{\mathrm{L}1} \sin \mathfrak{a} - V_{\mathrm{L}2} \cos \mathfrak{a} \\ 0 \end{bmatrix} \tag{3-75}$$

Again it is noted that all terms containing functions of $\mathfrak{a}$ vanish at the poles. Also, a wander azimuth strapdown formulation is obtained immediately by substituting $\boldsymbol{\omega}_{\mathrm{L}}$ for the reference angular rate in the preceding section.

EFFECTS OF INERTIAL INSTRUMENT BIAS

Just as the cross product terms in (3-73) were omitted from the performance analysis in Section 3.4.2, an approximate wander azimuth evaluation could be based on the first two terms on the right of (3-74). In fact, such an evaluation is immediately obtainable, at least for sufficiently slow azimuth torquing, by a reinterpretation of Section 3.4.2. Horizontal geographic axes are replaced by the wander azimuth reference directions, while "north" and "east" instrument errors are replaced by effective biases of the level instruments. Approximate performance of the wander azimuth system then conforms to the results already obtained, for durations up to an hour.

APPENDIX 3.A: GRAVITY AND THE EARTH'S ELLIPSOID MODEL

The direction for gravity as formulated in Section 3.4 is the geodetic vertical, introduced in the discussion accompanying (1-14). In many INS mechanizations the earth is modeled as an oblate spheroid whose north component of gravitation **G** just balances that of (3-28), so that **g** of (3-29) lies along $\mathbf{K}_{\mathrm{G}}$. This is a widely accepted approximation to the *geoid*, mathematically defined as a surface everywhere normal to the direction of gravity as experienced by a plumb-bob. The normal to the true geoid deviates from $\mathbf{K}_G$ by angles known as *vertical deflections*. Also, due to *gravity anomalies*, the magnitude of gravity actually experienced does not conform to the ellipsoid model. Adjustments could easily be added to (3-32) but, in any case, any uncompensated gravity vector uncertainty acts as a time-varying accelerometer error, considered in subsequent chapters, which extend the analysis of Section 3.4.2 (Ex. 3-41).

A common subtlety in kinematical convention will now be identified and resolved. Whereas (3-3) corresponds to an earth center position reference, (3-20) implies a reference point on the evolute (Ex. 3-42) of the ellipse (e.g., a point, denoted $\mathbb{Q}$, along the geodetic vertical line on Fig. 3-7). The explanation lies in two small but opposing approximations implicit in (3-29) and in (3-2), as applied to the earth's ellipsoid. If in this appendix $\boldsymbol{P}$ and $\boldsymbol{Q}$ denote displacements in geographic coordinates from earth center to the airborne INS and to the corresponding evolute base point, respectively, then $\mathbf{R}$ now represents the radius vector from $\mathbb{Q}$, and an accelerometer triad is sensitive to

$$\mathbf{A} = d^2\boldsymbol{P}/dt^2 - \mathbf{G} = d^2\mathbf{R}/dt^2 - \mathbf{G} + d^2\boldsymbol{Q}/dt^2 \tag{3-76}$$

With respect to a rigid earth ellipsoid, any given point on an evolute is fixed; thus

$$d^2\boldsymbol{Q}/dt^2 = \boldsymbol{\omega}_s \times (\boldsymbol{\omega}_s \times \boldsymbol{Q}) \tag{3-77}$$

With gravity force properly defined on the basis of an inertial reference,

$$\mathbf{g} = \mathbf{G} - \boldsymbol{\omega}_s \times (\boldsymbol{\omega}_s \times \boldsymbol{P}) \tag{3-78}$$

which combines with (3-76) to yield

$$d^2\mathbf{R}/dt^2 = \mathbf{A} + \mathbf{g} + \boldsymbol{\omega}_s \times (\boldsymbol{\omega}_s \times \mathbf{R}) \tag{3-79}$$

in agreement with the navigation equations of Section 3.4.

For completeness, the evolute is expressed here in terms of the standard ellipse parameters,

$$b_E = a_E(1 - e_E{}^2)^{1/2} \tag{3-80}$$

in the equation from Ref. 3-6, p. 421:

$$(a_E\, x)^{2/3} + (b_E\, y)^{2/3} = (a_E{}^2 e_E{}^2)^{2/3} \tag{3-81}$$

No similar development involving (3-21) is necessary, since sections of the ellipsoid parallel to the equator are circular.

APPENDIX 3.B: DYNAMICS OF PLATFORM ORIENTATION ERROR

Although (3-39) is a logical extension of (1-42), a derivation showing the added kinematical coupling and drift is instructive. Dynamics of $\boldsymbol{\psi}$ will first be described with disturbance from $\mathbf{n}_\omega$ alone, and the effect of velocity uncertainty will be added subsequently.

Consider a triad of gyros monitoring the absolute angular rate $\boldsymbol{\omega}_P$ of a gimballed platform,* and a fictitious coordinate frame $(\mathbf{I}_B, \mathbf{J}_B, \mathbf{K}_B)$ associated with an absolute angular rate $\boldsymbol{\omega}_B = \hat{\boldsymbol{\omega}}_P = \boldsymbol{\omega}_P - \mathbf{n}_\omega$ as follows: While the gyros actually experience the true angular rate $\boldsymbol{\omega}_P$, their outputs $\hat{\boldsymbol{\omega}}_P$ are processed as if they were mounted with IA directions along $(\mathbf{I}_B, \mathbf{J}_B, \mathbf{K}_B)$. The orthogonal transformation $\mathbf{T}_{B/P}$, corresponding to the orientation error, conforms to the dynamic equation

$$\dot{\mathbf{T}}_{B/P} = -\mathbf{T}_{B/P}\,\boldsymbol{\Omega}_P + \hat{\boldsymbol{\Omega}}_P\,\mathbf{T}_{B/P} \tag{3-82}$$

where, in analogy with (2-31), the $\boldsymbol{\Omega}$-matrices represent skew-symmetric operators containing the respective angular rate components in the off-diagonal positions. In the present example $(\mathbf{I}_B, \mathbf{J}_B, \mathbf{K}_B)$ is identified with the geographic frame so that, by combining the preceding relation with (2-15),

$$(\dot{\boldsymbol{\psi}} \times \ ) \doteq -\boldsymbol{\Omega}_P - (\boldsymbol{\psi} \times \ )\boldsymbol{\Omega}_P + \hat{\boldsymbol{\Omega}}_P + \hat{\boldsymbol{\Omega}}_P(\boldsymbol{\psi} \times \ ) \tag{3-83}$$

Through term-by-term association involving the off-diagonal elements of this expression, it readily follows that

$$\dot{\boldsymbol{\psi}} \doteq \mathbf{n}_\omega - \boldsymbol{\omega}_P \times \boldsymbol{\psi} \tag{3-84}$$

which conforms to (3-38) with $\mathbf{n}_\omega$ identified as the absolute time derivative in this example, and with $\boldsymbol{\omega}_P$ replaceable here by $\boldsymbol{\varpi}$ (Ex. 3-43). Instead of the full kinematical coupling in (3-38), azimuth torquing will be ignored in this near-equatorial analysis, and ψ_A will be assumed driven independently:

$$\frac{d\boldsymbol{\psi}}{dt} \doteq \begin{bmatrix} \dot{\psi}_N \\ \dot{\psi}_E \\ \dot{\psi}_A \end{bmatrix} + \begin{bmatrix} 0 & 0 & \varpi_2 \\ 0 & 0 & -\varpi_1 \\ 0 & 0 & 0 \end{bmatrix} \begin{bmatrix} \psi_N \\ \psi_E \\ \psi_A \end{bmatrix}, \qquad \lambda \doteq 0, \quad \psi_N \ll \psi_A, \quad \psi_E \ll \psi_A \tag{3-85}$$

This latter simplification is justified by relative performance of tilt and azimuth misorientation. Typically, ψ_N and ψ_E will have smaller magnitudes (and therefore less influence on ψ_A than this larger azimuth error can exert on tilt), due to the Schuler phenomenon, which will now be added.

* The procedure remains applicable in strapdown systems, but the overall considerations are somewhat more involved; see Section 4.3.4.

With imperfect velocity estimates at low latitude, (3-24) produces an actual platform command of essentially

$$\boldsymbol{\omega}_\mathrm{P} \doteq \begin{bmatrix} \omega_\mathrm{s} + \hat{V}_\mathrm{E}/\mathscr{R} \\ -\hat{V}_\mathrm{N}/\mathscr{R} \\ 0 \end{bmatrix}, \qquad \lambda \doteq 0 \tag{3-86}$$

so that the contribution of velocity uncertainty to $d\boldsymbol{\psi}/dt$ is

$$\begin{bmatrix} \omega_\mathrm{s} + \hat{V}_\mathrm{E}/\mathscr{R} \\ -\hat{V}_\mathrm{N}/\mathscr{R} \\ 0 \end{bmatrix} - \begin{bmatrix} \omega_\mathrm{s} + V_\mathrm{E}/\mathscr{R} \\ -V_\mathrm{N}/\mathscr{R} \\ 0 \end{bmatrix} = \frac{1}{\mathscr{R}} \begin{bmatrix} -\tilde{V}_\mathrm{E} \\ \tilde{V}_\mathrm{N} \\ 0 \end{bmatrix} \tag{3-87}$$

Combination with drift produces the total derivative $d\boldsymbol{\psi}/dt$; substitution into (3-85) provides (3-39) and (3-40).

EXERCISES

3-1 Accelerometers are sensitive to motion with respect to what set of axes? Specific force components are measured along which directions in a strapdown system? In a gimballed system?

3-2 An accelerometer, held stationary over the earth, is suddenly released. Essentially free fall conditions prevail initially, but a velocity-dependent drag force gradually retards the downward acceleration until a constant speed (terminal velocity) is reached. Sketch the time history of accelerometer output.

3-3 (a) Show that the components of $\boldsymbol{\omega}_\mathrm{E}$ are not changed by a transformation between earth coordinates and the $(\mathbf{I}_\mathrm{I}, \mathbf{J}_\mathrm{I}, \mathbf{K}_\mathrm{I})$ frame.

(b) Extend the results of Ex. 2-5d to show that the velocity, with respect to the earth center, of a helicopter hovering at fixed geographic position is $\boldsymbol{\omega}_\mathrm{E} \times \mathbf{R}_\mathrm{I}$, as expressed in inertial coordinates.

3-4 With $\mathbf{A}_\mathrm{A} = \mathbf{T}_{\mathrm{A/I}}\,\mathbf{A}_\mathrm{I}$ show that (3-6) is consistent with (3-3).

3-5 Gyros are sensitive to motion with respect to what set of axes? Observed rotation rates are resolved along which directions in a strapdown system? In a gimballed system?

3-6 Describe a three-dimensional attitude reference system involving (2-28) and Fig. 3-2. Extend this description to include computation of position relative to the earth center.

3-7 (a) On the basis of (3-13), what is the output of an error-free pickoff on a properly initialized east gyro torqued in accordance with (3-12)?

(b) Relate the vector in (3-14) to a transposed version of (1-28).

(c) In analogy with Fig. 3-4, sketch a plane normal to the geodetic pole, and apply the reasoning of part (a) to this sketch.

(d) Combine (3-15) with (1-12) and (1-13) to express approximate rotational rates of the geographic frame. Compute some typical values, for supersonic speed at 60° latitude. What are the rates at the south pole?

3-8 (a) An aircraft, initially heading north under straight and level flight conditions, is suddenly subjected to a 100°/sec angular rate about a level axis 1° to the right of $\mathbf{I}_A$. Due to static friction in platform gimbals, a set of instruments mounted on the inner element of Fig. 2-5 is instantaneously subjected to the aircraft rotation. What stabilizing commands are initially desired?

(b) Repeat part (a) for initial heading angles of 90° and 250°.

(c) Discuss the relation between Fig. 3-6 and a particular orthogonal transformation matrix.

3-9 Compare the torquing signals desired for an aircraft on a runway versus those for the helicopter in Ex. 3-3b.

3-10 An accelerometer triad and a gyro triad are mounted with sensitive axes along orthogonal directions that deviate from the aerodynamic reference by several degrees. Discuss the impact on the system considered in Ex. 3-6, and on a system that uses a geographic reference.

3-11 Show that (2-15) is consistent with (3-17), and that either can apply to a gimballed platform or a strapdown mechanization.

3-12 (a) Discuss (3-21) in relation to "ground east" velocity in (1-17). Why is there no similar distinction for the other two components of geographic velocity?

(b) Apply (3-24) and (3-25) to (2-35) for the helicopter in Ex. 3-3. When is $\boldsymbol{\omega}_s$ equal to $\boldsymbol{\varpi}$? What is the relation between $\boldsymbol{\omega}_s$ and $\boldsymbol{\omega}_E$?

3-13 (a) Show that the error introduced into (3-28) by $\mathscr{R}$ is of order $2(10)^{-5}g$. What level of error is produced in (3-27) for subsonic speeds at moderate latitudes?

(b) Verify (3-27) and discuss the direction of centripetal acceleration from sidereal rotation.

3-14 Verify (3-31).

3-15 Do the Euler angles (Φ, Θ, Ψ) appear in (3-32)? Does resolution of a platform gimbal pickoff control the accuracy of position and velocity? Is (3-32) dependent on an aerodynamic reference direction?

3-16 Show that the adjustment terms in (3-32) are typically a few milli-g at moderate latitudes. What would be the effect of ignoring these terms after five minutes? How does this relate to the opening discussion of Section 1.2?

3-17 (a) Show that, for an aircraft in a 4-g turn, the error introduced into (3-32) by 0.025 sec computational lag is of order $10^{-5}g$.

(b) Show that, for northbound flight at 350 kt, the effect of a 0.1-sec position computation lag is about 0.01 $\widehat{\text{min}}$.

(c) Recall the statement in regard to accelerometer pulse counting in Section 3.3. Consider a mechanization using a 40-Hz nav computer iteration rate, with digital integrating accelerometers on a gimballed platform, for which (3-33) is solved by accumulating increments of the form

$$\Delta\mathbf{V} = 0.025(\mathbf{g} + \mathbf{f}) + \upsilon \begin{bmatrix} \text{north accelerometer pulse count} \\ \text{east accelerometer pulse count} \\ \text{vertical accelerometer pulse count} \end{bmatrix}$$

Is each pulse count an algebraic sum? What directions are associated with positive and negative accelerometer pulses? When is each pulse count initialized? Has any consideration been given to small changes in **f** during a 0.025-sec interval?

3-18 (a) Compare (3-38) versus (2-33). Can finite rotations in three dimensions be represented by vectors? Is (3-38) exact?

(b) How is (3-2) related to the specific force components in (3-41)?

3-19 (a) Show that (1-42) conforms to (3-39) as a special case.

(b) Show that (1-45) conforms to (3-44) as a special case.

3-20 (a) Using (3-41), differentiate (3-44) and (3-45) and compare versus (3-42) and (3-43), respectively. Determine the initial values of these derivatives and verify their validity.

(b) Do (3-44) and (3-45) meet all requirements for initial conditions? Can any other solutions be obtained for (3-42) and (3-43) that meet all of these requirements?

3-21 (a) Gravitation in (3-1) can be represented by $\mu_E/\mathscr{R}^2$ where $\mu_E \doteq gR_E{}^2$ is the product, (earth mass) $\times$ (universal constant of gravitation). Using a first-order Taylor expansion, with $\tilde{h}$ denoting an altitude uncertainty, show that the *apparent* downward specific force can be approximated by

$$\hat{A}_V \doteq -d^2(h - \tilde{h})/dt^2 - (\mu_E/\mathscr{R}^2)(1 + 2\tilde{h}/\mathscr{R})$$

so that

$$d^2\tilde{h}/dt^2 \doteq 2W^2\tilde{h}, \qquad \hat{A}_V = A_V$$

(b) Compare the natural dynamics of INS errors in vertical and horizontal channels. Recall Ex. 1-8.

(c) Is altitude divergence a major threat if accurate updates are made at 5-min intervals in Fig. 1-14?

(d) Verify a vertical error sensitivity of order $10^{-7}g/\text{ft}$.

3-22 (a) Differentiate (3-51) and compare versus (3-48).

(b) Sketch a revised version of Fig. 3-8, with the signs of n_{a2} and $n_{\omega 1}$ reversed. What is the rate of steady error accumulation per 0.01°/hr total effective drift? How is this related to the connection between 1 $\widehat{\text{min}}$ and 1 nm over the earth?

3-23 (a) Use ***W*** to rework Ex. 3-13 as follows:

$$\tfrac{2}{3}(10)^{-2}[84/(24 \times 60)]^2 g \doteq 2(10)^{-5} g$$

By similar methods, show that comparable errors are introduced into (3-32) by (3-23).

(b) With errors an order of magnitude below the values in Table 3-2, would high precision requirements still allow (3-23) to be used in deriving the adjustment terms of (3-32)?

3-24 Write equations analogous to (3-47)–(3-51) for the other horizontal channel.

3-25 (a) Discuss allowable short-term drift in (3-53). Write the last term with a cubic coefficient $gd/6$.

(b) Add a quartic term to (3-52) for substitution into (3-53). Discuss short-term effects of the azimuth drift rate. If a system contained two accurate gyros and one poor quality gyro, how should they be oriented? How would the importance of azimuth gyro drift be affected by long-term unaided operation (i.e., navigation for several hours without fixes)?

3-26 (a) A strapdown system uses digital integrating accelerometers with a velocity resolution of $\frac{1}{32}$ fps, for a vehicle with a 6-g maximum acceleration. Compare the required DDA speed with that of the nav computer described at the end of Section 3.4.1. Give *two* reasons for separating the computations involved in (3-32) from those in (3-54).

(b) What associations are appropriate between directions and accelerometer pulse polarities, for the digital system considered in Section 3.5? Compare versus Ex. 3-17c.

(c) Would a change in mounting directions radically alter the performance of the strapdown system under consideration?

(d) If $tA_{\text{max}} < \upsilon$ could there be more than one pulse from any accelerometer during a DDA interval? If acceleration peaks are extremely infrequent, does it seem reasonable to let t exceed $v/A_{\max}$ and change pulse weight when the peaks occur?

3-27 Describe the substitution of velocity increments obtained from (3-54) into a mechanized integration of (3-6) and of (3-32). Identify the transformation matrix used and the velocity vector computed in each case. Make a comparison with Ex. 3-17c.

3-28 (a) Does $\mathbf{\Omega}_{\mathrm{A}}$ commute with its time integral for rotation about a fixed axis? Can an axis be considered fixed for infinitesimal durations? Nearly fixed for short durations? Assuming commutativity, differentiate (3-55) and compare with (2-31). Is this an extension of (2-68)?

(b) Should both the nonrotating and the geographic reference computation be fully developed in Section 3.5, or is one a special case of the other?

(c) How does this exercise relate to Ex. 2-9a,b?

(d) Can the same iteration rate be used for both attitude and velocity updating? Is this essential?

3-29 For DDA intervals with no gyro pulses detected, (2-31) is replaced by a transposed form of (2-29), with aircraft axes instantaneously nonrotating while the reference frame has an angular rate $\boldsymbol{\varpi}$. Note that these rates do not change much under general flight conditions and, if they did not change at all, rotational increments for (3-59) could be equated to

$$\begin{bmatrix} v_1 \\ v_2 \\ v_3 \end{bmatrix} = \begin{bmatrix} \varpi_1(t - t_i) \\ \varpi_2(t - t_i) \\ \varpi_3(t - t_i) \end{bmatrix}$$

Show that, as updated by (3-59), $\hat{T}_{11}$ could then be written in the form

$$\hat{T}_{11} = \hat{T}_{11(t_i)} + \hat{T}_{21(t_i)} v_3 - \hat{T}_{31(t_i)} v_2$$

with a derivative that conforms *exactly* to the applicable differential equation. Repeat for all elements of the direction cosine matrix. Since initial conditions are also satisfied, is (3-59) a unique solution for the case considered? Can (3-59) be an accurate approximation when this case is generalized to allow slow variations in reference rates?

3-30 Express the solution just obtained in a form similar to (2-73), for the special case $\varpi_2 = \varpi_3 = \omega_X = 0$. Is the form of the result familiar, after simplification?

3-31 (a) Use (2-8) with an equation in the form of (3-17), to show that an angular error of order ϕ^2 rad can be incurred in one DDA interval. Compare this degradation with the truncation error.

(b) Consider a fixed angular resolution just slightly greater than $\omega_{max} t$ and then imagine a design change whereby t is halved without changing ϕ. If (3-58) could be updated at the center of the interval shown in Fig. 3-9, would (3-61) be affected? In general, should significant improvement be expected from reduction of t below the requirement of (3-57)?

3-32 (a) In a strapdown system, are gyro torquers used to counteract rotations of the reference frame, or rotations experienced by the inertial instruments? Compare with a gimballed platform.

(b) What associations are appropriate between directions and gyro pulse polarities for the system considered in Section 3.5?

(c) Compare the amount of fast computation required for the geographic and the nonrotating reference. How would the scheme of Section 3.5 be modified by choice of a nonrotating reference frame?

3-33 Draw an analogy between (3-63) and (2-31). Compare the rates.

3-34 (a) Derive (3-65) from (3-18) and (3-22).

(b) Show that

$$\frac{1}{\mathrm{R_M} + h} \doteq \frac{1}{\mathrm{R_P} + h} + \frac{r}{(a_\mathrm{E} + h)^2}$$

Discuss the accuracy, given (3-19).

(c) What is the maximum value of the ratio r/a_E?

(d) Combine part (b) with (3-64) and (3-67), to derive (3-66).

(e) Express (3-24) as a special case of (3-64) and (3-66). Are the first two components valid at any latitude?

(f) Show that, at $\alpha = -\pi/2$, ω_{L1} and ω_{L2} agree with geographic frame rates about axes pointing east and south, respectively.

3-35 (a) Show that, for the wander azimuth system of Section 3.6, indicated heading is the azimuth angle obtained from the platform minus the computed wander angle.

(b) Use (1-40) and (3-67) to express the ground track angle as a function of the wander angle and components of $\mathbf{V}_\mathrm{L}$.

3-36 Obtain (3-68)–(3-70) from (3-62). Must the magnitudes of both inverse functions be computed in (3-69) and (3-70)? The signs?

3-37 (a) Apply the principles of (2-47) to write the quaternion for geographic velocity, with components resolved along the nonrotating frame, in the form

$$\boldsymbol{B}_\mathrm{L} \boldsymbol{V}_\mathrm{L} \boldsymbol{B}_\mathrm{L}^{-1} = \boldsymbol{B}_\mathrm{E} \boldsymbol{V}_\mathrm{E} \boldsymbol{B}_\mathrm{E}^{-1}$$

from which a quaternion representing the transformation between $(\mathbf{I}_E, \mathbf{J}_E, \mathbf{K}_E)$ and $(\mathbf{I}_L, \mathbf{J}_L, \mathbf{K}_L)$ frames can be obtained,

$$\boldsymbol{H} = \boldsymbol{B}_E^{-1}\boldsymbol{B}_L$$

(b) Show that the last element of one of the above operators is $\sin(\Xi/2)$. Which operator is it, and what are its other elements?

(c) Use the above results to obtain (3-71).

3-38 Show that (3-73) and (3-32) are equivalent.

3-39 (a) Why is gravity not subscripted in (3-74)?

(b) How does (3-54) apply to a wander azimuth strapdown system?

(c) Are there circumstances permitting or forbidding omission of the last term in (3-75) from the cross product in (3-74), at *low* latitudes? Is this question related to Ex. 3-23b?

3-40 Use (3-67) to express (3-74) in the geographic frame. Compare with (3-73).

3-41 (a) In a vertical channel system of Fig. 1-14, with updates accurate to 100 ft taken every 5 min, is there any pressing need to correct for gravity anomalies of order $3 \times 10^{-5}g$?

(b) With errors significantly lower than values in Table 3-2, could compensation for known vertical deflections of order 20 $\widehat{\text{sec}}$ be useful?

3-42 (a) Is the direction of $\mathbf{K}_G$, in combination with the definition of R_M, consistent with the concept of an evolute? Sketch the evolute in a meridian plane such as Fig. 3-7.

(b) Show that the ordinate of the y-intercept for the evolute is

$$(a_E + b_E)(a_E - b_E)/b_E \doteq \tfrac{2}{3} \times 10^{-2} a_E$$

Is any point on the evolute farther from the center?

(c) Assume that the fourth quadrant cusp of the evolute corresponds to the elliptical arc in the first quadrant of Fig. 3-7. Then, at $\lambda = \pi/2$ and $\lambda = 0$, the radii of curvature should be

$$b_E + (\text{ordinate of } y\text{-intercept}) \doteq a_E(1 + \tfrac{1}{2}e_E^2)$$

and

$$a_E - (\text{abscissa of } x\text{-intercept}) = a_E(1 - e_E^2)$$

respectively. Show this to be true.

(d) Express (3-77) as an application of (2-37). As in Ex. 3-23a, show that this quantity is of order $2 \times 10^{-5}g$. Is this a source of error?

(e) Describe the position reference for (3-21) as the intersection of $\mathbf{K}_E$ with a certain plane. Would this description be likely if the geoid were not approximated by a solid of revolution?

3-43 (a) Perform the term-by-term association of off-diagonal elements described in the text accompanying (3-83). Note that $\hat{\mathbf{\Omega}}_P$ is replaceable by $\mathbf{\Omega}_P$ in the last matrix product, since products of errors are much smaller than terms involving $\boldsymbol{\psi}$ or $\mathbf{n}_\omega$ alone.

(b) Discuss possible violation of the requirement for vectors in a cross product to be expressed in the same coordinate frame.

(c) Should cross-axis coupling be expected when inertial sensors are used in a rotating frame? Justify the replacement of $\boldsymbol{\omega}_P$ by $\boldsymbol{\varpi}$ in (3-84).

(d) Does (3-85) account for effects of azimuth error on tilt? Effects of tilt on azimuth error?

(e) Is an actual platform stabilization command generated from true or apparent velocity? What is the desired stabilization command? Justify the use of (3-23) in (3-86) by arguments similar to parts (a) and (c) of this exercise.

(f) Show that (3-87) is consistent with the sign convention of (2-15). Does velocity uncertainty exert an undesirable influence on tilt?

REFERENCES

3-1. Broxmeyer, C., *Inertial Navigation Systems.* New York: McGraw-Hill, 1964.

3-2. Fernandez, M., and Macomber, G. R., *Inertial Guidance Engineering.* Englewood Cliffs, New Jersey: Prentice-Hall, 1962.

3-3. O'Donnell, C. F. (ed.), *Inertial Navigation Analysis and Design.* New York: McGraw-Hill, 1964.

3-4. Kayton, M., and Fried, W. R. (eds.), *Avionics Navigation Systems.* New York: Wiley, 1969.

3-5. Clark, R. N., and Fosth, D. C., "Analysis and Test of a Precision Pulse-Rebalance Gyroscope," *AIAA JSR* **11** (4), April 1974, pp. 264–266.

3-6. *CRC Standard Mathematical Tables.* Cleveland, Ohio: Chemical Rubber Publ. Co., 1959.

Chapter 4

Advanced Inertial Navigation Analysis

4.0

A fitting topic for introducing the middle chapter of this book involves (1) how the preceding material will be placed into an integrated nav formulation, and (2) how this chapter will advance that effort. The illustration begins by expressing earlier kinematical relations in terms of a unified formalism for dynamic analysis. The equations of Section 3.4.2 conform to the relation (Ex. 4-1)

$$
\begin{bmatrix} \dot{\tilde{V}}_{\mathrm{N}} \\ \dot{\tilde{V}}_{\mathrm{E}} \\ \hdashline \\ \dot{\boldsymbol{\psi}} \\ \\ \end{bmatrix}
=
\left[\begin{array}{cc:ccc} 0 & 0 & 0 & -g & 0 \\ 0 & 0 & g & 0 & 0 \\ \hdashline 0 & -1/\mathscr{R} & & & \\ 1/\mathscr{R} & 0 & & (-\boldsymbol{\varpi}\times\ ) & \\ 0 & 0 & & & \end{array}\right]
\begin{bmatrix} \tilde{V}_{\mathrm{N}} \\ \tilde{V}_{\mathrm{E}} \\ \hdashline \\ \boldsymbol{\psi} \\ \\ \end{bmatrix}
+
\begin{bmatrix} n_{\mathrm{a}1} \\ n_{\mathrm{a}2} \\ \hdashline \\ \mathbf{n}_{\omega} \\ \\ \end{bmatrix}
\tag{4-1}
$$

With appropriate modifications of the matrix, this general form would remain applicable if the restrictions of (3-41) were removed and if position variables were added with effects of nonzero latitude included. Even the imperfections in $\hat{\mathbf{f}}$ and $\hat{\mathbf{g}}$, dropped from (3-37), can be included in long-term dynamic behavior. Furthermore, if inertial instrument errors are subdivided into effective fixed averages (denoted $\langle \mathbf{N} \rangle$) and rapidly varying random components (denoted $\mathbf{e}$),

$$\mathbf{n}_a = \langle \mathbf{N}_a \rangle + \mathbf{e}_a \tag{4-2}$$

$$\mathbf{n}_\omega = \langle \mathbf{N}_\omega \rangle + \mathbf{e}_\omega \tag{4-3}$$

then the original equations can be replaced by a higher order set driven by noise *alone*, e.g., (4-1) is easily replaced by

$$\dot{\mathbf{x}}_{10\times 1} = \left[\begin{array}{c:c} \mathbf{A}_{5\times 5} & \mathbf{I}_{5\times 5} \\ \hdashline \multicolumn{2}{c}{\mathbf{O}_{5\times 10}} \end{array}\right] \mathbf{x}_{10\times 1} + \left[\begin{array}{c} \mathbf{e}_{5\times 1} \\ \hdashline \mathbf{0}_{5\times 1} \end{array}\right] \tag{4-4}$$

where the first five and last five components of $\mathbf{x}_{10\times 1}$ are obtained from (4-1) and from $\langle \mathbf{N} \rangle$, respectively, while $\mathbf{e}_{5\times 1}$ contains random errors in two horizontal accelerations from (4-2) and three angular rate components from (4-3); $\mathbf{A}_{5\times 5}$ is the matrix in (4-1). As a final modifying step, the fixed average instrument errors can be replaced by slowly varying bias components with specified autocorrelation properties, driven by additional noise inputs replacing $\mathbf{0}_{5\times 1}$ in (4-4). All of these extensions conform to a unified dynamic representation, to be used for every mode of an integrated nav system. The necessary development, constituting the full estimation rationale noted at the end of Section 1.4, unfolds in subsequent chapters; this chapter presents an extensive analysis of the errors in (4-2) and (4-3).

Two questions that arise at the outset are (1) how simple error models, such as those in (4-2) and (4-3) or straightforward modifications thereof, can be useful for realistic system representation in the presence of complex motions, and (2) why much effort should be devoted to analyzing such simple models. The first question is answered by subdividing flight profiles into contiguous phases, each having its own characteristic static forces, turn rates, stationary vibration spectra, etc., so that environment-sensitive errors also acquire reasonably constant mean values and stationary statistics for each separate phase. Total errors within a flight phase are simply obtained by superposition, and terminal conditions of each phase (with possible adjustments to account for sudden maneuvers) initiate the next, in cascaded succession. Efforts to assess these errors are easily justified by their influence on performance prediction and nav mode sequencing.

The above-mentioned sensitivity of inertial instrument degradations to the dynamic environment, which may vary widely during a flight, could in principle be a major complicating factor in performance analysis. Collecting multiple error sources in a succession of phases affords a systematic and accurate (but not exact) representation. Fortunately the estimation scheme described in subsequent chapters can tolerate inexact models. Both theory and experience provide adequate guidelines for characterizing biases, narrow- and wideband component errors, and gravity uncertainties, with imperfect knowledge of autocorrelation times and other parameters, in the presence of aerodynamic forces and gusts. All of these factors are realistically taken into account, in a formulation sufficiently flexible to allow adaptation to any type of sensor* or mechanization. In fact, much of the analysis in this chapter is motivated by the profound influence of motion on strapdown system error.

A general INS performance assessment unfolds here as a sequence of basic physical principles applicable to the inertial instruments, degradation in the presence of complex motions, and system errors resulting from component degradations. The evaluation uses approaches from Refs. 4-1 and 4-2, although this material is now presented with some differences in purpose, notation, and viewpoint (i.e., in the context of an integrated system, without constraints on tutorial explanations).

4.1 ACCELEROMETERS IN GRAVITATIONAL FIELDS

While a terrestrial reference for aircraft navigation is conceptually pleasing, it should be realized that the earth's orbit deviates significantly from a constant velocity trajectory. Earth translation is characterized approximately by a circular path of radius $9 \times 10^7 \times 5280$ ft, making a full revolution every $365.25 \times 24 \times 3600$ sec. This corresponds to a centripetal acceleration of roughly $0.0006g$, definitely not a negligible quantity (Ex. 4-2). Nevertheless, a terrestrial INS customarily uses an earth position reference without mechanizing any provisions to account for translation of the earth itself; an explanation follows.

Translation of a star through space can be characterized by a velocity vector that is substantially constant over any duration of interest here. Let an inertial coordinate frame be defined by three orthogonal axes of fixed orientation, intersecting at some chosen origin that remains in a fixed location with respect to the sun. The "spherical earth" example corresponding to (3-2) can now be generalized as follows: $\mathbf{G}_I$ is now supplemented by the gravitation $\mathbf{G}_S$ from the remote sun, and the accelerometer triad position vector relative to the inertial frame origin is denoted $\boldsymbol{\rho}_I$. Then

$$\mathbf{A}_I = d^2\boldsymbol{\rho}_I/dt^2 - \mathbf{G}_I - \mathbf{G}_S \tag{4-5}$$

At the same time, however, $\mathbf{G}_S$ is substantially the same gravitational

* Inertial components analyzed here are elementary devices based on mechanical operating principles. Results are applicable in general, irrespective of floated versus nonfloated mechanization, types of bearings, etc., or the methodology can be applied to more sophisticated devices such as laser or electrostatic gyros.

acceleration experienced by the earth itself; ignoring any small differences (e.g., due to displacement between earth and accelerometers)

$$\mathbf{G}_S = d^2\boldsymbol{\rho}_E/dt^2 \tag{4-6}$$

where $\boldsymbol{\rho}_E$ denotes the position vector of the earth in the previously chosen inertial frame. Combination of the last two equations yields

$$\mathbf{A}_I = d^2\mathbf{R}_I/dt^2 - \mathbf{G}_I, \quad \mathbf{R}_I \triangleq \boldsymbol{\rho}_I - \boldsymbol{\rho}_E \tag{4-7}$$

so that (3-2) is acceptable for position vectors $\mathbf{R}_I$ referenced to an earth origin. By straightforward extension of this analysis it can be shown that this equation is likewise accurate in the presence of lunar and all other existing gravitational fields, provided that extreme precision is not required (Ex. 4-3).

4.2 GYRODYNAMICS

This section provides a derivation for (3-7) under the conditions stated, applicable to Fig. 3-1, and subsequently generalizes the torque equations for three-dimensional rotation. With components of angular rate experienced by the gyro gimbal equal to $\dot{\phi}$, $\dot{\theta}$, 0 (about the axes $\mathbf{I}_g$, $\mathbf{J}_g$, $\mathbf{K}_g$ in Fig. 3-1, respectively), a condition applicable to (3-7) is

$$\theta = 0 \tag{4-8}$$

Gyro behavior under these restrictions will now be developed.

4.2.1 Basic Gyro Torque Derivation

A coordinate frame ($\mathbf{I}_R$, $\mathbf{J}_R$, $\mathbf{K}_R$) will be defined rigidly attached to the rotor such that $\mathbf{K}_R = \mathbf{I}_R \times \mathbf{J}_R$ coincides with the SA and ($\mathbf{I}_R$, $\mathbf{J}_R$) are orthogonal diametral lines, somewhat arbitrarily chosen. With respect to these coordinates the inertia matrix of the rotor is not only time-invariant but diagonal; also, the first two elements are equal, so that the matrix can be written here as

$$\mathscr{I}_R = \begin{bmatrix} \mathscr{I} & 0 & 0 \\ 0 & \mathscr{I} & 0 \\ 0 & 0 & \mathscr{J} \end{bmatrix} \tag{4-9}$$

Now suppose that the gyro gimbal is instantaneously rotating at a rate $\omega_x = \dot{\phi}$ about $\mathbf{I}_g$ in Fig. 3-1. In combination with the relative motion $\dot{\theta}$

about $\mathbf{J}_g$ and a spin rate denoted Ω rad/sec, the total angular rate of the rotor is expressed in its own coordinate frame as

$$\boldsymbol{\omega}_R = \begin{bmatrix} 0 \\ 0 \\ \Omega \end{bmatrix} + \mathbf{T}_{R/g} \begin{bmatrix} \dot{\phi} \\ \dot{\theta} \\ 0 \end{bmatrix} \tag{4-10}$$

Since $\mathbf{T}_{R/g}$ is a z-axis rotation it is permissible to choose $(\mathbf{I}_R, \mathbf{J}_R)$ to coincide with $(\mathbf{I}_g, \mathbf{J}_g)$ at zero reference time, so that

$$\boldsymbol{\omega}_R = [\Omega t]_z \begin{bmatrix} \dot{\phi} \\ \dot{\theta} \\ \Omega \end{bmatrix} \tag{4-11}$$

and, by product differentiation with a fixed spin rate,

$$\dot{\boldsymbol{\omega}}_R = [\Omega t]_z \begin{bmatrix} \ddot{\phi} \\ \ddot{\theta} \\ 0 \end{bmatrix} + \begin{bmatrix} 0 & \Omega & 0 \\ -\Omega & 0 & 0 \\ 0 & 0 & 0 \end{bmatrix} [\Omega t]_z \begin{bmatrix} \dot{\phi} \\ \dot{\theta} \\ \Omega \end{bmatrix} \tag{4-12}$$

By substitution of (4-11) and (4-12) into a vector form of (2-83), with the A (aircraft) subscript replaced by R for the rotor,

$$\mathbf{L}_R = \mathscr{I}_R \dot{\boldsymbol{\omega}}_R + \boldsymbol{\omega}_R \times \mathscr{I}_R \boldsymbol{\omega}_R \tag{4-13}$$

while noting that the two 3×3 matrices in the last term of (4-12) commute,

$$\mathbf{L}_R = \mathscr{I}_R [\Omega t]_z \begin{bmatrix} \ddot{\phi} + \Omega\dot{\theta} \\ \ddot{\theta} - \Omega\dot{\phi} \\ 0 \end{bmatrix} + [\Omega t]_z \begin{bmatrix} \dot{\phi} \\ \dot{\theta} \\ \Omega \end{bmatrix} \times \left\{ \mathscr{I}_R [\Omega t]_z \begin{bmatrix} \dot{\phi} \\ \dot{\theta} \\ \Omega \end{bmatrix} \right\} \tag{4-14}$$

It is now convenient to re-express this torque, *acting on the rotor*, in the gyro gimbal coordinate frame; premultiplying all vectors in this expression by $[\Omega t]_z^T$, and noting that the similarity transformation of $\mathscr{I}_R$ about the rotor axis leaves the inertia matrix unchanged,

$$\mathbf{L}_g = \begin{bmatrix} \mathscr{I}\ddot{\phi} + \mathscr{J}\Omega\dot{\theta} \\ \mathscr{I}\ddot{\theta} - \mathscr{J}\Omega\dot{\phi} \\ 0 \end{bmatrix} \tag{4-15}$$

Of interest here is the OA torque; in particular, the term proportional to $\dot{\phi}$ has a coefficient of magnitude $H = \mathscr{J}\Omega$, as in (3-7) (Ex. 4-4).

Instantaneously before this developed torque takes effect on rotational displacements, (4-8) remains valid. For a restraining torque $B\dot{\theta}$, OA response is now obtained by applying (2-83) to the gyro gimbal. With its total axial moment of inertia (including the rotor diametral inertia $\mathscr{I}$) denoted J, and its other two principal moments tentatively assumed equal, the equation for OA rotation is (Ex. 4-5)

$$J\ddot{\theta} - H\dot{\phi} + B\dot{\theta} = 0 \tag{4-16}$$

In Laplace operator notation (4-16) is equivalent to

$$\frac{\theta(s)}{\phi(s)} = \frac{H/B}{1 + (J/B)s} \tag{4-17}$$

interpreted as follows: The physical motion is dominated by frequencies below $B/(2\pi J)$ Hz, so that gyro output is essentially proportional to ϕ in the spectral range of primary concern.

4.2.2 Generalization of Gyro Torque Equations

Removal of the restrictions in (4-8) affects the intrinsic rotational dynamics of a gyro in various ways, which will now be described in succession. Consider first a generalized angular motion about $\mathbf{J}_g$, so that the *absolute* OA angular rate of the gyro gimbal is $\dot{\theta} + \omega_y$; (4-16) is then replaced by*

$$(B/H)\dot{\theta} = \omega_x - \gamma_y(\ddot{\theta} + \dot{\omega}_y) \tag{4-18}$$

where J/H is denoted γ_y.

The next generalization to be considered is *anisoinertia* of the gyro gimbal (i.e., a difference j in moments of inertia about $\mathbf{I}_g$ and $\mathbf{K}_g$), in the presence of nonzero ω_z. With (4-16) modified by this effect alone, the extended equation could be written in the form (Ex. 4-6)

$$B\dot{\theta} = H\dot{\phi} - (J\ddot{\theta} + j\omega_x\omega_z) \tag{4-19}$$

Immediately this expression should be extended to include anisoinertia of the *rotor* as well; with nonzero ω_z, (4-11) is replaced here by

$$\boldsymbol{\omega}_R = [\Omega t]_z \begin{bmatrix} \dot{\phi} \\ \dot{\theta} \\ \Omega + \omega_z \end{bmatrix} \tag{4-20}$$

When this angular rate is used in the derivation of Section 4.2.1 (Ex. 4-7), the resulting expression can be put into the form

$$(B/H)\dot{\theta} = \dot{\phi} - (\gamma_y\ddot{\theta} + \gamma_{zx}\omega_x\omega_z) \tag{4-21}$$

where γ_{zx} includes the anisoinertia effects of the rotor and the gyro gimbal.

One more rotational effect remains to be considered here, i.e., the displacement angle θ. In general, the total angular rate of the rotor includes

* ω_x, ω_y, and ω_z are defined as instrument *case* rotation rates.

a spin velocity $\Omega \mathbf{K}_g$, the OA components $(\dot{\theta} + \omega_y)\mathbf{J}_g$, plus case rotations ω_x and ω_z about *displaced* axes $(\mathbf{I}_g \cos\theta + \mathbf{K}_g \sin\theta)$ and $(\mathbf{K}_g \cos\theta - \mathbf{I}_g \sin\theta)$, respectively, so that

$$\boldsymbol{\omega}_R \doteq [\Omega t]_z \begin{bmatrix} \dot{\phi} - \theta\omega_z \\ \dot{\theta} + \omega_y \\ \Omega + \omega_z + \theta\dot{\phi} \end{bmatrix} \tag{4-22}$$

The only terms not already analyzed are those involving θ. Of these, when (4-22) is substituted into (4-13), the resulting product containing $\theta\omega_z\Omega$ is by far the most significant (Ex. 4-8); the procedure of Section 4.2.1 then yields the indicated angular rate,

$$(B/H)\dot{\theta} \doteq \omega_x - \{\gamma_y(\ddot{\theta} + \dot{\omega}_y) + \gamma_{zx}\omega_x\omega_z + \theta\omega_z\} \tag{4-23}$$

which has a very simple physical interpretation. In addition to the terms derived previously, (4-23) includes the interaction of ω_z with an instantaneous misalignment of the gyro gimbal. With (3-10) used to approximate this last error source, (4-23) can be rewritten

$$(B/H)\dot{\theta} \doteq \omega_x - \left\{ \gamma_y(\ddot{\theta} + \dot{\omega}_y) + \gamma_{zx}\omega_x\omega_z + (H/B)\omega_z \int \omega_x \, dt \right\} \tag{4-24}$$

but the time integral used here is valid only in a midspectral region determined as follows: For vibratory angular rates with oscillation frequencies beyond cutoff in (4-17), (4-23) cannot be replaced by (4-24) because (3-10) is inapplicable. For rotations slow enough to be counteracted by platform servos (Fig. 3-3) or large enough to be rebalanced (Fig. 3-2), the *closed loop* dynamics of θ, and not (3-10), will be applicable to (4-23).

Since platform stabilization and rebalancing commands are mechanized on the basis of OA rotation, these last two equations demonstrate how generalized conditions produce unwanted additions to ω_x. There are also other errors, *not* originating from the Eulerian dynamics. The next section identifies these degradations, expresses each in the form of (4-3), and provides a systematic error analysis for INS components in general.

4.3 COMPONENT IMPERFECTIONS

For detailed INS performance analysis and for realistic error model parameters in nav systems using estimation, the development in Section 3.4.2 can be extended. INS degradations vary with the dynamic environment, which typically includes complex motions. Fortunately, expressions for motion-dependent instrument errors can be reduced to simple formulations, such as (4-2) and (4-3), for resultant inaccuracies. In strapdown, there is more to the "effective fixed" values $\langle \mathbf{N} \rangle$ than averages of time-varying instrument discrepancies. For example, strapdown accelerometer outputs containing zero mean oscillatory errors, multiplied in (3-54) by direction cosines driven with similar oscillations, produce additional bias effects through *rectification* (i.e., generation of constant components from products of equal-frequency sinusoids; see Ex. 4-9). This situation, demonstrating undesirable interaction between instrument error and stabilizing computations, is not the only source of rectification to be described in this chapter. More familiar to many inertial systems analysts are rectifications occurring wholly within an instrument [e.g., consider in-phase sinusoids for ω_x and ω_z in (4-21)]. A general analysis is presented here, wherein random vibrations are represented by bands of sinusoids, potentially rectifying whenever variables with overlapping spectra are multiplied.

To systematize the approach, absolute angular rates and specific forces *experienced by the IMU* are written as

$$\boldsymbol{\omega} = \boldsymbol{\kappa} + \mathbf{w} \tag{4-25}$$

$$\mathbf{B} = \mathbf{C} + \mathbf{a} \tag{4-26}$$

where $\boldsymbol{\kappa}$ represents a *profile rate* (i.e., the constant or slowly varying component of $\boldsymbol{\omega}$) and $\mathbf{C}$ represents a steady specific force; $\mathbf{a}$ and $\mathbf{w}$ denote oscillatory accelerations and angular rates, respectively. Thus, for a gimballed platform, $\mathbf{B} \doteq \mathbf{A}$ and $\boldsymbol{\omega} \doteq \boldsymbol{\varpi}$; for the strapdown system previously described, $\boldsymbol{\omega} = \boldsymbol{\omega}_{\mathrm{A}}$ and $\mathbf{C}$ in cruise consists of a 1-g upward lift (Appendix 4.A). Oscillations are represented here by narrowband noise waveforms, with center frequencies f_{a}, f_{w} defined such that

$$\frac{d}{dt} a_i(t) \doteq -2\pi f_{\mathrm{a}} \breve{a}_i \tag{4-27}$$

$$\int_0^t a_i(\mathrm{r})\, d\mathrm{r} \doteq \frac{1}{2\pi f_{\mathrm{a}}} \breve{a}_i \tag{4-28}$$

$$\frac{d}{dt} w_i(t) \doteq -2\pi f_{\mathrm{w}} \breve{w}_i \tag{4-29}$$

$$\int_0^t w_i(\mathrm{r})\, d\mathrm{r} \doteq \frac{1}{2\pi f_{\mathrm{w}}} \breve{w}_i \tag{4-30}$$

in which the overhead arc ($\breve{\ }$) denotes a 90° phase lag (Ex. 4-10). Series of the form $S_{\mathrm{a}1}$ and $S_{\mathrm{w}1}$, linearly proportional to these oscillatory waveforms, are more fully analyzed in Appendix 4.B. For immediate purposes it is sufficient to cite the last term of (4-21) as an example of rectifiable error; when components from (4-25) are substituted for ω_x and ω_z of the jth gyro, a product is obtained with the form

$$w_j w_{zj} = \langle w_j w_{zj} \rangle + S_{w2} \tag{4-31}$$

where the trigonometric series of the type denoted S_{w2}, spectrally centered about $2f_w$, is a zero-mean (generally non-Gaussian) random variable. Its detailed form, and the exact nature of **w**, need not be scrutinized. The main contributor to gyro degradation in this example is the *covariance* $\langle w_j w_{zj} \rangle$, which depends only on rms rotational oscillations and the extent of coupling between nominal IA and SA components. This notation can also be used to form other products such as

$$\breve{w}_j w_{zj} = \langle \breve{w}_j w_{zj} \rangle + \breve{S}_{w2} \tag{4-32}$$

$$a_j a_{zj} = \langle a_j a_{zj} \rangle + S_{a2} \tag{4-33}$$

$$\breve{a}_j a_{zj} = \langle \breve{a}_j a_{zj} \rangle + \breve{S}_{a2} \tag{4-34}$$

and similar expressions hold if OA oscillation components (subscripted with y) are inserted.

The above approach will be used throughout this chapter, to reduce various error sources into simple general classes. Total bias (constant or spectrally concentrated near zero frequency) is amenable to simplified analysis, such as that in Section 3.4.2. Narrowband noise, though non-cumulative in itself, contributes to total bias (through rectification), to $\mathbf{e}_a$ (e.g., in spectral regions near f_a and $2f_a$), and to $\mathbf{e}_\omega$ (e.g., in spectral regions near f_w and $2f_w$). These *wideband* errors have growth properties exhibited as follows: Suppose that, over a few minutes duration ($0 \le t \le \mathcal{T}$), all kinematical coupling for a gimballed platform can be ignored. For a north gyro drift rate consisting of white noise $e_\omega(t)$, and no initial platform tilt,

$$\psi_N = \int_0^{\mathcal{T}} e_\omega(t)\, dt \tag{4-35}$$

so that (Ex. 4-11)

$$\psi_N{}^2 = \int_0^{\mathcal{T}} \int_0^{\mathcal{T}} e_\omega(t) e_\omega(\mathrm{r})\, d\mathrm{r}\, dt \tag{4-36}$$

and, from a scalar version of (II-61),

$$\langle \psi_N{}^2 \rangle = \int_0^{\mathcal{T}} \int_0^{\mathcal{T}} E_\omega(t)\, \delta(t - \mathrm{r})\, d\mathrm{r}\, dt = \int_0^{\mathcal{T}} E_\omega(t)\, dt \tag{4-37}$$

where δ represents the Dirac delta function and E_ω denotes spectral density.

When E_ω is time-invariant,

$$\langle \psi_N^2 \rangle = E_\omega \mathscr{T} \tag{4-38}$$

so that rms tilt from this source grows as the square root of time. While this is more gradual than cumulative effects of constant drift rates, random errors can exert a dominating influence on some nav system operations if spectral densities are sufficiently large. The disposition of random errors, including approximations and conservative assumptions permissible with band limiting or other spectral shaping, is addressed in subsequent chapters. For the remainder of this chapter, error characterization using (4-31)–(4-34) or similar standard expressions is adequate; covariances will be treated here as known quantities (or quantities for which conservative models can be prescribed), constant within each flight phase.

Methods just discussed will now be applied to error analysis of INS components (including the inertial instruments *and* the means of mechanical or computational stabilization), with particular attention given to resultant biases.

4.3.1 Gyro Degradations

This section is not necessarily exhaustive, but most significant drift sources (imperfect null, scale factor, misalignment, and various additional environment-sensitive effects) are taken into account. Clarity is enhanced by description of individual sources before combined effects are considered. Explicit relations are thus given for each separate axis of a triad (e.g., for the jth SDF gyro). As suggested by comparing the $\mathscr{x}$ and $\mathscr{y}$ gyroscopic moments in (4-15), however, much of this analysis can be applied to 2DF gyros as well. Instead of the OA constraint on the rotor in Fig. 3-1, a 2DF ("free") gyro has another viscous restraint; this influences the coefficients in applicable expressions for drift, while often leaving the form of these expressions unchanged.

The subscripts $\mathscr{x}$, $\mathscr{y}$, $\mathscr{z}$ used in Section 4.2.2, as applied to coefficients and to angular rates or specific force components experienced, denote inertial instrument axes (e.g., $\omega_j = \omega_{\mathscr{x}j}$ here represents angular rate sensed by the jth gyro under error-free conditions, while $C_{\mathscr{z}j}$ is the steady component of specific force along its nominal spin axis direction). All coefficients, as well as profile rates $\boldsymbol{\kappa}$, static forces $\mathbf{C}$, and covariances [e.g., first term on the right of (4-31)–(4-34)], are treated as constant within a flight phase. Equations in Section 4.2.2, expressed in the form of apparent angular rate about the jth axis,

$$\hat{\omega}_j = \omega_j - n_{\omega j} \tag{4-39}$$

are then immediately adaptable to the general error analysis now to be presented.

DEGRADATIONS ORIGINATING FROM EULERIAN DYNAMICS

With OA rotation sensitivity as the only gyro imperfection considered, the terms in (4-18) would correspond to those of (4-39), in the same sequence. Since angular rate profiles do not include appreciable rotational accelerations sustained about any axis, average error *observed in the gyro output* would then be negligible. As will be shown in Section 4.4, however, the computed attitude reference would nevertheless drift steadily in a strapdown system.

The last term in (4-21), by reason of (4-25) and (4-31), contributes to both steady drift and oscillatory error:

$$\gamma_{\ast x}\,\omega_j\,\omega_{\ast j} = \gamma_{\ast x}(\kappa_j\kappa_{\ast j} + \langle w_j w_{\ast j}\rangle) + \gamma_{\ast x}(\kappa_j w_{\ast j} + \kappa_{\ast j} w_j + S_{w2}) \qquad (4\text{-}40)$$

The dominant steady drift from this source, significant in strapdown applications, is contributed by angular vibration rate covariances (Ex. 4-12).

By re-expressing the last term of (4-24) in the form of (4-40), a drift term of the type $\check{\gamma}_{\ast x}\langle \check{w}_j w_{\ast j}\rangle$ is obtained (Ex. 4-13). This effect, which can arise in both gimballed platform and strapdown mechanizations, is often referred to as *coning* drift. The name can be explained by an example already presented, in Appendix 2.A.2. With two sinusoidal rates in phase quadrature [(2-94) and (2-95)], the third axis describes a circular cone. More generally, a cone of arbitrary cross section can be formed by two irregular orthogonal angular rates containing components with cross-axis quadrature correlation. Sensitivity to this covariance $\langle \check{w}_j w_{\ast j}\rangle$ depends heavily on the *control* of gimbal displacement. With pulse-torquing in Fig. 3-2, for example, the last term of (4-24) is accurate only for angular oscillation amplitudes below the strapdown resolution ϕ. For gimballed platforms, the stabilization provisions (Fig. 3-5) considerably attenuate rotational displacements at frequencies below the servo bandwidth. Also, with either mechanization, gyro gimbal displacements cannot follow input oscillations at frequencies beyond cutoff in (4-17). The general representation $\check{\gamma}_{\ast x}\langle \check{w}_j w_{\ast j}\rangle$ is valid in all situations, provided the coefficient $\check{\gamma}_{\ast x}$ is a properly defined function of dynamic conditions.*

Reference 4-3 analyzes the above effect in geometric terms, and declares its presence "even in a theoretically perfect instrument." More precisely,

* As suggested by (4-30), $\check{\gamma}_{\ast x}$ is inversely proportional to oscillation frequency for a mid-band between cutoffs of platform stabilization and gyro response, or for subthreshold oscillations below gyro response cutoff in strapdown. For other conditions, the coefficient is reduced in accordance with the stabilization characteristics.

gimbal rotation of a gyro with finite angular momentum and imperfect stabilization can produce drift through interaction with ω_x, even if no other error sources are present.

ADDITIONAL ROTATION-SENSITIVE ERRORS

An unambiguous definition for gyro IA direction will now be chosen, prompted by consideration of misalignments. In Fig. 3-1, the SA coincides with the *spin reference axis* (SRA) and the IA points along $\mathbf{I}_g$ at rest. More generally, the IA here is normal to both OA and SRA. This definition, rather than an alternative (IA coincident with $\mathbf{I}_g$, used in some sources including Ref. 4-3), is convenient for the reference axis conventions of Section 3.3. Deviations of gyro gimbal axes from IA directions are then associated with degradations just analyzed, and IA misalignments are constant mounting imperfections. It follows that (4-25) and (4-26) denote inertial instrument *case*-referenced quantities (Ex. 4-14), while unintentional IA projections $\mathsf{y}_{\omega j}$, $\mathsf{z}_{\omega j}$ are fixed cross-axis sensitivities of the jth gyro, producing errors of the form

$$\mathsf{y}_{\omega j}\omega_{yj} + \mathsf{z}_{\omega j}\omega_{zj} = (\mathsf{y}_{\omega j}\kappa_{yj} + \mathsf{z}_{\omega j}\kappa_{zj}) + (\mathsf{y}_{\omega j}w_{yj} + \mathsf{z}_{\omega j}w_{zj}) \qquad (4\text{-}41)$$

A similar breakdown into bias and oscillatory components characterizes the effect of imperfect calibration, e.g.,

$$\mu_{\omega j}\omega_j = (\mu_{\omega j}\kappa_j) + (\mu_{\omega j}w_j) \qquad (4\text{-}42)$$

where $\mu_{\omega j}$ denotes first-order scale factor error of the jth gyro. Quadratic and higher order terms (e.g., proportional to powers of angular rate components) can be added, if appreciable nonlinear calibration errors are present.

Whereas many angular rate-dependent degradations take time to affect INS performance appreciably, misalignment and imperfect calibration of strapdown gyros take effect immediately upon maneuvering (Ex. 4-15).

The last rotation-dependent error to be considered, of particular importance in strapdown systems, is mismatch in effective response of gyros in a triad. If all instrument and data processing lags were equal (or equalized through intentionally delaying each information source by its departure from the longest lag in the system), imperfect response could be analyzed simply as band limiting [e.g., (4-17)]. Suppose, however, that the jth gyro output leads that of the slowest gyro in the triad by an uncompensated interval $\tau_{\omega j}$. Rotational excursion accumulated during this interval constitutes an error, with an approximate rate (Ex. 4-16)

$$d(\omega_j \tau_{\omega j})/dt \doteq -2\pi f_w \tau_{\omega j} \check{w}_j \qquad (4\text{-}43)$$

By expressing pulse-torqued gyro quantization in term of timing differentials,

Section 4.3.4 uses this last equation to derive a strapdown computational drift rate.

"g-SENSITIVE" AND "g^2-SENSITIVE" ERRORS

Undesired torques resulting from mass unbalance along any direction normal to $\mathbf{J}_g$ in Fig. 3-1 can produce drift as follows: Components of $\mathbf{B}$ interact with pendulous mass m at lever arm $\mathbf{r}$ off the $\mathbf{J}_g$ axis to produce an unwanted torque of the form $m\mathbf{r} \times \mathbf{B}$. Components can be resolved along $\mathbf{I}_g$ and $\mathbf{K}_g$ directions, allowing definition of coefficients γ_{xj} and γ_{zj} so that

$$\gamma_{xj} B_j + \gamma_{zj} B_{zj} = (\gamma_{xj} C_j + \gamma_{zj} C_{zj}) + (\gamma_{xj} a_j + \gamma_{zj} a_{zj}) \qquad (4\text{-}44)$$

Now consider an *instantaneous* displacement of the gyro gimbal, due to unequal elastic displacements in different directions, under the influence of $\mathbf{B}$. In effect this is an *instantaneous* mass unbalance proportional to the appropriate component of $\mathbf{B}$, producing an unwanted torque equal to this instantaneous unbalance times another (perpendicular) component of $\mathbf{B}$. Again, only OA torque components are of interest, and the error (Ex. 4-17) can be expressed as

$$\gamma_{xzj} B_j B_{zj} = \gamma_{xzj}(C_j C_{zj} + \langle a_j a_{zj} \rangle) + \gamma_{xzj}(C_j a_{zj} + C_{zj} a_j + S_{a2}) \qquad (4\text{-}45)$$

As a straightforward extension, suppose that the aforementioned displacement of the gyro gimbal, due to mechanical lags, does not quite correspond with the instantaneous value of $\mathbf{B}$. Instantaneous vibratory displacement can then be decomposed, identifying one component that correlates with $\mathbf{B}$ (producing the effect just described, with γ_{xzj} accounting for only the "in-phase" correlation) and another component that correlates with $\breve{\mathbf{B}} = \breve{\mathbf{a}}$ (producing a "cylindrical drift" with a coefficient denoted here as $\breve{\gamma}_{xzj}$; see Ex. 4-18). Effects of cylindrical motion on gyro drift can also be found in Ref. 4-4, along with several other topics treated here.

COMBINATION OF INDIVIDUAL ERROR SOURCES

For purposes such as manufacturing tolerance specification, certain gyro error coefficients may be treated as random variables in a statistical description of total drift. For performance analysis of interest here, however, a mission is flown with a particular INS having *deterministic* (i.e., nonrandom) parameters. An imperfect null $\nu_{\omega j}$ is then added to the bias sources already described, and steady drift contributions are combined algebraically to produce a resultant value, e.g., for gimballed platforms (Ex. 4-19),

$$\begin{aligned}\langle n_{\omega j(\text{platf})} \rangle \doteq\ & \nu_{\omega j} + \mu_{\omega j} \kappa_j + y_{\omega j} \kappa_{yj} + z_{\omega j} \kappa_{zj} + \breve{\gamma}_{zx} \langle \breve{w}_j w_{zj} \rangle + \gamma_{xj} C_j \\ & + \gamma_{zj} C_{zj} + \gamma_{xzj}(C_j C_{zj} + \langle a_j a_{zj} \rangle) + \breve{\gamma}_{xzj} \langle \breve{a}_j a_{zj} \rangle \end{aligned} \qquad (4\text{-}46)$$

while the oscillatory contributions are included with the superimposed error $\mathbf{e}_\omega$.

4.3.2 Accelerometer Errors

Ground rules for the development to follow match those stated for gyro drift analysis; all error coefficients are deterministic and time-invariant for any given flight phase. In the event that compensating commands are present in any form, the coefficients herein are defined as the uncompensated residual portion. By analogy with (4-39), the apparent value, true value, and uncertainty for the jth component of specific force are related as

$$\hat{B}_j = B_j - n_{aj}, \qquad 1 \leq j \leq 3 \tag{4-47}$$

Contributions to n_{aj} from null bias v_{aj}, misalignments y_{aj}, z_{aj}, and proportionality constant error μ_{aj} are now self-explanatory (Ex. 4-20). For platform applications the total accelerometer error is often characterized as the resultant of these three sources combined with an offset effect analogous to anisoelastic drift in gyros (Ex. 4-21),

$$n_{aj(\text{platf})} \doteq \mathsf{v}_{aj} + \mu_{aj} B_j + \mathsf{y}_{aj} B_{yj} + \mathsf{z}_{aj} B_{zj} + \beta_{xzj} B_j B_{zj} \tag{4-48}$$

The following considerations, noted in the preceding gyro error analysis, are applicable here also:

(1) Scale factor error analysis is readily extended to include a power series.

(2) Since the z axis of each instrument can be chosen along either of two directions, there is an opportunity to select orientations giving the lowest systematic error. For strapdown systems no instrument z axis should be along yaw; for gimballed platforms no instrument z axis should be along the azimuth axis. For the instrument with its IA along yaw (strapdown) or azimuth (platform), the direction of higher thrust should be avoided for z-axis selection.

(3) Equations (4-26), (4-33), and (4-48) can be combined (Ex. 4-22) to produce an expression similar to (4-46).

It is now instructive to recall the mounting alignment convention of Section 3.3. The #2 accelerometer IA (along $\mathbf{J}_P$ or $\mathbf{J}_A$ for a gimballed platform or a strapdown system, respectively) has no projection on either of the other two reference directions; also the #1 accelerometer IA has no projection along the instrument-based z direction. Therefore,

$$\mathsf{y}_{a2} \equiv \mathsf{z}_{a2} \equiv \mathsf{y}_{a1} \equiv 0 \tag{4-49}$$

Furthermore, in view of the order of largest acceleration components [item (2)], this convention provides greater benefit than other instrument references (e.g., if the #3 accelerometer had been chosen as a reference axis by definition, the other two IA's would not necessarily be insensitive to the largest static force). This benefit is actually a matter of proper definition: nav performance is a function of instrument alignments relative to *each other*.

Three reference variables can be set by choice, and the remaining degrees of freedom (in the present case, $\mathbf{z}_{a1}$, $\mathbf{y}_{a3}$, $\mathbf{z}_{a3}$, and all gyro IA misalignments) are expressed as departures from the conventions imposed.

Accelerometer errors remaining to be discussed here are of interest primarily for strapdown applications. First, in close analogy with (4-43), a relative timing error $\boldsymbol{\tau}_{aj}$ produces an error

$$d(B_j \boldsymbol{\tau}_{aj})/dt \doteq -2\pi f_a \boldsymbol{\tau}_{aj} \check{a}_j \tag{4-50}$$

Also inherent in any finite-sized accelerometer triad is a degradation due to noncoincident location. Unlike angular rates, which would be uniform throughout any rigid assembly of instruments, accelerations are position-dependent. Let an IMU reference point be chosen (e.g., equidistant from the accelerometer centroids) and apply (2-37) to the fixed position vector $\mathbf{r}_j = (r_{j1}, r_{j2}, r_{j3})$ for each of these centroids. The difference between the theoretical outputs and the true specific force at the IMU reference point in a strapdown system is expressed by (Ex. 4-23)

$$\begin{bmatrix} \mathbf{I}_A \cdot d^2\mathbf{r}_1/dt^2 \\ \mathbf{J}_A \cdot d^2\mathbf{r}_2/dt^2 \\ \mathbf{K}_A \cdot d^2\mathbf{r}_3/dt^2 \end{bmatrix} = \begin{bmatrix} (\boldsymbol{\omega}_A \cdot \mathbf{r}_1)\omega_{A1} - r_{11}|\boldsymbol{\omega}_A|^2 \\ (\boldsymbol{\omega}_A \cdot \mathbf{r}_2)\omega_{A2} - r_{22}|\boldsymbol{\omega}_A|^2 \\ (\boldsymbol{\omega}_A \cdot \mathbf{r}_3)\omega_{A3} - r_{33}|\boldsymbol{\omega}_A|^2 \end{bmatrix} + \begin{bmatrix} \dot{\omega}_{A2} r_{13} - \dot{\omega}_{A3} r_{12} \\ \dot{\omega}_{A3} r_{21} - \dot{\omega}_{A1} r_{23} \\ \dot{\omega}_{A1} r_{32} - \dot{\omega}_{A2} r_{31} \end{bmatrix} \tag{4-51}$$

From the now familiar decomposition of motion components and formation of rectification products, it follows that the centripetal term [first vector on the right of (4-51)] produces a bias of order $[|\boldsymbol{\omega}_A|^2 \times$ (mounting separation distance)]. The last term, when resolved along reference coordinates (Section 4.4), produces a "rectified tangential acceleration" error at this same nominal level, by the mechanism already described in Ex. 4-9.

Conceivably, numerous additional degradations could be discussed here. Many of the remaining error sources, however, pertain to specific types of accelerometers (e.g., OA angular acceleration sensitivity of the form $\boldsymbol{\beta}_y \dot{\omega}_{yj}$ and anisoinertia effects expressible as $\boldsymbol{\beta}_{xx} \omega_j \omega_{xj}$, in analogy with corresponding gyro drifts previously discussed, are applicable to so-called pendulous accelerometers). Rather than attempting to catalog every possible type of degradation, this subsection can terminate with the concept of overall resultant error separable into steady and oscillatory constituents as already described.

4.3.3 Platform Servo Imperfections

Gimballed platform servo response is conveniently expressed in terms of degradations added to gyro drift rates. Clearly it makes little difference, from the standpoint of overall nav system performance, whether a given misorientation resulted from erroneous dynamic commands (actual gyro drift) or imperfect response to those commands. The latter will now be analyzed under simplifying assumptions, after which a more general class of conditions will be discussed.

This development begins by characterizing the platform as a rotating body subjected to a net torque $\mathbf{L}_P$ and responding in a manner analogous to (2-83) [i.e., with all A subscripts replaced by P, provided that the triad ($\mathbf{I}_P$, $\mathbf{J}_P$, $\mathbf{K}_P$) corresponds to principal axes of inertia]. As defined prior to (2-19), $\boldsymbol{\omega}_P$ represents the total angular rate of the platform; its departure from the instantaneously commanded rate is simply obtained by subtraction. When this departure is formulated under conditions of equal moments of inertia about ($\mathbf{I}_P$, $\mathbf{J}_P$, $\mathbf{K}_P$), the result for any axis appears in the simple form,

$$\mathsf{J}\ddot{\varphi} = \mathsf{L}_E \tag{4-52}$$

where φ represents a contribution to one component of $\boldsymbol{\psi}$ resulting from platform servo disturbances; J, of course, denotes the corresponding moment of inertia; and

$$\mathsf{L}_E = \mathsf{L}_D - \mathsf{L}_F \tag{4-53}$$

is an error torque due to incomplete cancellation of a disturbance L_D by a feedback torque L_F. The transfer function relating L_F and φ is the product of a feedback transfer function F multiplied by a gyro transfer function G, From this, in combination with the last two equations (Ex. 4-24), the transfer function relating undesired platform angular rate $w = \dot{\varphi}$ to disturbance torque is

$$w(s)/\mathsf{L}_D(s) = s/(\mathsf{J}s^2 + \mathsf{FG}) \tag{4-54}$$

A symbolic representation appears in Fig. 4-1a; all pertinent servo theory can be found in standard texts on control systems. This development first considers the simplified case in which $\mathsf{G} = H/B$ (see Section 3.2), and F can take a simple lead-lag form. By appropriate design the open loop transfer function of Fig. 4-1b can then be obtained (and can be maintained in this common form when further design additions are adopted, such as compensating networks for increased low-frequency stiffness). Moreover, a closed-loop function of the type shown in Fig. 4-1c can be obtained that suitably approximates the response when the disturbance L_D is rate-dependent gimbal friction. With the same transfer function in all three axes, the overall response can be expressed in terms of an effective proportionality factor K, damping coefficient ζ, and natural frequency ω_n; thus

$$\mathbf{w}(s) = \left[\mathsf{K}\frac{s/\omega_n}{1 + 2\zeta s/\omega_n + (s/\omega_n)^2}\right]\mathbf{u}(s) \tag{4-55}$$

where $\mathbf{u}(s)$ here denotes the Laplace transform of aircraft rate in platform

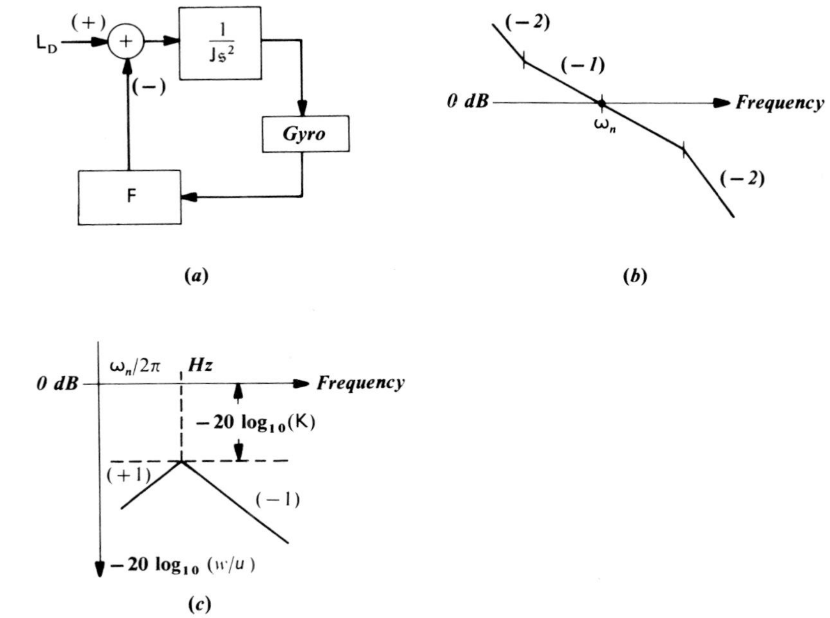

FIG. 4-1. Platform servo error; (a) single axis rotation error, (b) open- and (c) closed-loop transfer function.

axes, $\mathbf{T}_{P/A}\boldsymbol{\omega}_A$. This is the type of behavior described in Ref. 4-5 for the direct drive case (i.e., without gears).

In addition to band limiting via (4-17), this development ignores a host of smaller effects. Among these are cross-axis coupling, nonorthogonality of the platform gimbals, static as well as dynamic unbalance, various misalignments (including that of the azimuth resolver mentioned in Section 3.2), nonrigidity (including anisoelasticity of the platform itself), and resonances (Ref. 4-6). These degradations can be collectively represented by a conservative assessment of their contribution to $\mathbf{e}_\omega$.

In many analyses the effects of platform servo transients are viewed as delayed, distorted, differentiated miniature replicas of aircraft rotation, via (4-55), with the superimposed random degradations just described. These effects are often less serious than *gyro* drifts associated with imperfect stabilization. Suppose, for example, that the peak response of Fig. 4-1c occurs at 5 Hz, while J/B is 0.001 sec. Then a 10-Hz angular oscillation is well within the bandwidth of (4-17), but causes appreciable stabilization error in (4-55). Reference 4-3 postulates a gyro whose $\mathbf{I}_g$ axis describes a circular cone with 10-mr vertex angle at this 10-Hz rate, while the $\mathbf{K}_g$ axis undergoes similar motion in phase quadrature. The SA *rate* ω_z is then correlated with angular *displacement* of the gyro gimbal (Ex. 4-25), and the average value of the last term in (4-23) is 0.18°/sec, orders of magnitude beyond any acceptable drift rate. This emphasizes the importance of the closing statement in Ref. 4-3, recommending that shock mounts be designed to disallow conical motions in certain frequency bands.

4.3.4 Strapdown Data Processing Errors

The configuration defined at the beginning of Section 3.5 generates an attitude drift from inexact updating (Ex. 4-26). Data processing degradations are often subdivided into roundoff, truncation, and commutation errors. Since the direction cosine adjustments from (3-58) and (3-59) are equally prone to high and low roundoff, this discrepancy* can be included in the zero-average $\mathbf{e}_\omega$. Series truncation error in those two equations, as already explained in Section 3.5, can be ignored. A significant drift, however, can be traced to processing of rotation data just before computation, i.e., sampling or quantizing of gyro information. For any given time history of vehicle rates, complete over any specified time interval, there is a unique set of outputs that would be seen over this interval by a triad of perfect gyros (Ex. 4-28). When these outputs are sampled or quantized, however, the converse is not true; for any triplet of *processed* gyro outputs, there must exist an infinity of vehicle rate time histories that could have generated it. Within this infinite set, the range of possible angular orientations at the end of the interval is a measure of drift error. This error can be attributed to possible subthreshold angular excursions in a DDA mechanization, or spectral aliasing associated with signals of finite duration (and therefore infinite bandwidth) sampled at finite rates in a system using numerical integration. In any case, the dominant processing error arises from combining multiaxis rotational data, with some of the time-history "fine structure" missing, while commutativity holds only in the infinitesimal limit.

During the early years of strapdown development, propagation characteristics of commutation error were largely unknown. Short duration examples as typified by (3-61) were of course recognized, and for larger excursions situations shown in Fig. 4-2 could be postulated, leading to an orientation error (Ex. 4-29)

$$[\phi]_y[\pi]_x[-\pi]_x\{[\pi]_x[\phi]_y[-\pi]_x\}^{-1} = [2\phi]_y \tag{4-56}$$

which produces a much greater error than that indicated by (3-61) but over a significantly longer time. The available examples, however, did not

* After a large number K of active iterations with a DDA word length of $\mathscr{W}$ bits, cumulative effects of roundoff produce an angular orientation error proportional to $2^{-\mathscr{W}}\sqrt{K}$ rad for typical motion patterns (Ex. 4-27).

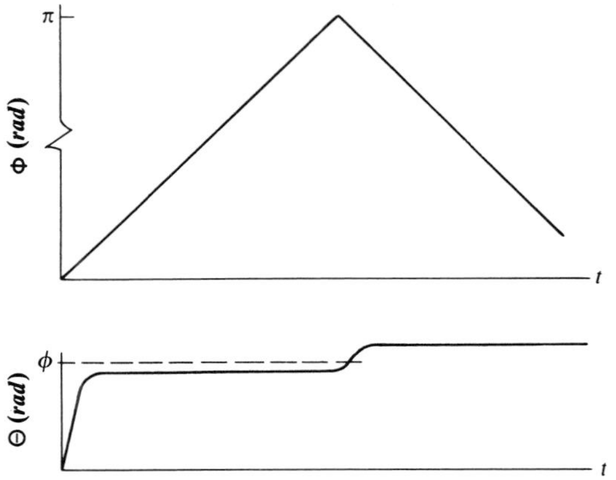

FIG. 4-2. Reversal of pitch axis direction before threshold crossing of pitch displacement.

provide a systematic approach for predicting general error growth in the presence of random angular rates, fully or partially correlated across axes, with or without coning, lasting for millions of attitude computer iterations (quite realistic for flight phases of several minutes duration, with less than 1-msec iteration time). To fill this need Ref. 4-1 characterized the quantization effect by timing errors directly proportional to resolution and inversely proportional to angular rates, for substitution into (4-43) and (3-82) reinterpreted as follows: In the discussion preceding (3-82), the vehicle angular rate $\boldsymbol{\omega}_A$ can be substituted in place of $\boldsymbol{\omega}_P$ while $\boldsymbol{\omega}_B$ can be redefined for strapdown as $\hat{\boldsymbol{\omega}}_A = \boldsymbol{\omega}_A - \mathbf{n}_\omega$ (Ex. 4-30). Then, with the notation $\boldsymbol{\xi}$ used to denote strapdown attitude error, (3-82) is replaced by

$$\dot{\mathbf{T}}_{B/A} = \frac{d}{dt}\begin{bmatrix} 0 & -\xi_3 & \xi_2 \\ \xi_3 & 0 & -\xi_1 \\ -\xi_2 & \xi_1 & 0 \end{bmatrix} = -\mathbf{T}_{B/A}\boldsymbol{\Omega}_A + \hat{\boldsymbol{\Omega}}_A \mathbf{T}_{B/A} \tag{4-57}$$

By a derivation paralleling that in Appendix 3.B,

$$\dot{\boldsymbol{\xi}} = \mathbf{n}_\omega - \boldsymbol{\omega}_A \times \boldsymbol{\xi} \tag{4-58}$$

As a close approximation to the solution of this equation, the time integral of the drift rate can be substituted into the vector product for the last term (Ex. 4-31); since the present analysis is concerned primarily with averages,

$$\langle \dot{\boldsymbol{\xi}} \rangle \doteq \langle \mathbf{n}_\omega \rangle - \left\langle \boldsymbol{\omega}_A \times \int \mathbf{n}_\omega \, dt \right\rangle \tag{4-59}$$

which explicitly illustrates how rectification can occur in the presence of rotation-dependent drift. In the case of unequal delays in (4-43), the contribution from rectification becomes essentially

$$\left\langle -\mathbf{w} \times \begin{bmatrix} \tau_{\omega 1} & 0 & 0 \\ 0 & \tau_{\omega 2} & 0 \\ 0 & 0 & \tau_{\omega 3} \end{bmatrix} \begin{bmatrix} w_1 \\ w_2 \\ w_3 \end{bmatrix} \right\rangle = -\begin{bmatrix} (\tau_{\omega 3} - \tau_{\omega 2})\langle w_2 w_3 \rangle \\ (\tau_{\omega 1} - \tau_{\omega 3})\langle w_3 w_1 \rangle \\ (\tau_{\omega 2} - \tau_{\omega 1})\langle w_1 w_2 \rangle \end{bmatrix} \tag{4-60}$$

where, through (4-25), $\mathbf{w}$ is identified with $\boldsymbol{\omega}_A$ for strapdown. Immediately this result is applicable to timing errors due to quantization, which may be half the interval between successive ϕ-threshold crossings. If timing error for the jth gyro due to resolution may be approximated by the ratio $\phi/(2w_{j(\text{rms})})$, there is a steady component of commutation error,

$$-\frac{\phi}{2}\begin{bmatrix} \left(\dfrac{1}{w_{3_r}} - \dfrac{1}{w_{2_r}}\right)\langle w_2 w_3 \rangle \\ \left(\dfrac{1}{w_{1_r}} - \dfrac{1}{w_{3_r}}\right)\langle w_3 w_1 \rangle \\ \left(\dfrac{1}{w_{2_r}} - \dfrac{1}{w_{1_r}}\right)\langle w_1 w_2 \rangle \end{bmatrix} = \frac{\phi}{2}\begin{bmatrix} \rho_{23}(w_{3_r} - w_{2_r}) \\ \rho_{31}(w_{1_r} - w_{3_r}) \\ \rho_{12}(w_{2_r} - w_{1_r}) \end{bmatrix} \tag{4-61}$$

(where ρ_{ij} represents the normalized cross-correlation coefficient between w_i and w_j, and the subscript r denotes rms value), *if* all three axes have sufficient angular rates. However, when the excursion about any axis is less than ϕ, the other two components of commutation drift vanish (Ex. 4-32), and all commutation error vanishes if any two axes are inactive. Thus (4-61) is consistent with all previously known commutation error properties; it is also consistent with accumulation (Ex. 4-33) of short-term errors from (3-61).

Just as gyro drift dominates long-term INS error in Section 3.4.2, commutation error is the principal long-term strapdown data processing degradation (Ex. 4-34). Accelerometer quantization (Ex. 4-35), however, can degrade accuracy in the short term.

FURTHER DATA PROCESSING OPTIONS

The foregoing model of commutation error is based on a simple concept: uncertainty in timing of a resolution threshold crossing, characterized by choosing the center of the time elapsed between successive crossings, can reach a value of half that elapsed period. Since an important drift component could be predicted on the basis of this insight, it seems reasonable to postulate a mechanized correction based on (4-61). Actually a variety of error sources could be similarly compensated, to the extent that motion properties (mean and rms levels of specific force and angular rate components, correlation data, etc.) are correctly known. That topic has already been mentioned in this chapter, and the error analyses herein are in terms of uncompensated levels remaining after corrective measures, if any, are taken. The subject is briefly revived here, however, in order to *discourage* expectations for increased sophistication in attitude algorithms. Specifically, the optimal estimation method discussed in subsequent chapters, so crucial to mixing navigation data from different sources, is *not* a candidate for replacement of (3-58) and (3-59). Not only would the high-speed computational requirements be markedly increased, but (see Chapter 5) a model of the dynamics would be essential—the same information that, if available, would enable the previously described commutation error correction! Since this is the dominant processing error, an estimation scheme would offer little theoretical advantage as a strapdown attitude algorithm. Moreover, even this theoretical improvement would vanish in the presence of inevitable dynamic model imperfections. Further consideration of general estimation model requirements must await the next chapter but, for the present topic, the following should be kept in mind:

(1) Compensation of a conventional DDA algorithm by (4-61) is easily mechanized and need not be exact, e.g., if the rotational characteristics were partly known a partial correction would be obtained.

(2) An estimation algorithm, in contrast, requires either a correct model or added provisions to counteract the model imperfections. These imperfections otherwise become a severe threat to computational stability.

It might be expected from the preceding material that alternative parametrizations would not remove the degradations introduced by quantization of finite rotations. This is true, whether the chosen parameter set consists of quaternion elements [e.g., generalization of (2-41) analogous to (2-31), followed by approximating the result in analogy with (3-58) and (3-59)], Euler angles [e.g., comparable representation of rotation based ultimately upon (2-22), with restricted pitch excursion], or any other means of characterization. A similar statement holds for arithmetical procedures, such as replacement of the second-order expression in (3-58) by a reversible sequential first-order algorithm (Ref. 4-7). While various algorithms may be reasonable candidates for implementation, they cannot of themselves be expected to remove commutation error introduced ahead of the DDA computation. It follows that an approach should be selected on the basis of overall computing requirements. For attitude alone, the quaternion parametrization would hold an advantage. This is compromised, however, in view of subsequent coordinate resolution requirements [recall Ex. 2-9 and note that a similar conclusion holds for other related algorithms, such as that described in Ref. 4-8. Note also that (2-48) is equivalent to Eq. (8) of Ref. 4-8].

The preceding paragraph offers an explanation for the emphasis given to direction cosine algorithms here. While certainly not the only candidate approach, (3-58) and (3-59) afford efficiency in implementation and, for any well chosen DDA computation, the foregoing analysis provides a general performance assessment. Note that neither this assessment nor the mechanization description includes consideration of Appendix 2.B. The footnote on the last page of Ref. 4-1, which describes the orthogonalization, explains that it is unnecessary in this application.

Much of the discussion thus far has been slanted toward pulse rebalancing with DDA computation. For attitude at least, various numerical integration algorithms are often considered as alternatives. Suppose, for example, (3-55) were abandoned in favor of a Runge–Kutta solution to (2-31), with vehicle rotation data taken from gyro outputs in analog waveform (unquantized) signal formats. In effect, the number $\mathcal{N}$ of derivatives taken into account would be determined by the order of the numerical integration algorithm. Simple algorithms would call for intervals dictated by desired rotational resolution and angular rates encountered, while higher values of $\mathcal{N}$ would allow larger time steps; thus overall computational complexity would not be markedly greater or less than that required by the DDA scheme. Nor would the performance for any level of complexity differ greatly between the pulse torque/DDA mechanization and the analog gyro/numerical integration approach, in the presence of general rotational oscillations [i.e., containing correlations defined by both (4-31) and (4-32)]. The reason is traceable to an analogy between finite quantization in one mechanization versus finite sampling rates in the other. While the plurality of possible numerical integration schemes precludes a single all-encompassing formula, a general synopsis can be offered, as follows:

Absence of information regarding rotational derivatives of order beyond $\mathcal{N}$ is tantamount to loss of some fine detail regarding the precise time history of angular motions. Any component of sudden rotation involving these missing derivatives will not be properly accounted for in a numerical integration; when this occurs about multiple axes a commutation error can result. For fixed-step integration schemes, correlation of algorithm error with angular rates or their Hilbert transforms (90°-shifted waveforms) will depend on whether $\mathcal{N}$ is even or odd. This in turn will determine whether the scheme is primarily sensitive to correlations from (4-31) or (4-32).

With a well-chosen algorithm then, computational errors should be influenced more heavily by environment and system parameters than by selection between alternative computing schemes of comparable complexity. Attitude drift arises from processing of gyro data that takes place ahead of the computation.

4.4 GENERALIZED ERROR PROPAGATION ANALYSIS

The ensuing development extends Section 3.4.2 to include inertial instrument rotations and the general component errors just described. Performance is demonstrated in a geographic frame, for three SDF pulse-torqued rate integrating gyros and digital integrating accelerometers, with IA orientation nominally parallel to aircraft axes; application to other configurations and conditions is straightforward.* As in the preceding analysis, error dynamics are expressed in terms of fixed covariances, static forces, system parameters, and initial conditions applicable to a flight phase.

With specific force in aircraft coordinates denoted

$$\mathbf{F} = \mathbf{T}_{A/G}\,\mathbf{A} \tag{4-62}$$

the basis of this mechanization, from (3-33), is

$$\dot{\mathbf{V}} = \mathbf{T}_{G/A}\,\mathbf{F} + \mathbf{g} + \mathbf{f} \tag{4-63}$$

while the apparent derivative of geographic velocity is

$$\dot{\hat{\mathbf{V}}} = \hat{\mathbf{T}}_{G/A}\,\hat{\mathbf{F}} + \hat{\mathbf{g}} + \hat{\mathbf{f}} \tag{4-64}$$

in which the computed transformation matrix is expressible as

$$\hat{\mathbf{T}}_{G/A} = \hat{\mathbf{T}}_{G/I}\,\hat{\mathbf{T}}_{I/A} = \mathbf{T}_{\hat{G}/G}\,\mathbf{T}_{G/I}\,\mathbf{T}_{I/A}\,\mathbf{T}_{A/\hat{A}} \tag{4-65}$$

where

$$\mathbf{T}_{A/\hat{A}} = \mathbf{I} - (\boldsymbol{\xi} \times\ ) \tag{4-66}$$

while the accumulated effect of uncertainty $\tilde{\boldsymbol{\varpi}}$ in reference frame rotation rate is represented by $\boldsymbol{\psi}_G$ such that

$$d\boldsymbol{\psi}_G/dt = \tilde{\boldsymbol{\varpi}} \tag{4-67}$$

and

$$\mathbf{T}_{\hat{G}/G} = \mathbf{I} - (\boldsymbol{\psi}_G \times\ ) \tag{4-68}$$

In combination with the total drift effect, re-expressed in geographic coordinates,

$$\boldsymbol{\psi}_F = \mathbf{T}_{G/A}\,\boldsymbol{\xi} \tag{4-69}$$

* The development is easily applied to a wander azimuth reference, for reasons given at the end of Section 3.6. Reference 4-2 used an inertial frame to obtain similar results. Other instrument mounting geometries are readily accommodated, by inserting appropriate transformations for IA directions. Equations in this section can also characterize gimballed platform performance, when various rotation-sensitive errors are ignored.

the total misorientation vector is (Ex. 4-36)

$$\boldsymbol{\psi} = \boldsymbol{\psi}_F + \boldsymbol{\psi}_G \tag{4-70}$$

This formulation is fully compatible with error analysis in both the preceding chapter and the preceding section (Ex. 4-37), for misorientation dynamics governed by

$$\dot{\boldsymbol{\psi}}_G = \tilde{\boldsymbol{\varpi}} - \boldsymbol{\varpi} \times \boldsymbol{\psi}_G \tag{4-71}$$

and

$$\dot{\boldsymbol{\psi}}_F = \dot{\mathbf{T}}_{G/A}\boldsymbol{\xi} + \mathbf{T}_{G/A}\dot{\boldsymbol{\xi}} = -\boldsymbol{\varpi} \times \boldsymbol{\psi}_F + \mathbf{T}_{G/A}(\dot{\boldsymbol{\xi}} + \boldsymbol{\omega}_A \times \boldsymbol{\xi}) \tag{4-72}$$

since, from the last three equations and (4-58),

$$\dot{\boldsymbol{\psi}} + \boldsymbol{\varpi} \times \boldsymbol{\psi} = \tilde{\boldsymbol{\varpi}} + \mathbf{T}_{G/A}\,\mathbf{n}_\omega \tag{4-73}$$

This last equation enables immediate identification of rectified drifts for addition to (4-46). With $\mathbf{T}_{G/A}$ expressed (Ex. 4-38) as an oscillatory rotational displacement superimposed on mean heading and pitch angles $\langle \Psi \rangle$ and $\langle \Theta \rangle$, respectively,

$$\mathbf{T}_{G/A} \doteq [\langle \Psi \rangle]_z^T [\langle \Theta \rangle]_y^T \left\{ \mathbf{I} + \frac{1}{2\pi f_w} (\breve{\mathbf{w}} \times \quad) \right\} \tag{4-74}$$

the interaction between $\breve{\mathbf{w}}$ and oscillatory components of $\mathbf{n}_\omega$ in (4-73) causes drift. Equivalently, the time integral of $\mathbf{n}_\omega$ is rectified by the derivative of $\breve{\mathbf{w}}$, as already expressed in (4-59). Contributions to $\langle \mathbf{N}_\omega \rangle$ from expressions paralleling (4-60) can include (Ex. 4-39) the last term in (4-18), the last terms in (4-41) and (4-42) in the presence of general coning motion, and translation/rotation interactive drifts from (4-44) and (4-45). Drift sources associated with rotational oscillations are summarized in Table 4-1, expressed in terms of motion correlations defined in Appendix 4.B.

Generalization of (3-37) follows readily from combining (4-65), (4-66), and (4-68), with small error product terms omitted, to form the matrix

$$\hat{\mathbf{T}}_{G/A} \doteq \mathbf{T}_{G/A} - (\boldsymbol{\psi}_G \times \quad)\mathbf{T}_{G/A} - \mathbf{T}_{G/A}(\boldsymbol{\xi} \times \quad) \tag{4-75}$$

and substitution into (4-64), subtracted from (4-63), again with error products omitted:

$$\dot{\tilde{\mathbf{V}}} = (\boldsymbol{\psi}_G \times \quad)\mathbf{T}_{G/A}\mathbf{F} + \mathbf{T}_{G/A}(\boldsymbol{\xi} \times \mathbf{F}) + \mathbf{T}_{G/A}(\mathbf{F} - \hat{\mathbf{F}}) + \mathbf{g} - \hat{\mathbf{g}} + \mathbf{f} - \hat{\mathbf{f}} \tag{4-76}$$

With (4-62), (4-69), and (4-70), and the identification of accelerometer error in this system as

$$\mathbf{n}_a = \mathbf{F} - \hat{\mathbf{F}} \tag{4-77}$$

TABLE 4-1

DRIFT RATE RECTIFICATION PRODUCTS INVOLVING ROTATIONAL OSCILLATIONS

Rectification of rotational error from	Combined with oscillations in	In the presence of motion defined by Table 4-3, Appendix 4.B, case
Finite gyro gimbal rotation		5
Anisoinertia		4
OA angular acceleration		4, 5[a]
Scale factor	angular rate	5
Misalignment		5
Response time differential		4
Computational lag		4
Mass unbalance		6, 8
(Anisoelasticity) × (steady specific force)	specific force	6, 8
(Cylindrical drift sensitivity) × (steady specific force)		7, 9

[a] Possible rectification in the presence of generalized coning would arise directly from squaring individual angular rate components, rather than from quadrature correlation. This phenomenon could also occur with uncoupled rotation.

the preceding expression reduces to (Ex. 4-40)

$$\dot{\tilde{\mathbf{V}}} = \boldsymbol{\psi} \times \mathbf{A} + \mathbf{T}_{G/A}\,\mathbf{n}_a \tag{4-78}$$

where imperfections in $\hat{\mathbf{f}}$ and $\hat{\mathbf{g}}$ are again deferred to Chapter 6. Just as the last term in (4-73) produces drift through rectification, the last term of (4-78) contributes to $\langle \mathbf{N}_a \rangle$ via (4-74). This is the same phenomenon introduced at the beginning of Section 4.3 and demonstrated in Ex. 4-9. Most important effects of this type are listed in Table 4-2, but those interested in a more complete assessment of rectification are referred to Appendix 4.B.

Since $\mathbf{n}_a$ and $\mathbf{n}_\omega$ are defined in the inertial instrument-based reference frame, while $\langle \mathbf{N}_a \rangle$, $\langle \mathbf{N}_\omega \rangle$, $\mathbf{e}_a$, and $\mathbf{e}_\omega$ are expressed in the nav reference, the relation

$$\begin{bmatrix} \mathbf{T}_{G/A}\,\mathbf{n}_a \\ \mathbf{T}_{G/A}\,\mathbf{n}_\omega \end{bmatrix} = \begin{bmatrix} \langle \mathbf{N}_a \rangle \\ \langle \mathbf{N}_\omega \rangle \end{bmatrix} + \begin{bmatrix} \mathbf{e}_a \\ \mathbf{e}_\omega \end{bmatrix} \tag{4-79}$$

replaces (4-2) and (4-3) for strapdown systems. Substitution of these expressions into (4-73) and (4-78) reduces strapdown system error propagation to the form

$$\begin{bmatrix} \dot{\tilde{\mathbf{V}}} \\ \hline \dot{\boldsymbol{\psi}} \end{bmatrix} = \left[\begin{array}{c:c} \mathbf{O}_{3\times 3} & (-\mathbf{A}\,\times\ \ ) \\ \hdashline \partial \boldsymbol{\varpi}/\partial \mathbf{V} & (-\boldsymbol{\varpi}\,\times\ \ ) \end{array}\right] \begin{bmatrix} \tilde{\mathbf{V}} \\ \hline \boldsymbol{\psi} \end{bmatrix} + \begin{bmatrix} \langle \mathbf{N}_a \rangle + \mathbf{e}_a \\ \hline \langle \mathbf{N}_\omega \rangle + \mathbf{e}_\omega \end{bmatrix} \tag{4-80}$$

TABLE 4-2

RECTIFICATION PRODUCTS OBTAINED FROM INTERACTION BETWEEN ROTATIONAL VIBRATION AND OSCILLATORY ACCELEROMETER ERROR

Rectification of acceleration error from	Combined with oscillation in	In the presence of motion defined by Table 4-3, Appendix 4.B, case
Tangential component of accelerometer separation error	angular rate	4, 5[a]
OA angular acceleration	angular rate	4, 5[a]
Scale factor	specific force	8
Misalignment	specific force	6, 8
Response time differential	specific force	9
(g-squared sensitivity) × (steady specific force)	specific force	6, 8
(Quadrature g-squared sensitivity) × (steady specific force)	specific force	7, 9

[a] Possible rectification in the presence of generalized coning would arise directly from squaring individual angular rate components, rather than from quadrature correlation. This phenomenon could also occur with uncoupled rotation.

which is the *same* relation used to analyze gimballed platform error dynamics. Therefore, with components of $\langle \mathbf{N}_a \rangle$ and $\langle \mathbf{N}_\omega \rangle$ substituted for biases in Section 3.4.2, the solutions obtained therein characterize major system errors in that analysis (valid for one or two hours). In addition, (4-80) provides a basis for further extensions of that analysis, applicable to both gimballed platform and strapdown systems.

APPENDIX 4.A: DYNAMIC ENVIRONMENT OF AIRBORNE INERTIAL INSTRUMENTS

Dependence of gyro and accelerometer degradations upon forces and angular rates calls for a basic description of translation and rotation experienced. Typical motions are described here for conventional aircraft with aileron, elevator, and rudder control only (i.e., no consideration is given to wing slats, flaps, spoilers, speed brakes, etc.). The presentation begins with general principles from classical dynamics.

Fundamental concepts of rigid body mechanics involve distinction between translational and rotational motion, as well as the extent of their interaction or their independence. It must first be realized that *complete* separation of translation and rotation is meaningful only for points lying on the axis of rotation; every other point in a rotating body will experience translation which, as already illustrated in Section 2.2.2, will vary with its location relative to the rotational axis. Since this axis is in general time-varying, further clarification is required for an overall formulation. This need has of course been met and clearly summarized in Ref. 4-9. Rigid body motion is defined by translation *of* the mass center, combined with rotation *about* the mass center. Not surprisingly, position is fully specified by three degrees of freedom for translation (e.g., Cartesian components for mass center location) and three degrees of freedom for orientation (e.g., the Euler angles defined in Chapter 2). Given a full specification of velocities or accelerations and angular rates during any time interval, a complete time history of position for that interval can be expressed in terms of kinematic differential equations in the preceding chapters.

Due to translation experienced at any point off a rotation axis, the concept of independence between the two fundamental motions is qualified thus: If the dynamics governing translation of the mass center can be expressed independently of rotational variables and vice versa, the translation and rotation are uncoupled. Many high-altitude earth satellites, and indeed the earth itself, very nearly satisfy this condition. A frequently cited counter-example is a ball rolling on a surface. There are also intermediate cases in which one, but not both, sets of equations can be considered independent; e.g., this can result from coupling that is not inherent in the homogeneous system definition, but enters via either translation- *or* rotation-dependent forcing functions.

In introducing the example of earth satellites above, care was taken to stipulate the high-altitude condition. With appreciable atmospheric densities, aerodynamic effects influence translational motion and also produce torques on a rigid body that is not completely symmetrical. It is easy to understand how the forces and torques both depend upon rotation; it is also widely recognized that they depend upon translational velocity. Furthermore, for aircraft the translational motion is kinematically coupled to rotation via the angular rates of the $(\mathbf{I}_A, \mathbf{J}_A, \mathbf{K}_A)$ frame. Finally, the control surfaces (rudder, ailerons, and elevators) are driven to take advantage of predictable interactions between the rotations induced and translational adjustments obtained. It follows that aircraft fall decidely within the class of vehicles whose translational and rotational equations of motion are coupled.

The ensuing description of aircraft dynamics will take the form of simultaneous differential equations, to be solved together for translation and rotation variables. Sets of equations at several levels of complexity are obtainable (Ref. 4-10) but, as already explained, a simple approximate representation is adequate for this book. The first idealization, already justified in Section 3.3, is to use the axes $(\mathbf{I}_A, \mathbf{J}_A, \mathbf{K}_A)$ as defined therein for the present analysis [thus ignoring various imperfections in mechanical assembly, insofar as they affect motion characterization; recall Ex. 4-14c]. This idealization permits the $(\mathbf{I}_A, \mathbf{J}_A, \mathbf{K}_A)$ triad to be identified with the aircraft principal axes of inertia for application of (2-83), while offering a simple translational velocity formulation as follows: Let the *true airspeed* vector be denoted V_A, as expressed in aircraft coordinates. Then its projections along $\mathbf{J}_A$ and $\mathbf{K}_A$ are essentially $u\beta$ and $u\alpha$, respectively, where u represents true airspeed (TAS) and α, β respectively denote *true angle of attack* and *sideslip* as customarily defined (see Fig. 4-3). These conventions

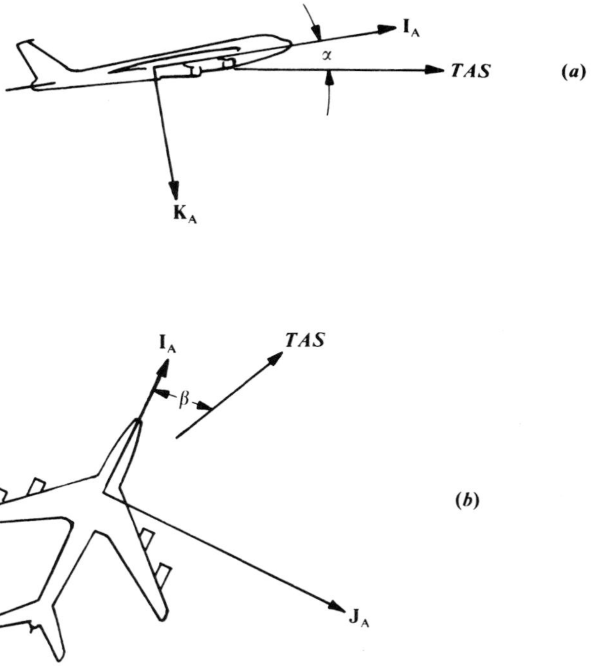

FIG. 4-3. (a) Angle of attack and (b) sideslip (TAS, true airspeed).

will now be substituted into the standard Newtonian and Eulerian dynamics, with specialized notation common to flight analysis. For simplicity the case of zero wind velocity will be considered first, and discussion of generalizations will follow.

FLIGHT EQUATIONS

It is convenient to use an *initial* ground track coordinate frame $(\mathbf{I}_0, \mathbf{J}_0, \mathbf{K}_0)$ defined so that, at the starting time t_0 of a short interval (e.g., a duration of a few minutes chosen for aerodynamic analysis), the $(\mathbf{I}_0\text{-}\mathbf{K}_0)$ plane contains the initial velocity vector and $\mathbf{K}_0$ points downward. To a close approximation, the $(\mathbf{I}_0, \mathbf{J}_0, \mathbf{K}_0)$ frame is inertial and the transformation from this frame to aircraft coordinates can be characterized as (Ex. 4-41)

$$\mathsf{T}_{A/0} = [\Phi]_x[\Theta]_y[\Psi - \zeta_0]_z \tag{4-81}$$

where ζ_0 is the initial value of the ground track angle in (1-40). In essence this transformation conforms to the same type of kinematic relation as (2-27),

$$\dot{\mathsf{T}}_{A/0} = \mathbf{\Omega}_A \mathsf{T}_{A/0} \tag{4-82}$$

with initial conditions obtained by substituting into (4-81) the initial values of roll, pitch, and drift angle, respectively [recall (1-39)]. In the $(\mathbf{I}_0, \mathbf{J}_0, \mathbf{K}_0)$ frame, Newton's second law can be applied; to a close approximation,

$$\int (\mathsf{T}_{0/A} F + \mathbf{g})\, dt = \mathsf{T}_{0/A} \begin{bmatrix} u \\ u\beta \\ u\alpha \end{bmatrix} \tag{4-83}$$

where F denotes specific force at the mass center of the aircraft, expressed in the $(\mathbf{I}_A, \mathbf{J}_A, \mathbf{K}_A)$ frame. The present analysis ignores variations in the airspeed vector magnitude; α and β are amenable to small angle approximations (e.g., if these angles were each $\frac{1}{10}$ rad, the root sum square of airspeed vector components would be affected by less than 1%). By combination of the last two equations, followed by re-expression in the aircraft coordinate frame (Ex. 4-42),

$$\dot{\alpha} - q = -\beta p + \frac{g}{u}\cos\Phi + F_3/u \tag{4-84}$$

$$\dot{\beta} + r = \alpha p + \frac{g}{u}\sin\Phi + F_2/u \tag{4-85}$$

under the condition of reasonably small pitch angle (i.e., $\cos\Theta \doteq 1$). In characterizing the forcing functions F_3/u, F_2/u a trim condition $\alpha = \alpha_t$ is defined such that, in static straight and level flight, aircraft weight is just balanced by *lift*. The complete force F_3 will, in general, also contain terms dependent upon the pitch rate q and a control surface displacement (in this case, the elevator deflection, denoted δ_e). For small deviations from some chosen nominal set of conditions these latter terms can be approximated by linear functions with coefficients denoted z_q and $z_{\delta e}$, respectively. Equation (4-84) is then rewritten as

$$\dot{\alpha} = \left(1 + \frac{z_q}{u}\right) q - \beta p - \frac{g}{u}(1 - \cos\Phi) + z_w(\alpha - \alpha_t) + \frac{z_{\delta e}}{u}\delta_e \tag{4-86}$$

where z_w, by the same procedure described above, is treated as a *stability derivative* for the angle of attack, applicable near the chosen nominal conditions. The method of stability derivatives will be used here to obtain sideslip forcing functions and to drive the moment equations (differential equations for components of $\boldsymbol{\omega}_A$) also; the former will include terms with coefficients labeled y, subscripted by β, p, r, and δ_r to denote sensitivities to sideslip, roll rate, yaw rate, and off-trim rudder deflection δ_r, respectively. Thus (4-85) becomes

$$\dot{\beta} = \left(\frac{y_r}{u} - 1\right) r + \left(\frac{y_p}{u} + \alpha\right) p + \frac{g}{u}\sin\Phi + y_\beta \beta + \frac{y_{\delta r}}{u}\delta_r \tag{4-87}$$

In the moment equations, stability derivatives are commonly labeled ℓ, m, n for roll, pitch, and yaw torques, respectively. The roll and yaw equations contain terms for sideslip, roll and yaw rate, rudder deflection, and aileron deflection δ_a; the pitch equation includes terms for $(\alpha - \alpha_\ell)$, q, and δ_e. Therefore, (2-83) provides the relations

$$\dot{p} = \left(\frac{\mathscr{I}_{A2} - \mathscr{I}_{A3}}{\mathscr{I}_{A1}}\right) qr + \ell_\beta \beta + \ell_p p + \ell_r r + \ell_{\delta a}\delta_a + \ell_{\delta r}\delta_r \tag{4-88}$$

$$\dot{q} = \left(\frac{\mathscr{I}_{A3} - \mathscr{I}_{A1}}{\mathscr{I}_{A2}}\right) rp + m_\alpha(\alpha - \alpha_\ell) + m_q q + m_{\delta e}\delta_e \tag{4-89}$$

$$\dot{r} = \left(\frac{\mathscr{I}_{A1} - \mathscr{I}_{A2}}{\mathscr{I}_{A3}}\right) pq + n_\beta \beta + n_p p + n_r r + n_{\delta a}\delta_a + n_{\delta r}\delta_r \tag{4-90}$$

The last five equations constitute a common representation for constant speed flight dynamics. Many of the coefficients are functions of flight conditions (speed, altitude, angle of attack, etc.). In integrating the equations this dependence is accounted for by using specific sets of tabulated coefficients for narrow ranges of variables and/or by programming additional relations for functional dependence of stability derivatives. Obviously, several modifications and extensions of (4-86)–(4-90) are possible. Additional terms and/or equations could be added by relaxing some of the underlying assumptions (e.g., allowing larger pitch angles and speed variations), while other terms are sometimes negligibly small (e.g., those involving z_q, y_p, y_r, $y_{\delta r}$, and ℓ_r). Stability derivative definitions could be enhanced by expressing them in terms of physical aircraft parameters (lift and drag coefficients, etc.). For near-trim conditions (2-22) and (4-86)–(4-90) can be simplified by omitting all products of variables while introducing small angle approximations ($\sin \Phi \doteq \Phi$; $\cos \Phi \doteq 1$) for roll as well as pitch. Interestingly, the resulting linear equations would then separate into a *longitudinal channel* (for angle of attack and pitch) and a *lateral channel* (for sideslip, roll, and yaw), with each channel exhibiting its own characteristic modes and excursions. Alternatively, the aerodynamic behavior could be rederived in terms of interacting transfer functions in the frequency domain. Relaxation of the rigidity constraint would allow coupled elastic deformations to be taken into account. A wind velocity V_W as resolved into aircraft coordinates can be introduced such that total aircraft velocity is expressible as

$$\mathbf{V}_A = V_A + V_W \tag{4-91}$$

This decomposition of $\mathbf{V}_A$ is, of course, acceptable for a static situation but, in general, (4-91) requires a more involved interpretation: The vector V_A contains the attenuated and delayed response to the wind oscillations (gust), while the relatively quiescent portion of the wind velocity effect is contained in V_W. The gust response is customarily obtained by modifying the preceding aerodynamic equations to include effects of a standard shaped noise spectrum that corresponds to a three-dimensional turbulent field model.

ILLUSTRATIVE EXAMPLE

The present analysis only scratches the surface of aerodynamic theory. A much broader treatment of the pertinent topics is available (e.g., Refs. 4-11 and 4-12), but it suffices here to describe time histories for a loosely piloted maneuver. Suppose the preceding equations were integrated simultaneously with (2-22) under short-term conditions, so that aircraft orientation is referred to a geographic frame assuming inertially fixed directions. In a trim condition during perfect cruise with the roll and sideslip angles, vehicle rates, and control surface deflections initially zero, all derivatives on the left of (4-86)–(4-90) would remain quiescent until receipt of a command or disturbance. The initial pitch angle would then be equal in magnitude to the trim angle α_ℓ, so that the flight path angle defined in (1-37) is zero for constant altitude. A typical turn maneuver would be initiated by a sequence of aileron and rudder deflections, the latter being employed to enhance stability and coordination.* Because the ensuing roll

* An indication of successful coordination is the absence of specific force components in both the lateral direction and parallel to the airspeed vector during the maneuver described here. Departures from this ideal goal are discussed in the material that follows.

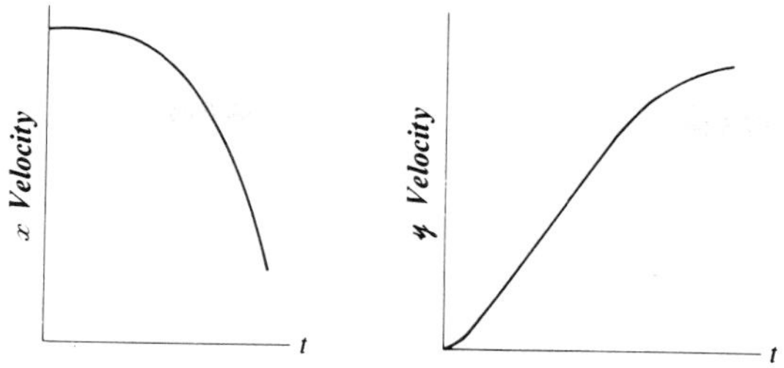

FIG. 4-4. Horizontal velocity components vs. time.

rotation upsets the lift–gravity balance in (4-86) and also induces pitch variations through (4-89) and (2-22), elevator control is generally used to govern the rate of ascent or descent (e.g., to hold altitude constant; this is the turn condition adopted for present discussion). The effect of the manual or automatic sequence of control surface deflections on the aircraft trajectory, velocity, specific forces experienced, and rotation can be described in terms of accompanying diagrams as follows: Let the $(\mathbf{I}_0, \mathbf{J}_0, \mathbf{K}_0)$ frame origin be chosen directly below the initial aircraft location. Then, after a small delay (due to lags in responding to the control surface commands), the x and y components of aircraft velocity will begin to change, as illustrated in Fig. 4-4; corresponding position dynamics will, of course, conform to Fig. 4-5 (abcissa scales for all aerodynamic time histories may be taken to represent several seconds). Typically, these motions are accompanied by variations in altitude and vertical velocity, with excursion amplitudes dependent upon tightness of elevator control. Angular orientation time

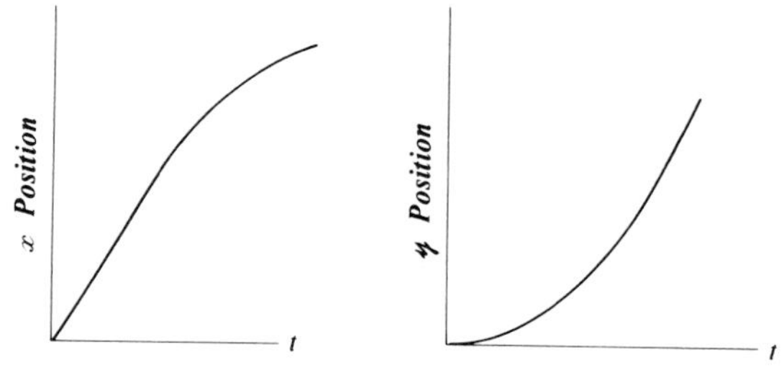

FIG. 4-5. Horizontal position components vs. time.

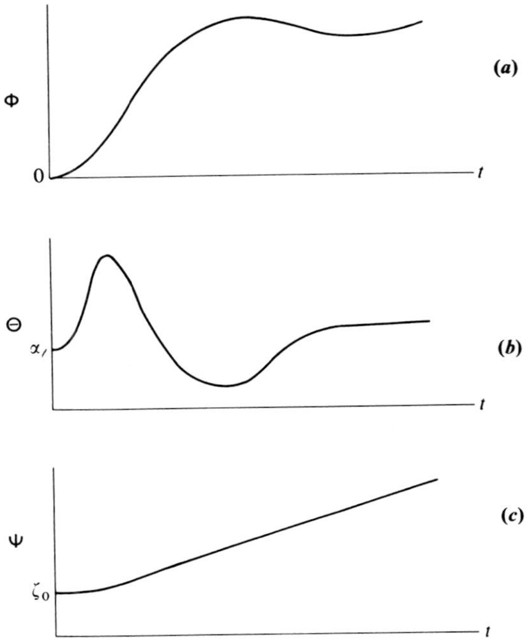

FIG. 4-6. (a) Roll, (b) pitch, and (c) heading angles vs. time.

histories, approximated in Fig. 4-6, are also readily explained. The roll angle behavior somewhat resembles the step response of a conventional servo. The pitch angle undergoes a transient and settles at a level that, in general, differs from its initial value. The purpose of this difference is to counteract any unintended change in flight path angle, which would otherwise result from angle of attack shifts necessitated by lift–gravity considerations. Heading angle remains near the initial ground track angle for a short lag interval and assumes the desired near-linear variation in the turn. Corresponding components of ζ appear in Fig. 4-7. Roll rate is dominated by a transient and, in view of (2-21), approaches a value $(-\dot{\Psi} \sin \Theta)$. Pitch and yaw rates progress toward levels approximately equal to the resolved components (proportional to the sine and cosine of the roll angle, respectively) of the heading rate $\dot{\Psi}$ (Ex. 4-43).

Dynamic descriptions for α and β are also in order here. However, the most interesting parts of their time history curves resemble the dynamic patterns of specific force components. This will now be illustrated under conditions of constant altitude as well as constant speed and zero wind

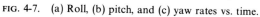

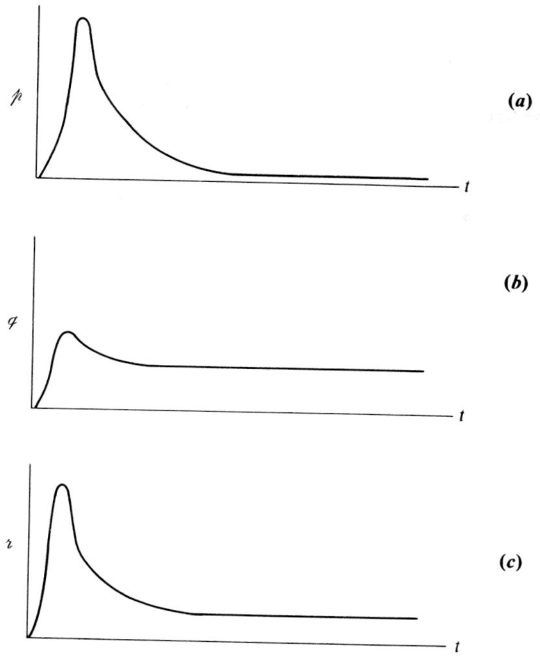

FIG. 4-7. (a) Roll, (b) pitch, and (c) yaw rates vs. time.

with small pitch displacements, while the rudder force is ignored (Ex. 4-44) and small stability derivatives are set to zero:

$$\begin{aligned} &\boldsymbol{V}_{W} = \boldsymbol{0}, \qquad \cos\Theta \doteq 1, \qquad \sin\Theta \doteq \Theta \\ &z_q = y_p = y_r = 0, \qquad y_{\delta_r}\,\delta_r \ll u y_\beta \beta \end{aligned} \tag{4-92}$$

Specific force in aircraft coordinates is obtained from (4-83) as

$$\boldsymbol{F} = u\begin{bmatrix} 0 \\ \dot{\beta} \\ \dot{\alpha} \end{bmatrix} + u\begin{bmatrix} q\alpha - r\beta \\ r - p\alpha \\ p\beta - q \end{bmatrix} - g\begin{bmatrix} -\Theta \\ \sin\Phi \\ \cos\Phi \end{bmatrix} \tag{4-93}$$

From (4-87), under conditions of (4-92),

$$F_2 = u y_\beta \beta \tag{4-94}$$

i.e., lateral specific force is directly proportional to sideslip. For steady state conditions in a coordinated turn,

$$\beta = 0 \text{ rad}, \qquad \dot{\alpha} = \dot{\beta} = \dot{\Phi} = \dot{\Theta} = 0 \text{ rad/sec} \tag{4-95}$$

and, in combination with (4-93) and (2-21) neglecting steady state roll rate (Ex. 4-45),

$$F_{(\text{turn})} \rightarrow u \begin{bmatrix} \alpha\dot{\Psi} \sin \Phi \\ \dot{\Psi} \cos \Phi \\ -\dot{\Psi} \sin \Phi \end{bmatrix} - g \begin{bmatrix} -\Theta \\ \sin \Phi \\ \cos \Phi \end{bmatrix} \tag{4-96}$$

From the last three expressions, steady roll angle in a coordinated turn is

$$\Phi_{(\text{turn})} \rightarrow \arctan(u\dot{\Psi}/g) \tag{4-97}$$

The "load factor" (sec Φ) is used to denote wing force, in g-units. Inertial force, as already shown, is intended to be normal to $\mathbf{J}_A$. It will now be demonstrated that this force should also be normal to $\mathbf{V}_A$ in this case. The flight path angle, by reason of previous definitions, conforms to the relation

$$\begin{bmatrix} 1 \\ \beta \\ \alpha \end{bmatrix}^{\mathrm{T}} \left\{ \mathbf{T}_{A/0} \begin{bmatrix} 0 \\ 0 \\ -1 \end{bmatrix} \right\} \doteq \Theta - \alpha \cos \Phi - \beta \sin \Phi \tag{4-98}$$

For constant altitude at zero sideslip,

$$\Theta - \alpha \cos \Phi \rightarrow 0 \tag{4-99}$$

so that, by (4-96),

$$\begin{bmatrix} u \\ u\beta \\ u\alpha \end{bmatrix}^{\mathrm{T}} F_{(\text{turn})} \rightarrow 0 \tag{4-100}$$

because, in this restrictive case, longitudinal specific force is

$$F_1 \rightarrow \alpha(u\dot{\Psi} \sin \Phi + g \cos \Phi) = -\alpha F_3 \tag{4-101}$$

At this point the curves in Fig. 4-8 are readily understandable. Vertical specific force begins at $-1g$ for lift in cruise and proceeds toward a level dictated by the third component of (4-96). Longitudinal and lateral behavior are dominated by influence from the angle of attack and sideslip, respectively, which experience changes upon turn command and proceed toward quiescent levels (a new value for lift balance in the former and a null for the latter).

GENERALIZED SPECIFIC FORCE AT IMU LOCATION

Extended consideration is now given to the curves just described, for the case in which gusts are present. Irregular oscillations would be superimposed, especially on Figs. 4-7 and 4-8; due to integration these oscillations would be less pronounced in position, velocity, and angular orientation time histories. More generally, all of the conditions of (4-92) would be relaxed,

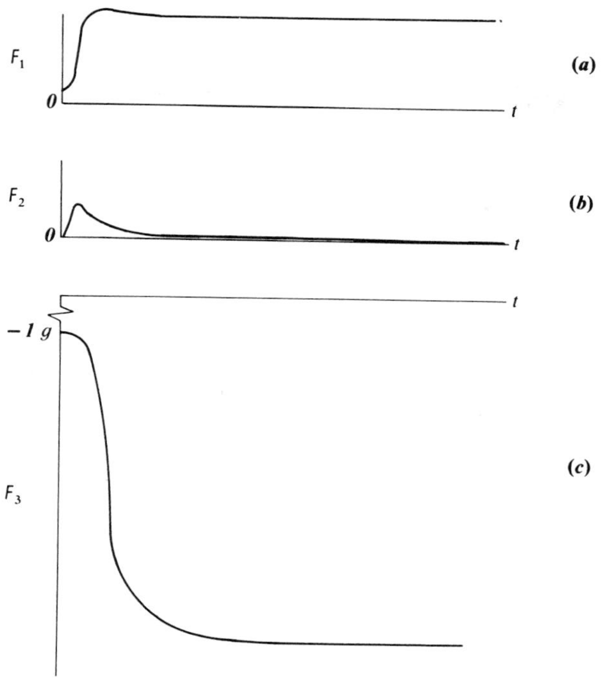

FIG. 4-8. (a) Longitudinal, (b) lateral, and (c) downward components of specific force vs. time.

and nonzero long-term average wind as well as gusts can be present; furthermore, a general flight path will include speed and altitude variations. To discuss the forces experienced at a fixed displacement $\mathbf{d}_{C/N}$ from the mass center, aircraft velocity $\mathbf{V}_A$ of (4-91) will be combined with $\boldsymbol{\omega}_A \times \mathbf{d}_{C/N}$ for application of Newton's second law in the $(\mathbf{I}_0, \mathbf{J}_0, \mathbf{K}_0)$ frame:

$$\int (\mathbf{T}_{0/A}\mathbf{F} + \mathbf{g})\,dt = \mathbf{T}_{0/A}(\mathbf{V}_A + \boldsymbol{\omega}_A \times \mathbf{d}_{C/N}) \tag{4-102}$$

This expression is then reduced to

$$\mathbf{F} = \dot{\mathbf{V}}_A + \boldsymbol{\omega}_A \times \mathbf{V}_A + \dot{\boldsymbol{\omega}}_A \times \mathbf{d}_{C/N} + \boldsymbol{\omega}_A \times (\boldsymbol{\omega}_A \times \mathbf{d}_{C/N}) - \mathbf{T}_{A/0}\,\mathbf{g} \tag{4-103}$$

for a rigid structure (Ex. 4-46). When shock mounts are introduced, $\mathbf{d}_{C/N}$ is no longer constant and $\boldsymbol{\omega}_A$ no longer represents true angular rate at the IMU location. In all cases, however, dynamic behavior is heavily influenced by the relations just presented, and all motions can be collected into total static components with superimposed oscillations, as in Section 4.3 and the next appendix.

APPENDIX 4.B: ANALYSIS OF RANDOM OSCILLATIONS

Standard analytical techniques, applicable to **a** and **w** introduced in Section 4.3, will now be demonstrated. Let the ith component of oscillatory angular rate **w** be represented by a series of the form S_{w1}:

$$w_i = \sum_k (\mathsf{C}_{ik} \cos 2\pi f_{ik} t + \mathsf{D}_{ik} \sin 2\pi f_{ik} t) \tag{4-104}$$

from which several properties are readily apparent:

(1) The amplitude and phase angle of the kth spectral component are $(\mathsf{C}_{ik}^2 + \mathsf{D}_{ik}^2)^{1/2}$ and $\arctan(\mathsf{D}_{ik}/\mathsf{C}_{ik})$, respectively (Ex. 4-47).

(2) When there are many spectral components with unrelated phase angles, w_i conforms to the standard definition for a random waveform.

(3) The rms value of w_i is

$$w_{ir} = \left\{\frac{1}{2} \sum_k (\mathsf{C}_{ik}^2 + \mathsf{D}_{ik}^2)\right\}^{1/2} \tag{4-105}$$

(4) When the center frequency, denoted f_w, in (4-104) far exceeds the spread (i.e., difference between the largest and smallest values of f_{ik}), w_i is a *narrowband* waveform. Differentiation and integration conform closely to (4-29) and (4-30) for narrowband waveforms, and a 90° phase lag nearly corresponds to a time shift of $1/(4f_w)$ in this example.

(5) For small amplitude oscillations, f_w is at least one order of magnitude greater than w_{ir}.

Much of the error analysis in this chapter involves products of oscillatory functions, such as the *covariance* $\langle w_i w_j \rangle$. Within any single flight phase, all spectral components in (4-104) are constant in this analysis (or can be bounded by fixed conservative estimates), and all random vibrations are *erogodic*, i.e., probabilistic ensemble averages can be obtained from time averages. Under these conditions (Ex. 4-48),

$$\langle w_i w_j \rangle = \frac{1}{2} \sum_k (\mathsf{C}_{ik} \mathsf{C}_{jk} + \mathsf{D}_{ik} \mathsf{D}_{jk}) \tag{4-106}$$

so that the normalized cross-correlation coefficient used in Section 4.3.4 is

$$\rho_{ij} = \frac{\sum_k (\mathsf{C}_{ik} \mathsf{C}_{jk} + \mathsf{D}_{ik} \mathsf{D}_{jk})}{\{\sum_k (\mathsf{C}_{ik}^2 + \mathsf{D}_{ik}^2)\}^{1/2} \{\sum_k (\mathsf{C}_{jk}^2 + \mathsf{D}_{jk}^2)\}^{1/2}} \tag{4-107}$$

When the sum of all positive products on the right of (4-106) is counterbalanced by the resultant negative contribution, w_i and w_j are uncorrelated. At the other extreme, ρ_{ij} has its maximum absolute value of unity when all of these products have the same sign and every ratio $\mathsf{C}_{ik}/\mathsf{C}_{jk}$, $\mathsf{D}_{ik}/\mathsf{D}_{jk}$

is uniform for all spectral components (Ex. 4-49). Therefore, the covariance

$$\langle w_i w_j \rangle = \rho_{ij} w_{ir} w_{jr}, \qquad \langle w_i \rangle = \langle w_j \rangle = 0 \tag{4-108}$$

is bounded here by

$$-w_{ir} w_{jr} \leq \langle w_i w_j \rangle \leq w_{ir} w_{jr} \tag{4-109}$$

since

$$-1 \leq \rho_{ij} \leq 1 \tag{4-110}$$

For correlation coefficients of ± 1, w_i and w_j are said to be *tightly coupled.* When w_i and w_j are only *partially* correlated ($0 < |\rho_{ij}| < 1$), the instantaneous axis of rotation has a time-varying direction. Immediately this raises the possibility of quadrature correlation, producing a covariance

$$\langle \breve{w}_i w_j \rangle = \frac{1}{2} \sum_k (\mathsf{C}_{ik} \mathsf{D}_{jk} - \mathsf{D}_{ik} \mathsf{C}_{jk}) \tag{4-111}$$

interpreted geometrically as follows: Any triad of axes (e.g., $\mathbf{I_g}$, $\mathbf{J_g}$, $\mathbf{K_g}$, rigidly attached to a gyro gimbal; an instrument case reference IA, OA, SRA; platform principal axes of inertia; aircraft axes $\mathbf{I_A}$, $\mathbf{J_A}$, $\mathbf{K_A}$), as well as its instantaneous axis of rotation, can be visualized in terms of figures traced on the surface of a unit sphere centered at the intersection of the triad. Each figure may be an irregular self-intersecting open or closed curve that may or may not exhibit a growing number of net encirclements about a fixed point. When the direction of the instantaneous axis of rotation for a triad is time-varying, and the time history of accumulated clockwise or counterclockwise net encirclements contains a linearly time-varying constituent, the motion can be referred to as cumulative coning; this can only result from correlations (full or partial) between phase-quadrature components of angular rates in correspondence with (4-111). Just as the concept of a cone is not restricted to a right circular solid of revolution, the concept of coning motion is not limited to the regular precession demonstrated in Appendix 2.A.2. In general, w_j may contain (1) a component fully correlated with w_i, (2) a component fully correlated with $\breve{w}_i$, and (3) a component independent of w_i and $\breve{w}_i$. Superposition of these three components covers the full range of coupling relationships possible between the two waveforms.

Similar analyses can be applied to translational vibration, and to coupling between rotation and translation. Table 4-3 provides a geometric description for each coupling relationship (Ex. 4-50).

TABLE 4-3

GEOMETRIC INTERPRETATION OF COUPLED OSCILLATIONS FOR i AND j AXES $(1 \leq i, j \leq 3)$[a]

	a_i	a_j	$\breve{a}_i$	$\breve{a}_j$	w_i	w_j	$\breve{w}_i$	$\breve{w}_j$
a_i	1	2	0	3	7	9	6	8
a_j		1	3	0	9	7	8	6
$\breve{a}_i$			1	2	6	8	7	9
$\breve{a}_j$				1	8	6	9	7
w_i					1	4	0	5
w_j						1	5	0
$\breve{w}_i$							1	4
$\breve{w}_j$								1

[a] Interpretation: (i) $i \neq j$, (ii) table is symmetrical, (iii) table entries correspond to notes enumerated below.

0. Inherent decoupling; (SIN)(COS) products of equal frequencies, as well as products of trigonometric functions of unequal frequencies, average to zero.
1. Inherent full coupling.
2. Translational vibration along an inclined axis in i-j plane.
3. Generalized cylindrical translation in i-j plane (e.g., circular for a single sinusoid).
4. Rotational oscillation about an inclined axis in i-j plane.
5. Coning motion [e.g., rotation characterized by right circular cone for a single sinusoid; Eqs. (2-94) and (2-95)].
6. Reciprocating corkscrew motion.
7. Reciprocating translation accompanied by delayed coaxial rotation.
8. Vibration experienced at a point at some separation distance from a center of rotation.
9. Component of vibration described in (8) with delay (e.g., due to nonrigidity of connecting arm).

FORMULATION OF INDEPENDENT ERROR FORCING FUNCTIONS

Tables 4-1 and 4-2 adequately characterize rectification errors only if further correlations between **A** and the random forcing functions are taken into account. For example, components of $\mathbf{e}_\omega$ from angular vibrations introduce similar oscillations, in phase quadrature, into $\boldsymbol{\psi}$; interaction with translational vibration can produce a *sculling* effect.* The degradation considered here arises from rectification in $\boldsymbol{\psi} \times \mathbf{A}$, which is not included in (4-79). Yet for (4-80) and all further nav error analysis, it is highly desirable to characterize $\mathbf{e}_a$ and $\mathbf{e}_\omega$ *without* regard to other correlated variables. It will now be demonstrated that, to use an independent probabilistic model for these random forcing functions, additional rectification products are included in $\langle \mathbf{N}_a \rangle$.

Rigorous analysis of (4-80), a vector differential equation with randomly varying coefficients as well as random forcing functions, will not be attempted here. Enough insight is nevertheless gained by writing (4-80) in the form of (II-58), with the dynamics matrix and the forcing function replaced by (Ex. (4-52):

$$\mathbf{a}_1 = \left[\begin{array}{c|c} \mathbf{O}_{3\times 3} & (-\mathbf{A}\times\ ) \\ \hline \partial\boldsymbol{\varpi}/\partial\mathbf{V} & (-\boldsymbol{\varpi}\times\ ) \end{array}\right], \qquad \mathbf{u}_1 = \left[\begin{array}{c} \mathbf{n}_a \\ \hline \mathbf{n}_\omega \end{array}\right] \tag{4-112}$$

$$\mathbf{a}_2 = \left[\begin{array}{c|c} \mathbf{O}_{3\times 3} & \mathbf{T}_{G/A}(-\mathbf{F}\times\ )\mathbf{T}_{A/G} \\ \hline \partial\boldsymbol{\varpi}/\partial\mathbf{V} & (-\boldsymbol{\varpi}\times\ ) \end{array}\right], \qquad \mathbf{u}_2 = \left[\begin{array}{c} \mathbf{T}_{G/A}\mathbf{n}_a \\ \hline \mathbf{T}_{G/A}\mathbf{n}_\omega \end{array}\right] \tag{4-113}$$

for gimballed platform and strapdown systems, respectively. In either case, sculling and related effects are highlighted by a simple transformation; the unknown solution does not appear explicitly on the right of (II-60):

$$\left[\begin{array}{c} \dot{\tilde{\mathbf{V}}} \\ \hline \boldsymbol{\psi} \end{array}\right] = \mathbf{a}(t)\boldsymbol{\Phi}_a(t, t_0)\left[\begin{array}{c} \tilde{\mathbf{V}}_0 \\ \hline \boldsymbol{\psi}_0 \end{array}\right] + \mathbf{u}(t) + \mathbf{a}(t)\int_{t_0}^{t} \boldsymbol{\Phi}_a(t, \mathrm{r})\mathbf{u}(\mathrm{r})\, d\mathrm{r} \tag{4-114}$$

where the transition matrix $\boldsymbol{\Phi}_a$ conforms to the same dynamic relation as (II-55), initialized as the identity in (II-56). This latter condition assures that the first term in a series expansion for $\boldsymbol{\Phi}_a$ will be the identity matrix; thus a major contribution to the last term in (4-114), for gimballed platforms, is

$$\mathbf{a}_1(t)\int_{t_0}^{t} \mathbf{u}_1(\mathrm{r})\, d\mathrm{r} = \mathbf{a}_1(t)\left[\begin{array}{c} \langle \mathbf{N}_a \rangle (t - t_0) \\ \hline \langle \mathbf{N}_\omega \rangle (t - t_0) \end{array}\right] + \frac{1}{2\pi f_u}\mathbf{a}_1(t)\left[\begin{array}{c} \check{\mathbf{e}}_a(t) \\ \hline \check{\mathbf{e}}_\omega(t) \end{array}\right] \tag{4-115}$$

in which f_u is replaced by f_a and f_w for errors sensitive to specific force and to rotation, respectively. Total specific force **A** is then represented by (4-26)

* A sinusoidal analysis applicable to gimballed platform vibration appears on pp. 507–508 of Ref. 4-4. Application to strapdown as well as gimballed systems, in the presence of random vibrations, is addressed in Ex. 4-51.

and substituted into (4-112); a product proportional to $\mathbf{a} \times \breve{\mathbf{e}}_\omega$ is immediately obtained (Ex. 4-53). In addition to sculling, this product includes acceleration biases from *vibratory* mass unbalance, and certain other translation-dependent gyro degradations often overlooked in platform analysis. These effects are augmented by various rotation-sensitive errors in strapdown systems (Ex. 4-54), for which a summary appears in Table 4-4.

TABLE 4-4

RECTIFICATION PRODUCTS OBTAINED FROM INTERACTION BETWEEN OSCILLATORY ORIENTATION ERROR AND TRANSLATIONAL VIBRATION

Rectification of rotational error from	Combined with oscillations in	In the presence of motion defined by Table 4-3, case
OA angular acceleration	angular rate	7, 9
Scale factor	angular rate	8
Misalignment	angular rate	6, 8
Response time differential	angular rate	9
Computational lag	angular rate	9
Mass unbalance	specific force	3
(Anisoelasticity) × (steady specific force)	specific force	3
(Cylindrical drift sensitivity) × (steady specific force)	specific force	2, 3[a]

[a] Possible rectification in the presence of generalized cylindrical motion would arise directly from squaring individual acceleration components, rather than from quadrature correlation. This phenomenon could also occur with uncoupled translation.

EXERCISES

4-1 Show that (4-1) is consistent with (3-37) and the equations following it in Section 3.4.2. Show conformance with (II-58) as well. Is (4-4) consistent with (II-58) and the equations following it in Appendix II?

4-2 (a) Does the generalization mentioned after (3-2), allowing superposition of a fixed velocity vector, justify an earth reference for terrestrial inertial navigation?

(b) Sketch a *heliocentric* coordinate frame (i.e., with the sun at the origin), assuming an earth position at fixed radial distance and a constant rate of revolution. With this vector substituted for $\mathbf{R}$ in (2-37), verify the value of $6(10)^{-4}g$ for centripetal acceleration in Section 4.1.

(c) Show that, if this amount of acceleration appeared as a bias error, a velocity uncertainty of over 15 fps could result.

(d) How should *differences* in solar gravitation, as experienced by an accelerometer and the earth nearby, compare with the value of $6(10)^{-4}g$?

4-3 (a) The assumption implicit in (4-6) is that altitudes reached in terrestrial navigation are too low to produce appreciable differences between solar gravitation as experienced by the INS and by the earth itself. To test this assumption, let (4-6) establish the basic "two-body" relation for the earth's orbit. This basic motion (which ignores perturbing forces such as other gravitational fields) is planar and, in this case, slightly elliptical. Suppose that $|\mathbf{G}_S|$ varies inversely as the square of distance from the sun and, by Ex. 4-2, the attraction along the earth–sun line is approximately

$$|\mathbf{G}_S| = \frac{\mu_S}{(90 \times 10^6 \times 5280)^2}\ \text{fps}^2 \doteq 0.0006g$$

where μ_S denotes the product (solar mass) × (universal gravitational constant). Reduce the 90 million mile distance by 4000 miles, to find the difference in solar gravitation acting on a low altitude aircraft at noon time.

Hint: Write a Taylor approximation of the form

$$|\Delta\mathbf{G}_S| \doteq \frac{\mu_S}{(90 \times 10^6 \times 5280)^2}\left\{1 + 2\left(\frac{4 \times 10^3}{9 \times 10^7}\right) - 1\right\}$$

and use the preceding relation to eliminate μ_S.

(b) For a nominal lunar radius of $\frac{1}{4}R_E$ and a lunar surface gravitation of about $g/6$,

$$\mu_M/(\tfrac{1}{4}R_E)^2 \doteq \tfrac{1}{6}g$$

show that lunar gravitation at 238,000 miles can introduce a $10^{-7}g$ error into the INS, about double the value produced by the solar gravitation differential.

(c) If the earth's gravity were modeled to within $10^{-6}g$, and a triad of accelerometers had comparable accuracy, would solar or lunar gravitational differentials have to be taken into account? At what level of accuracy do these differentials become significant? Compare with related statements on p. 310 of Ref. 4-5.

(d) Are the conclusions just reached affected by ellipticity or perturbations of the earth's orbit? By the nonspherical shape of the earth?

4-4 (a) Re-express the rotor angular rate of (4-11) in gyro gimbal coordinates. Do the same for the corresponding angular momentum. Show that developed OA torque is *not* merely the cross product of these two vectors. What simplification, assumed for angular momentum, could force that cross product to yield the correct answer? Would such a procedure have any theoretical justification? Discuss the significance of the first term on the right of (4-14).

(b) Suppose that

$$A' = \mathbf{T}A, \qquad B' = \mathbf{T}B, \qquad C' = \mathbf{T}C, \qquad A' = B' \times C'$$

It follows that $A = B \times C$ and, therefore,

$$\mathbf{T}^{\mathrm{T}}A' = (\mathbf{T}^{\mathrm{T}}B') \times (\mathbf{T}^{\mathrm{T}}C')$$

Apply this reasoning to the second term on the right of (4-14).

(c) Recall the definition of $\boldsymbol{\omega}_{\mathrm{A}}$ at the beginning of Section 2.2.2, and of $\mathscr{I}_{\mathrm{A}}$ at the beginning of Appendix 2.A.2. These are the quantities applicable to Euler's dynamic law, in the form of (2-83), for a rigid aircraft. Do (4-9) and (4-10) conform to proper definitions for the gyro rotor under the conditions stated?

(d) Briefly discuss a cause-and-effect relation between moment and rotation, applicable to (4-15).

4-5 (a) Interpret (4-16) in terms of a self-contained assembly (Fig. 3-1), with no external torque applied.

(b) Interpret (3-10) as a special case of (4-17). Sketch a frequency response of (4-17).

4-6 Use (2-83) to generalize (4-16), so that (4-19) is obtained.

4-7 Extend the preceding exercise, using (4-20), to obtain (4-21). Noting that the axial moment of inertia for the rotor–gimbal assembly in Fig. 3-1 includes a diametral moment of inertia for the rotor, while a diametral moment of inertia for the assembly includes the polar moment of inertia for the rotor, describe γ_{xx} in terms of parameters defined in the text.

4-8 Use (4-22) to derive (4-23), and show that neglected quantities are small in comparison with the term involving $\omega_x \omega_x$ in (4-23).

4-9 Consider a strapdown system having exact knowledge of attitude with respect to inertial axes [i.e., perfect angular resolution, etc., such that $\hat{\mathbf{T}}_{\mathrm{I/A}} = \mathbf{T}_{\mathrm{I/A}}$ in (3-54)], in the presence of a roll oscillation $\boldsymbol{\omega}_{\mathrm{A}} = \mathbf{1}_1\omega_0 \cos \omega_X t$. Suppose that the #2 accelerometer, with OA along $+\mathbf{I}_{\mathrm{A}}$, has an error $\boldsymbol{\beta}_y \dot{\omega}_y$. Use (2-70) to show that, if $\mathbf{T}_{\mathrm{I/A}} = \mathbf{I}$ at $t_0 = 0$, the time integral of transformed specific force measurements would be

$$\int \left\{ \begin{bmatrix} 1 & 0 & 0 \\ 0 & \cos\left(\dfrac{\omega_0}{\omega_X} \sin \omega_X t\right) & -\sin\left(\dfrac{\omega_0}{\omega_X} \sin \omega_X t\right) \\ 0 & \sin\left(\dfrac{\omega_0}{\omega_X} \sin \omega_X t\right) & \cos\left(\dfrac{\omega_0}{\omega_X} \sin \omega_X t\right) \end{bmatrix} \begin{bmatrix} A_{\mathrm{A}1} \\ A_{\mathrm{A}2} + \boldsymbol{\beta}_y \omega_0 \omega_X \sin \omega_X t \\ A_{\mathrm{A}3} \end{bmatrix} \right\} d$$

Compare accelerometer error convention here versus (3-34); identify one similarity and one difference. Show that, for small amplitude oscillations, average error in $\hat{A}_{13}$ has a magnitude $\doteq \beta_y \omega_0^2/2$ (note the end of Ref. 4-2).

4-10 (a) Apply (4-25) to Ex. 4-9; identify w_1 and f_w. Show that $\breve{w}_1 = \omega_0 \sin \omega_X t$, in conformance with (4-29) and with the 90° phase lag definition.

(b) Add two more sinusoids of arbitrary amplitude, frequency, and phase angle to w_1 of part (a). Differentiate with respect to time. For *narrowband* oscillations (i.e., the bandwidth spanned by the three sinusoids is much less than the center frequency), does (4-29) closely characterize the derivative?

(c) Generalize the angular rate in Ex. 4-9, so that

$$\boldsymbol{\omega}_A = (\mathbf{1}_1 \omega_{01} + \mathbf{1}_3 \omega_{03}) \cos \omega_X t.$$

In (4-21) find the average roll gyro degradation if mounted such that nominal SA direction coincides with $\mathbf{K}_A$. Are there additional oscillatory errors at $(2\omega_X)$ rad/sec?

(d) Show that a 45° phase lag between ω_{A1} and ω_{A3} would reduce the drift in part (c) by a factor of $1/\sqrt{2}$. What would a 90° phase lag do to this error? In view of (4-30) and (4-32), how would this phase lag affect the last term of (4-24)? Express a phase lead in terms of a lag combined with a sign reversal.

(e) Let the three closely spaced frequencies of part (b) characterize both w_1 and w_3, such that their ratio is constant. With part (c) generalized accordingly, would the drift rate still be proportional to mean squared angular rate?

(f) If Ex. 4-9 were generalized to include three closely spaced frequencies, would the average acceleration bias remain proportional to mean squared angular rate?

(g) A linear operator commonly used in communication theory is the *Hilbert transform*, which, as applied to an arbitrary waveform, retards the phase angle of each spectral component by 90°. Relate this operator to differentiation and integration, as applied to narrowband functions.

4-11 (a) Show that (4-35) and (4-36) are equivalent.

(b) Show that (4-37) is a special case of the last equation in Appendix II. Express the derivation, from (4-35)–(4-37), in terms of the last six equations in Appendix II.

4-12 (a) Do steady drifts exert their main influence on short- or long-term degradations of an attitude reference? Can oscillatory errors affect both short- and long-term performance?

(b) Is (4-39) consistent with the attitude error convention in Appendix 3.B?

(c) Verify (4-40) and relate it to Ex. 4-10e.

(d) Does the long-term degradation from (4-40) differ substantially from the value it *would* have if the drift rate were fixed at its average level? How does this question relate to the definition of an average?

4-13 (a) Combine (4-25), (4-30), and (4-32) with the last term of (4-24). Does a rectification product materialize?

(b) Integrate (2-94) and express the result in terms of (2-95).

(c) Recognize the first matrix on the right of (2-97) as a solution to (2-28). For this example show that the x axis of the vehicle describes a cone in space.

(d) Show that if the waveforms used in part (b) are substituted into part (a), $\check{\gamma}_{\dot{x}x}$ varies inversely with a frequency. Express this coefficient in terms of frequency and two gyro parameters, under the condition that (4-24) is valid.

4-14 (a) Does the possibility of SDF gyro gimbal rotation, relative to the case, introduce a potential ambiguity into the definition of OA direction? Does Fig. 3-1 illustrate whether the IA is defined normal to SA or SRA?

(b) Does the derivation of (4-23) depend upon the gyro IA definition chosen?

(c) Note the directions associated with ω_x and $\omega_{\dot{x}}$ in the discussion preceding (4-22). Are these directions associated with the gyro case? Note that imperfect alignment would introduce a discrepancy in an error model based on specifications for aircraft dynamics; how would this discrepancy typically compare with the error itself? Extend this reasoning to the use of small angle approximations for $\boldsymbol{\theta}$ and for misalignments; possible effects of $\boldsymbol{\psi}$ on general characterization involving (4-25) and (4-26), departure of $(\mathbf{I}_A, \mathbf{J}_A, \mathbf{K}_A)$ from aerodynamic axes, etc.

4-15 (a) An aircraft cruising northward has a strapdown INS with an imperfectly calibrated #3 gyro ($\mu_{\omega 3} = 0.001$), and the anisoinertia coefficient $\gamma_{\dot{x}x}$ for each gyro in the triad is 0.001 (rad/sec)/(rad/sec)^2 = 0.001 sec. The #1 and #3 gyros are mounted with nominal SRA directions along $\mathbf{J}_A$, while the #2 gyro has its SRA parallel to $\mathbf{I}_A$. If angular vibrations about the pitch axis are independent of roll and yaw rates, are all SRA mounting directions wisely chosen? Discuss possible influence of anisoinertia and coning effects.

(b) Suppose that roll and yaw oscillations are *tightly coupled*, so that $\langle w_1 w_3 \rangle$ is equal to the product of rms rates 0.005 and 0.01 rad/sec about $\mathbf{I}_A$ and $\mathbf{K}_A$, respectively. Show that, if the #3 gyro SRA had been mounted along $\mathbf{I}_A$, a drift rate of order 0.01°/hr could have resulted.

(c) A gradual turn is initiated so that, after two minutes, the aircraft is flying eastward. Describe the change in profile rate, and a fairly abrupt change in azimuth error. Describe the sequence as two separate phases of flight, with an "instantaneous" adjustment of nav error between phases. How would tilt be affected by 0.1-mr IA projections of #1 and #2 gyros along $\mathbf{K}_A$? Would 0.1-mr alignment accuracy be unnecessarily stringent?

(d) Recall the profile rates at moderate latitudes in Ex. 3-7d. If the system considered in part (c) had been a gimballed platform, would the scale factor and misalignment errors have taken effect during the 2-min interval? If in addition the system had used a wander azimuth reference with $\omega_{L3} = 0$, how would INS performance be affected by level gyro misalignments and azimuth gyro calibration error?

4-16 Neglecting interaction between profile rates and timing mismatch, substitute $w_j(t - \tau_{\omega j})$ for *apparent* angular rate about the jth axis at time t, expressed as a narrow band of sinusoids with center frequency f_w. For timing errors much smaller than an oscillation period, verify (4-43) rigorously, and show correspondence to the sign convention of (4-39).

4-17 (a) Do certain rotation-sensitive drift coefficients such as γ_y, γ_{zx} seem likely to be equal for all gyros in a triad? If highly mismatched, could another subscript be added to them, in analogy with (4-44) and (4-45)?

(b) Suppose that, on the basis of laboratory tests, a system is equipped with mechanical or computational provisions to compensate mass unbalance. Can (4-44) still be applied, with coefficients redefined as an uncorrected effect remaining from imperfect compensation? Extend this reasoning to various degradations throughout a navigation system.

(c) For mechanical compensation, could estimated drift rates multiplied by H be used to adjust torquing commands in (3-16)?

(d) Recall the discussion preceding (4-35). If $f_a \neq f_w$, name *four* spectral regions containing oscillatory gyro error. Describe the sources of these errors.

(e) Let the gyros in Ex. 4-15 be sensitive to reciprocating motion, caused by engine vibrations transmitted through the aircraft structure to the IMU. Assume tightly coupled 10-fps^2 rms vibrational accelerations along IA and SRA directions during cruise, with g-sensitive and g^2-sensitive coefficients of order 0.03°/hr/g and 0.03°/hr/g^2. Re-express these coefficients per NOMENCLATURE (Appendix I) and find the maximum tilt that they could contribute during $\frac{1}{2}$-hr cruise. Can the errors in (4-44) and (4-45) be ignored during the 2-min turn interval, for the "instantaneous" approximation in Ex. 4-15c? Can this reasoning be generalized?

(f) If the system just described were replaced by a gimballed platform with different vibration levels along each aircraft axis, how would drift in the latter phase be affected by the turn?

4-18 (a) Suppose that, due to elastic deformation characteristics that vary with direction, the jth gyro gimbal experiences reciprocating motion along its SRA, causing an "effective mass unbalance" interacting with B_j. Suppose that this instantaneous mass unbalance is excited by, but lags 10° behind, a low-amplitude sinusoidal $\frac{1}{2}$-g rms acceleration, 30° off $\mathbf{K_g}$ in the $\mathbf{I_g}$-$\mathbf{K_g}$ plane. Express the resulting drift rate as a function of two coefficients denoted γ_{xzj} and $\check{\gamma}_{xzj}$.

(b) If a cylindrical motion existed independently of the phenomenon just described, could the overall result of quadrature-correlated translational vibrations be combined into one term with a coefficient $\check{\gamma}_{xzj}$?

(c) What is the net effect of interaction between C_j and the excitation described in part (a)? Discuss generalization of the waveforms and directions considered for acceleration components.

(d) Combining moments of inertia for the derivation of (4-21) implies an assumption that the rotor–gimbal assembly of Fig. 3-1 responds as a rigid structure to rotations about both IA and SRA. This is largely true for typical aircraft oscillations, but suppose that a high-frequency angular vibration of a gyro case about the SRA produced a slight modulation of *relative* angular rate between rotor and gyro gimbal. If a phase lag were applied to ω_z in (4-20), how would (4-21) be changed? Would a completely new form of drift be generated, or could the general case be covered by a redefinition of certain coefficients? Can an analogy be drawn with part (a) of this exercise?

4-19 (a) Does (4-46) completely describe the drift of a strapdown attitude reference? Can SRA orientations be chosen to minimize drift effects other than those considered in Ex. 4-15?

(b) Could a random rotor asynchronism, generated independently of all other motions, produce a contribution to (4-46) or to $\mathbf{e}_\omega$?

(c) It is sometimes convenient to include profile rate interactions with misalignment and scale factor errors in the definition of $\mathbf{v}_\omega$. Was that definition used in Section 4.3.1?

(d) Could performance of a gimballed platform be accurately represented by substituting (4-46) for drift biases in Section 3.4.2?

4-20 (a) Interpret the accelerometer IA projections y_{aj} and z_{aj} as SRA and OA rotations, respectively. Include algebraic signs.

(b) Recall the discussion preceding (4-35). Does oscillatory acceleration contribute to both gyro and accelerometer errors?

4-21 Some accelerometers are degraded by a force sensitivity proportional to a displacement in another direction normal to the OA. For a force-sensitive displacement with phase lag, generalize the last term of (4-48).

4-22 Extract the bias from (4-48) as suggested in the text. Relate y_{a1} in (4-49) to item (2) following (4-48).

4-23 Apply (2-37) to the derivation of (4-51). Compare this result with Ex. 4-20b, and discuss its relation to Ex. 4-9.

4-24 Show that the expression

$$\mathsf{J}\mathfrak{s}^2\boldsymbol{\varphi}(\mathfrak{s}) = \mathsf{L}_D(\mathfrak{s}) - \mathsf{F}(\mathfrak{s})\mathsf{G}(\mathfrak{s})\boldsymbol{\varphi}(\mathfrak{s})$$

is consistent with (4-52)–(4-54).

4-25 (a) Suppose that

$$\boldsymbol{\theta} = 0.01 \cos(20\pi t), \qquad \omega_z = (0.01)(20\pi)\cos(20\pi t)$$

Verify an average drift rate of 0.18°/sec. Although imperfect stabilization contributes to the generation of this error, is it nevertheless a gyro drift? Make a statement regarding substitution of narrowband noise for the sinusoidal oscillations.

(b) With ω_z defined as the angular rate about the SRA, part (a) contains an approximation. Relate this to Ex. 4-14c.

(c) Since strapdown gyros are subjected to aircraft rotation, they are provided with tight rebalancing. If $\phi = 10\ \widehat{\text{sec}}$ and the vertex angle in part (a) were $0.4(10)^{-4}$ rad, show that the coning drift rate would be $\sim$0.01°/hr. Could the vertex angle get much larger if $B = H$?

4-26 Recall the opening discussion of Section 4.3.3. Can analogous statements be made concerning strapdown data processing imperfections?

4-27 (a) Each direction cosine in (3-58) is stored in a register with $\mathscr{W}$ bits set aside for fractional magnitude. The ith active DDA iteration introduces an independent uncertainty ϵ_i into each direction cosine and, after K iterations, roundoff error is of order $(\epsilon_1 + \epsilon_2 + \cdots + \epsilon_K)$, where

$$|\epsilon_i| = 2^{-\mathscr{W}}, \qquad i = 1, 2, \ldots, K$$

Give an approximation to the squared error. If $\langle \epsilon_i \epsilon_j \rangle = 0,\ i \neq j$, what is its mean squared value? The rms value?

(b) Assume that the rms angular rate during a 15-min flight segment is roughly 0.1 rad/sec, and the number of active DDA iterations is about $(0.1)(900)/10^{-4}$, or roughly a million. What value of $\mathscr{W}$ produces an accumulated roundoff error of order 10^{-4} rad?

(c) Relate the roundoff error growth pattern to (4-38). Can roundoff error be included in $\mathbf{e}_\omega$?

4-28 (a) In the context of this book, a perfect gyro is defined by an unqualified ability to sense exact angular rate about its IA, regardless of dynamic environment. By this definition, can a perfect gyro be sensitive to imperfect stabilization? Can a gyro with finite angular momentum be perfect?

(b) Give an analogous definition for a perfect accelerometer. By this definition, can an accelerometer of finite size be perfect?

4-29 Verify (4-56). If $\omega_{\max} = \phi/t$ it takes at least how long to produce the rotational sequence in Fig. 4-2? How high a drift rate could then be associated with (4-56)? Show that this is comparable to, but less than, a maximum accumulation rate $\phi^2/(2t)$ from (3-61), with active iterations in all intervals.

4-30 In (4-57) identify rotations actually sensed along directions associated with $\boldsymbol{\omega}_{\mathrm{A}}$. Are gyro outputs processed as if they had IA directions along $\hat{\boldsymbol{\omega}}_{\mathrm{A}}$? Derive (4-58).

4-31 (a) In (II-58), replace $\mathbf{x}$ by $\boldsymbol{\xi}$, $\mathbf{u}$ by $\mathbf{n}_{\omega}$, and $\mathbf{A}$ by $\boldsymbol{\Omega}_{\mathrm{A}}$. Show that (4-58) is the result.

(b) Substitute the orthogonal matrix

$$\mathbf{T} = \exp\left\{\int \boldsymbol{\Omega}_{\mathrm{A}}(\mathfrak{r})\, d\mathfrak{r}\right\}$$

for $\boldsymbol{\Phi}$, along with the replacements just described, in (II-59). Obtain the relation

$$\dot{\boldsymbol{\xi}} = \boldsymbol{\Omega}_{\mathrm{A}}\, \mathbf{T}(t, t_0)\boldsymbol{\xi}_0 + \boldsymbol{\Omega}_{\mathrm{A}} \int_{t_0}^{t} \mathbf{T}(t, \mathfrak{r})\mathbf{n}_{\omega}(\mathfrak{r})\, d\mathfrak{r} + \mathbf{n}_{\omega}$$

Would the absence of this first term invalidate (4-58) if the solution vector $\boldsymbol{\xi}$ were properly initialized?

(c) Aside from the means of expressing initial conditions, show how the true derivative $\dot{\boldsymbol{\xi}}$ differs from the approximation

$$\dot{\boldsymbol{\xi}} \doteq \boldsymbol{\Omega}_{\mathrm{A}} \int_{t_0}^{t} \mathbf{n}_{\omega}(\mathfrak{r})\, d\mathfrak{r} + \mathbf{n}_{\omega}$$

when an expansion similar to (2-69) is applied to $\mathbf{T}$. Use (2-73) to justify (4-59) for small amplitude rotations. Is this justification restricted to sinusoidal oscillations?

4-32 (a) Assume sinusoidal oscillations for w_i and w_j in the expression

$$\langle w_i w_j \rangle = \rho_{ij} w_{i\mathrm{r}} w_{j\mathrm{r}}$$

in which case the correlation coefficient is identified as what function of the phase angle?

(b) When gyro bandwidth mismatch coexists with commutation error, can their drift rates be added vectorially?

(c) Both $\langle w_1 w_2 \rangle$ and $\langle w_1 w_3 \rangle$ vanish for an inactive roll channel in (4-61), while the timing error normally associated with $\phi/(2w_{1_r})$ becomes inoperative. What is the drift rate due to commutation error in that case? Generalize to two inactive channels.

4-33 (a) Sum the squares of vector components on the right of (4-61). Since the square of each parenthetical quantity is nonnegative, what values for ρ_{ij}^2, between zero and unity, will maximize this sum? Show that, for a fixed value of $|\mathbf{w}|$ in the expression

$$2\{|\mathbf{w}|^2 - (w_{1_r} w_{2_r} + w_{2_r} w_{3_r} + w_{3_r} w_{1_r})\}$$

a minimization of products $w_{i_r} w_{j_r}$ is desirable. Consider Ex. 4-32c with one angular rate essentially equal to ϕ/t while the other is just large enough to produce excursions of ϕ rad; show conformance to the maximum accumulation rate in Ex. 4-29. Note that this is an *effective* long-term accumulation rate, not necessarily indicating patterns similar to Fig. 3-9 repeated on every iteration pair.

(b) Do small errors of ϕ^2 rad per iteration pair effectively accumulate in a consistent direction, for any fixed direction of $\boldsymbol{\omega}_A$?

(c) Justify the inclusion of unrectified commutation error in $\mathbf{e}_\omega$.

(d) Could extended time histories based on Fig. 3-9 be postulated for a fixed rotation axis, with apparent but fictitious sharp changes in axial direction?

4-34 (a) Show that, in the context of Section 3.4.2 and Ex. 4-25c or 4-29, a 0.1-mr resolution can be somewhat crude.

(b) Would commutation error be absent from a strapdown gyro triad with infinitesimal quantization, infinite computing rates, and infinitely tight torquing in Fig. 3-2? Could these provisions also remove all drifts associated with OA gyro gimbal rotation? Is commutation error nevertheless a separately identifiable drift source?

(c) If $\omega_{max} > \phi/t$, could unrecognized rotation become an overriding degradation? Would safe design therefore dictate pessimistic assessment of maximum angular rates, for selection of computing speed? Beyond the safety margin, would further reduction of the DDA iteration interval significantly affect accuracy?

(d) Closed-form solutions for attitude and angular rates in Appendix 2.A.1 enabled an economical simulation in Ref. 4-1. Could expressions in Appendix 2.A.2 be used for similar purposes?

4-35 Let W represent the time integral of **B**, while integer truncation is denoted by an operator "Trun( )." Integrating accelerometer quantization error is then

$$\mathbf{e}_v = W - \upsilon \operatorname{Trun}(W/\upsilon)$$

Could this error be approximated, with the use of (4-50)? Define an approximate delay time for each accelerometer.

4-36 (a) The *actual* angular offset of a gimballed platform from $(\mathbf{I}_G, \mathbf{J}_G, \mathbf{K}_G)$ constitutes an error, whereas for a strapdown system only the *uncertainty* in orientation contributes directional error. Subdivide this uncertainty into two constituents, representing absolute uncertainty in $(\mathbf{I}_A, \mathbf{J}_A, \mathbf{K}_A)$ directions and imperfect knowledge of the nav reference frame orientation with respect to inertial axes. Relate this interpretation to (4-65).

(b) Relate (3-17), (3-36), and (3-54) to a *computational* frame $(\mathbf{I}_c, \mathbf{J}_c, \mathbf{K}_c)$ displaced from true geographic axes by

$$\begin{bmatrix} \mathbf{I}_c \\ \mathbf{J}_c \\ \mathbf{K}_c \end{bmatrix} = \{\mathbf{I} + (\boldsymbol{\psi}_F + \boldsymbol{\psi}_G) \times \ \} \begin{bmatrix} \mathbf{I}_G \\ \mathbf{J}_G \\ \mathbf{K}_G \end{bmatrix}$$

Transpose (3-17) and discuss the transformation direction.

(c) Recalling Ex. 2-5a, use (II-17) to express $\boldsymbol{\psi} \times \mathbf{A}$ in terms of $\boldsymbol{\psi}_F$ and $\boldsymbol{\psi}_G$. Discuss the relation of Ex. 2-1b to this example. Apply the principle of Ex. 4-4b to show that

$$(\boldsymbol{\psi} \times \ ) = (\boldsymbol{\psi}_G \times \ ) + \mathbf{T}_{G/A}(\boldsymbol{\xi} \times \ )\mathbf{T}_{A/G}$$

Demonstrate agreement with (3-17) and (4-75).

(d) Define $\boldsymbol{\varpi}_A = \mathbf{T}_{A/G}\,\boldsymbol{\varpi}$ and, in analogy with (2-19), $\boldsymbol{\zeta}_G \triangleq \boldsymbol{\omega}_A - \boldsymbol{\varpi}_A$. By analogy with part (c), use these quantities with (2-28) to derive (2-31).

4-37 (a) Identify $\tilde{\boldsymbol{\varpi}}$ in Sections 1.4 and 3.4.2.

(b) Show conformance of (4-71) to (4-67). Use (2-31), (2-32), and (4-69) to derive (4-72). Verify (4-73). Identify the total time derivative, $d\boldsymbol{\psi}/dt$.

(c) The reference cited in Ex. 4-3c validates the assumption of constant sidereal rate, even at drift rates of order 10^{-4} °/hr. If a significant unrecognized variation $\dot{\omega}_s$ *did* exist, could it be included in $d\boldsymbol{\psi}/dt$?

4-38 Truncate (2-69) and substitute $\check{\mathbf{w}}$ from (4-30), with the initial transformation matrix expressed by mean heading and pitch angles. Assume that roll is small or short-lived. Compare the results versus (4-74).

4-39 (a) The discussion following (4-24) defines an open-loop condition

$$\boldsymbol{\theta}_j = (H/B) \int \omega_j \, dt$$

Differentiate this expression and compare with (3-9). Differentiate again and combine with (4-16) to obtain a degradation of the form $(J/B)\dot{\omega}_j$. Apply the general interpretation of Ex. 4-14c here. Apply (4-59) to this degradation, assuming a triad of uniform gyros. Why does average drift from this degradation vanish, regardless of OA directions chosen?

(b) What happens when the example just considered is generalized as in (4-18)? Show that OA directions can be identified for which (4-59) contains squares, as well as products, of angular rate components. Explain the footnote for Table 4-1.

(c) Use (4-41), (4-42), and (4-59) to determine drift from misalignment and scale factor errors, in the presence of angular rates with cross-axis quadrature correlation.

(d) Use (4-44), (4-45), and (4-59) to identify drift sources associated with certain vibratory motions.

(e) Use (4-43), (4-73), and (4-74) to obtain an expression involving the product

$$\frac{1}{2\pi f_w}\,\breve{\mathbf{w}} \times \begin{bmatrix} \mathsf{T}_{\omega 1} & 0 & 0 \\ 0 & \mathsf{T}_{\omega 2} & 0 \\ 0 & 0 & \mathsf{T}_{\omega 3} \end{bmatrix} (-2\pi f_w)\breve{\mathbf{w}}$$

and show conformance to (4-60). Does it matter whether drift components in Table 4-1 are derived from (4-59) or (4-73) and (4-74)?

4-40 (a) Use (II-34), (4-62), (4-69), and a vector identity to express $\boldsymbol{\psi}_F \times \mathbf{A}$ as

$$\begin{bmatrix} (\mathbf{T}_2 \cdot \boldsymbol{\xi})(\mathbf{T}_3 \cdot \mathbf{F}) - (\mathbf{T}_2 \cdot \mathbf{F})(\mathbf{T}_3 \cdot \boldsymbol{\xi}) \\ (\mathbf{T}_3 \cdot \boldsymbol{\xi})(\mathbf{T}_1 \cdot \mathbf{F}) - (\mathbf{T}_3 \cdot \mathbf{F})(\mathbf{T}_1 \cdot \boldsymbol{\xi}) \\ (\mathbf{T}_1 \cdot \boldsymbol{\xi})(\mathbf{T}_2 \cdot \mathbf{F}) - (\mathbf{T}_1 \cdot \mathbf{F})(\mathbf{T}_2 \cdot \boldsymbol{\xi}) \end{bmatrix}$$

Use orthogonality conditions as in Ex. 2-8a to reduce this quantity to $\mathbf{T}_{A/G}^{T}(\boldsymbol{\xi} \times \mathbf{F})$.

(b) In Ex. 4-9, show that average error is perpendicular to both the angular rate and the measured acceleration uncertainty. Interpret this in connection with combination of (4-74) and (4-78).

(c) Recall Ex. 4-35. Can accelerometer quantization error be rectified in strapdown systems? Is much of this effect likely to arise in a gimballed platform?

(d) How does mean pitch and heading orientation act on the errors in Table 4-2?

4-41 (a) Show that (4-81) conforms to the pertinent angle definitions and conventions given previously.

(b) Show that (4-83) is a statement of Newton's second law. What approximations have been adopted?

4-42 (a) Derive (4-84) and (4-85) as indicated in the text. Illustrate a lift–gravity balance $F_3 = -g$ for cruise (no roll angle or sideslip and negligible angular rates).

(b) In (4-88)–(4-90), how does the usage of components of ζ from (2-19) relate to the assumption that $(\mathbf{I}_0, \mathbf{J}_0, \mathbf{K}_0)$ is an inertial frame?

4-43 Can q remain positive in Fig. 4-7 despite negative slope of pitch angle in Fig. 4-6? Explain in terms of (2-21). Use this same equation to explain steady state angular rates in a coordinated turn.

4-44 (a) Control surfaces considered in Appendix 4.A exert their predominant *direct* effect on aircraft rotation; translational adjustments then evolve from the flight dynamics described by (4-86)–(4-90). How is this related to the inequality in (4-92)?

(b) In (4-94) let $\mathscr{Y}_\beta = C_{y\beta}\mathscr{F}/(um_A)$, where $C_{y\beta} \doteq 0.7$ is a coefficient for a given aircraft of mass m_A, and $\mathscr{F}$ represents the product (dynamic pressure) × (wing area). At 5000-ft altitude (dry air density of approximately 0.0636 lb/ft^3) and an airspeed of 900 fps, dynamic pressure is essentially

$$\tfrac{1}{2}(0.0636/g)(900)^2 \doteq 800 \text{ lb/ft}^2$$

Show that for a 20-ton aircraft with 500-ft^2 wing area, sideslip can be approximated under the above conditions as

$$\beta \doteq \frac{m_A}{\mathscr{F} C_{y\beta}} F_2 \doteq 0.1 \frac{F_2}{g C_{y\beta}} \text{ rad}$$

What sideslip angle corresponds to an instantaneous 0.1-g lateral accelerometer output?

4-45 (a) From (4-97) show that the roll angle for a coordinated turn in Ex. 4-15c would be about 10° at 250 kt. What bank angle corresponds to a 2-g turn? Would a pilot in a coordinated 2-g turn sense an unusual specific force direction? Interpret the expression, "A pilot flies by the seat of his pants."

(b) Let $C_{L\alpha}$ represent a lift coefficient for the aircraft described in Ex. 4-44b. Consider flight at constant 5000-ft altitude, but with airspeed variations that invalidate (4-101). If, under somewhat simplified conditions,

$$\alpha \doteq -\frac{m_A}{\mathscr{F} C_{L\alpha}} F_3 \doteq -0.1 \frac{F_3}{g C_{L\alpha}} \text{ rad}$$

and $C_{L\alpha} \doteq 3.5$ at 5000 ft for speeds near 900 fps, what is the angle of attack during a level, coordinated, nominal 2-g, 4.1°/sec, turn?

4-46 (a) What slow rotation is ignored in (4-102) and (4-103)? Show that, under the conditions assumed,

$$\mathbf{F} = \boldsymbol{F} + \dot{\boldsymbol{\omega}}_{\mathrm{A}} \times \mathbf{d}_{\mathrm{C/N}} + \boldsymbol{\omega}_{\mathrm{A}} \times (\boldsymbol{\omega}_{\mathrm{A}} \times \mathbf{d}_{\mathrm{C/N}})$$

(b) Use (4-103) to describe $\boldsymbol{W}$ of Ex. 4-35; also define **A** from (4-62).

(c) Show that, with no motion of the air mass and no displacement of the IMU from aircraft mass center, (4-102) reduces to (4-83).

4-47 (a) Apply the composite angle formula to

$$(\mathsf{C}^2 + \mathsf{D}^2)^{1/2} \cos\{2\pi f t + \arctan(\mathsf{D}/\mathsf{C})\}$$

and relate to (4-104).

(b) Relate the inequality ($f_{\mathrm{w}} \gg w_{ir}$) to Ex. 4-9.

(c) Use the property,

$$\mathrm{av}\{\cos^2 2\pi f t\} = \mathrm{av}\{\sin^2 2\pi f t\} = \tfrac{1}{2}$$

to obtain (4-105) from

$$\lim_{t \to \infty} \frac{1}{t} \int_0^t w_i^2 \, d\tau$$

4-48 (a) Derive (4-106) as in the preceding exercise. Before averaging, what type of series appears in the product $w_i w_j$? What is its center frequency and where has it appeared before in this chapter?

(b) With nonzero frequencies throughout (4-104), what is $\langle w_i \rangle$? Add constant angular rate components κ_i, κ_j, and show that

$$\langle \omega_i \omega_j \rangle = \kappa_i \kappa_j + \langle w_i w_j \rangle$$

Re-express the correlation coefficient of (4-107) in terms of $\langle \omega_i \omega_j \rangle$, κ_i, κ_j, and rms values ω_{ir}, ω_{jr}.

4-49 (a) Consider a fixed vibration axis 30° off an x axis in an x-y plane, such that

$$\mathsf{C}_{ik}/\mathsf{C}_{jk} = \mathsf{D}_{ik}/\mathsf{D}_{jk} = \sqrt{3}$$

for every spectral component in (4-104). Show tight coupling for this case. Show that the same ratio characterizes w_i/w_j for all time in this example.

(b) Replace $\sqrt{3}$ in part (a) by $-1/\sqrt{2}$ and interpret the result.

4-50 Sketch a fulcrum at the left of a horizontal connecting rod, attached to a vibrating IMU on the right. With an angular vibration axis normal to the plane of the sketch passing through the fulcrum, show correlation between the rotation angle and the translational displacement perpendicular to the connecting rod. For small angular excursions use (4-27) and (4-28) to approximate the translational displacement as $-a_j/(2\pi f_a)^2$, and use (4-30) to represent this motion by case 8 of Table 4-3, with

$$f_a = f_w, \qquad \langle \check{w}_i a_j \rangle = w_{ir} a_{jr}$$

4-51 (a) Suppose that, due to oscillation frequencies beyond the servo bandwidth in Section 4.3.3, a gimballed platform experiences the type of motion described in Ex. 4-50. Assume that the rotation occurs about the vertical axis, while level accelerometers correctly measure the vibratory translational acceleration. Recalling the opening statement of Ex. 4-36a, describe an effective bias obtained from $\boldsymbol{\psi} \times \mathbf{A}$.

(b) Modify part (a) to assess the interaction between gyro scale factor error and translational vibration in a strapdown system. Extend this example to include gyro misalignment, and nonrigid connecting arms.

4-52 (a) Identify $\partial\boldsymbol{\varpi}/\partial\mathbf{V}$ for Sections 3.4.2 and 4.4. Does it present added possibilities for rectification, necessitating further analysis? How does its interaction with forcing functions differ from that of the operator $(-\mathbf{A} \times \quad)$?

(b) Relate the analysis of Section 3.4.2 to a transition matrix

$$\begin{bmatrix} \cos Wt & 0 & 0 & 0 & -u \sin Wt & 0 \\ 0 & \cos Wt & 0 & u \sin Wt & 0 & 0 \\ 0 & 0 & \cosh \sqrt{2}Wt & 0 & 0 & 0 \\ 0 & \dfrac{-\sin Wt}{u} & 0 & \cos Wt & 0 & 0 \\ \dfrac{\sin Wt}{u} & 0 & 0 & 0 & \cos Wt & 0 \\ 0 & 0 & 0 & 0 & 0 & 1 \end{bmatrix}$$

What is the value of this matrix at zero time?

4-53 (a) Relate the terms on the right of (4-114) to instrument error and to cross products in (4-73) and (4-78).

(b) Consider (4-26), (4-62), and (4-74) for an added role of specific force in generating rectification products as follows: A static force component of order $\langle \check{\mathbf{w}} \times \mathbf{a} \rangle/(2\pi f_w)$ is present but, for small angular excursions with $|\mathbf{a}| < |\mathbf{C}|$, this added component is outweighed by other effects such as lift.

(c) Describe the role of (4-44) and (4-45) in the generation of translation-dependent rectification products from $\mathbf{a} \times \check{\mathbf{e}}_\omega$. Discuss degradations from typical angular vibration (e.g., roll/yaw, with IMU location off the aurcraft mass center), regardless of chosen SA orientations.

4-54 (a) Using (4-26) and (4-62), describe an explicit interaction between $\mathbf{a}$ and $\mathbf{n}_\omega$ in a strapdown system.

(b) A common approach in characterizing solutions of differential equations is to assume the presence of terms having the same frequencies as the forcing functions. Relate this procedure to Ref. 4-2 and to the end of Appendix 4.B.

REFERENCES

4-1. Farrell, J. L., "Performance of Strapdown Inertial Attitude Reference Systems," *AIAA J. of Spacecraft and Rockets* **3** (9), Sept. 1966, pp. 1340–1347.

4-2. Farrell, J. L., "Analytic Platforms in Cruising Aircraft," *AIAA Journal of Aircraft* **4** (1), Jan.–Feb. 1967, pp. 52–58.

4-3. Goodman, L. E., and Robinson, A. R., "Effect of Finite Rotations on Gyroscopic Sensing Devices," *ASME Trans., J. Appl. Mech.* **25** (2), June, 1958, pp. 210–213.

4-4. Fernandez, M., and Macomber, G. R., *Inertial Guidance Engineering.* Englewood Cliffs, New Jersey: Prentice-Hall, 1962.

4-5. Kayton, M., and Fried, W. R. (eds.), *Avionics Navigation Systems.* New York: Wiley, 1969.

4-6. Parvin, R. H., *Inertial Navigation.* Princeton, New Jersey: Van Nostrand, 1962.

4-7. Wiener, T. F., "Theoretical Analysis of Gimballess Inertial Reference Equipment Using Delta-Modulated Instruments," MIT Instrumentation Lab., Sc.D. Thesis T-300, March 1962.

4-8. Bortz, J. E., "A New Mathematical Formulation for Strapdown Inertial Navigation," *IEEE Trans.* **AES-7** (1), Jan. 1971, pp. 61–66.

4-9. Goldstein, H., *Classical Mechanics*, London: Addison-Wesley, 1950, pp. 93–96, 143–144.

4-10. Rhoads, D. W., and Schuler, J. M., "A Theoretical and Experimental Study of Airplane Dynamics in Large-Disturbance Maneuvers," *J. Aero Sci.* **24** (7), July, 1957, pp. 507–526.

4-11. Etkin, B., *Dynamics of Flight*, New York: Wiley, 1959.

4-12. Bisplinghoff, R. L., Ashley, H., and Halfman, R. L., *Aeroelasticity*, Cambridge, Massachusetts: Addison-Wesley, 1955.

Chapter 5

Estimation with Discrete Observations

5.0

Estimation is an attempt to determine a set of variables (e.g., Cartesian components of position, velocity, etc.) from *observations* (synchronized or sequential measurements) related to those variables. The observations may be

incomplete: related to some, but not all, of the variables to be estimated,
indirect: expressed as some known function of the variables to be estimated, rather than identified as the variables themselves,
intermittent: taken at irregularly spaced instants of time, or
inexact: corrupted by biases and by sequentially uncorrelated random errors.

In many applications the variables to be estimated are *dynamic*, i.e., they are components of the solution $\mathbf{x}$ to a vector differential equation in the general form,

$$\dot{\mathbf{x}} = \boldsymbol{f}(\mathbf{x}, t) \tag{5-1}$$

which for the case of a *linear* functional relationship, reduces to [Appendix

II, Section II.3]*

$$\dot{\mathbf{x}} = \mathbf{A}\mathbf{x} + \mathbf{e} \tag{5-2}$$

where $\mathbf{A}$ is a matrix of coefficients describing the system dynamics and $\mathbf{e}$ a forcing function with specified characteristics; $\mathbf{A}$ and $\mathbf{e}$ may be functions of time, but not of $\mathbf{x}$. For dynamic systems estimation subdivides into three operations, i.e.,

(1) *Extrapolation:* the solution $\mathbf{x}_m$ of (5-1) or (5-2) at any time t_m is estimated from observations taken prior to t_m.

(2) *Filtering:* $\mathbf{x}_m$ is estimated from observations taken at and before t_m.

(3) *Smoothing:* $\mathbf{x}_m$ is estimated from observations including those taken after t_m (e.g., in postflight data reduction and analysis).

These operations are also categorized as to

(1) whether the dynamic process characteristics are known, partially known, or largely unknown; and

(2) whether the applicable equations are linear or nonlinear. The former case is characterized by linearity in the dynamics [e.g., (5-2)] and in the equations relating $\mathbf{x}$ to the observations; e.g., a scalar measurement at the discrete instant t_m could have an indicated value denoted by

$$z_m = \mathbf{H}_m \mathbf{x}_m + \epsilon_m \tag{5-3}$$

where $\mathbf{H}$ expresses the proportionality between the measurement and each component of $\mathbf{x}$, while ϵ denotes measurement error.

For estimators using knowledge of the dynamic process that generates $\mathbf{x}$, the dynamic equations in combination with observation equations [e.g., (5-3)] constitute a system *model*. Many practical applications of estimation theory involve fitting data to an imperfect model, based on equations in nearly linear form. This chapter specifies guidelines for these applications, through exposition of fundamentals (Section 5.1), optimal estimator derivation under ideal conditions (Section 5.2), extensions to include systematic error sources (Section 5.3) and practical degradations (e.g., nonlinearities, model imperfections; Section 5.4), and application to navigation (Section 5.5). The appendices to this chapter then enhance the overall interpretation. This presentation of estimation theory is slanted toward navigation, and simple examples are used to demonstrate various principles as the need arises.

* As with all other chapters, familiarity with Appendix II is assumed here. Note that applications of (5-2) have already been introduced at the outset of Chapter 4.

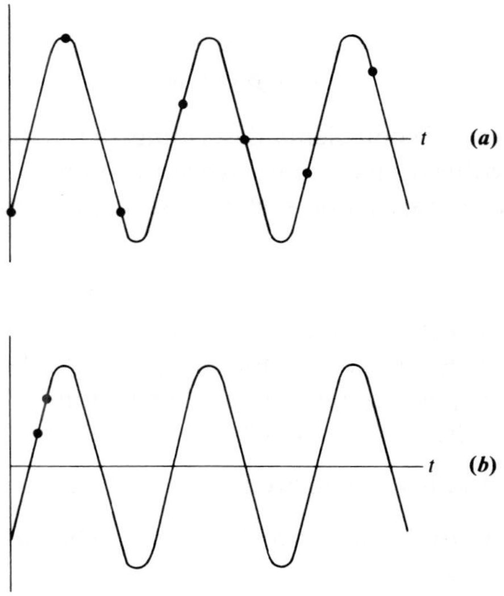

FIG. 5-1. Construction of continuous time histories from discrete data samples; (a) high sample density, (b) minimum sample input, given (5-4).

5.1 BASIC CONCEPTS IN LINEAR ESTIMATION

Dynamic estimation using discrete observations is exemplified in Fig. 5-1. A fixed transfer function filter would reconstruct a time history from the samples in Fig. 5-1a, using a spectral characterization as discussed in Section 1.3.3. However, when a dynamic process of the form

$$d^2x/dt^2 + \omega^2 x = 0 \tag{5-4}$$

is known to govern the waveform generation, then x could be specified for all time from the two samples in Fig. 5-1b, if the samples are measured exactly (Ex. 5-1). More generally, measurement errors impose requirements for more observations, though not as many as indicated in Fig. 5-1a when (5-4) is given.

The fitting of imperfect data to a model is approached by a method that has gained wide acceptance through a host of successful applications. As a first step in the approach, *any* linear dynamical system can be described by some number N of linearly independent* first-order differential equations, re-expressible in the form (5-2) for a linear N-dimensional vector differential

* That is, none of the N equations can be obtained from a linear transformation of the others. The equations can still be *coupled*, however, since $\mathbf{A}$ can have nonzero off-diagonal elements.

equation. Its solution $\mathbf{x}_m$ at any time t_m is called the *state vector* or the *state*, while its components are called the *state variables*, or simply, the *states*. Thus $\mathbf{x}_m$ can represent the "initial conditions" applicable to the dynamic differential equation at time t_m, from which it follows that (Ex. 5-2)

(1) The state is *unambiguous*; a complete set of values for the state variables at any time t_m would enable a complete and unique determination of $\mathbf{x}$ for all time in (5-2), if $\mathbf{A}$ and $\mathbf{e}$ were fully specified; and

(2) The state is *nonredundant*; no fewer than N variables could enable the full determination of dynamic behavior just described.

VERTICAL CHANNEL APPLICATION

The general linear formulation is readily demonstrated in terms of altitude h and vertical velocity V_V, since

$$\begin{bmatrix} -\dot{h} \\ \dot{V}_V \end{bmatrix} = \begin{bmatrix} 0 & 1 \\ 0 & 0 \end{bmatrix} \begin{bmatrix} -h \\ V_V \end{bmatrix} + \begin{bmatrix} 0 \\ Z_V \end{bmatrix} \tag{5-5}$$

where Z_V denotes total downward acceleration (Ex. 5-3). If Z_V were fully specified, knowledge of h and V_V at any instant would enable their determination for all time from (5-5). In a system with imperfections, however, apparent* vertical channel dynamics could be characterized by

$$\begin{bmatrix} -\dot{\hat{h}} \\ \dot{\hat{V}}_V \end{bmatrix} = \begin{bmatrix} 0 & 1 \\ 0 & 0 \end{bmatrix} \begin{bmatrix} -\hat{h} \\ \hat{V}_V \end{bmatrix} + \begin{bmatrix} 0 \\ \hat{Z}_V \end{bmatrix} \tag{5-6}$$

and, by subtraction of these last two equations,

$$\begin{bmatrix} -\dot{\tilde{h}} \\ \dot{\tilde{V}}_V \end{bmatrix} = \begin{bmatrix} 0 & 1 \\ 0 & 0 \end{bmatrix} \begin{bmatrix} -\tilde{h} \\ \tilde{V}_V \end{bmatrix} + \begin{bmatrix} 0 \\ \tilde{Z}_V \end{bmatrix} \tag{5-7}$$

It follows that discrepancies in initialization and in measured acceleration $\hat{Z}_V$ produce the result (Ex. 5-4)

$$\begin{bmatrix} -\tilde{h} \\ \tilde{V}_V \end{bmatrix} = \begin{bmatrix} 1 & t - t_m \\ 0 & 1 \end{bmatrix} \begin{bmatrix} -\tilde{h}_{(t_m)} \\ \tilde{V}_{V(t_m)} \end{bmatrix} + \int_{t_m}^{t} \begin{bmatrix} (t - \mathrm{r})\tilde{Z}_V(\mathrm{r}) \\ \tilde{Z}_V(\mathrm{r}) \end{bmatrix} d\mathrm{r} \tag{5-8}$$

which demonstrates how these imperfections produce position uncertainties

* *Apparent* in this context means *observed*, *estimated*, or *obtained from* observed and/or estimated quantities, all denoted by the circumflex ($\hat{}$). A tilde ($\tilde{}$) denotes *uncertainty* or error in an apparent value, e.g., $\tilde{V}_V = V_V - \hat{V}_V$.

that grow with time. A method of counteracting these effects, already illustrated in Fig. 1-14, involves an updating cycle as follows:

(1) Continuously monitored vertical acceleration is used to integrate (5-6) between discrete observations.
(2) Each observation is used for sudden adjustment of $\hat{h}$ and $\hat{V}_V$.
(3) The adjusted estimates initiate a new integration time segment, which lasts until the next observation.

The updating can be infrequent and irregularly timed, so long as the accumulated effects of inexact derivatives in (5-6) stay below an allowable altitude uncertainty. These principles will now be generalized, for application to all modes of navigation under consideration.

"CONTINUOUS–DISCRETE" ESTIMATION FORMALISM

Continuous dynamic equations and discrete updates here involve *zero-mean random vectors* $\mathbf{e}$ and $\boldsymbol{\epsilon}$, respectively, defined such that each component is a zero-mean random variable. A *sequentially uncorrelated* property is expressed for $\boldsymbol{\epsilon}$ in terms of covariance matrices $\mathbf{R}_m$ [Appendix II, development following (II-30)], specified at discrete instants t_m such that

$$\langle \boldsymbol{\epsilon}_m \boldsymbol{\epsilon}_n^{\mathrm{T}} \rangle = \begin{cases} \mathbf{O}, & m \neq n \\ \mathbf{R}_m, & m = n \end{cases} \tag{5-9}$$

and for $\mathbf{e}$ in terms of a *process covariance matrix* $\mathbf{E}$ with the Dirac delta function δ, such that

$$\langle \mathbf{e}(t)\mathbf{e}^{\mathrm{T}}(\mathrm{r}) \rangle = \mathbf{E}(t)\,\delta(t - \mathrm{r}) \tag{5-10}$$

in which the elements of $\mathbf{E}$ are *white noise* spectral densities (Ex. 5-5). While allowed to change with time, $\mathbf{R}$ and $\mathbf{E}$ are subject to restrictions of *positive definiteness* and *positive semidefiniteness*, respectively, i.e., for any nonzero vector $\boldsymbol{\alpha}$ of conformable dimension,

$$\boldsymbol{\alpha}^{\mathrm{T}}\mathbf{R}\boldsymbol{\alpha} > 0, \qquad \boldsymbol{\alpha} \neq \mathbf{0} \tag{5-11}$$

$$\boldsymbol{\alpha}^{\mathrm{T}}\mathbf{E}\boldsymbol{\alpha} \geq 0, \qquad \boldsymbol{\alpha} \neq \mathbf{0} \tag{5-12}$$

The random vector $\mathbf{e}(t)$ influences the dynamics of $\mathbf{x}(t)$ in accordance with the state transition equation of Section II.3,

$$\mathbf{x}(t) = \boldsymbol{\Phi}(t, t_{m-1})\mathbf{x}_{m-1} + \int_{t_{m-1}}^{t} \boldsymbol{\Phi}(t, \mathrm{r})\mathbf{e}(\mathrm{r})\, d\mathrm{r} \tag{5-13}$$

while the random vector $\boldsymbol{\epsilon}_m$ degrades the observations as follows: A *data vector*, consisting of M_m simultaneous scalar measurements, has an error-free value of

$$\mathbf{y}_m = \mathbf{H}_m \mathbf{x}_m \tag{5-14}$$

and an observed value given by the M_m-dimensional generalization of (5-3),

$$\mathbf{z}_m = \mathbf{H}_m \mathbf{x}_m + \boldsymbol{\epsilon}_m \tag{5-15}$$

where $\mathbf{H}_m$ is the $M_m \times N$ matrix expressing the proportionality between each *observable* (i.e., each measured quantity) and each state variable. A continuous–discrete estimation problem may now be defined as follows:

Let a sequence of noise-corrupted data vectors, characterized by (5-15), be used to construct estimates $\hat{\mathbf{x}}_m$ for the state of a dynamic process defined by (5-2) and (5-13), under conditions of known $\mathbf{A}(t)$, $\mathbf{E}(t)$, $\mathbf{H}_m$, and $\mathbf{R}_m$. The next section derives a *minimum variance* unbiased estimate, i.e., an estimate that minimizes the variance $\langle \tilde{x}_j^2 \rangle$ of every component $\tilde{x}_j$ of the state uncertainty $\tilde{\mathbf{x}}$ defined as

$$\tilde{\mathbf{x}} \triangleq \mathbf{x} - \hat{\mathbf{x}} \tag{5-16}$$

5.2 MINIMUM VARIANCE LINEAR ESTIMATION

Expressions will now be derived relating state estimates to observations under conditions just given. The covariance matrix of estimation error is also presented. Processing of all observations in a block is first considered, followed by description of repeated updating as observations appear in succession.

5.2.1 Linear Estimation by Block Processing

A special case of the linear estimation problem just stated follows from the condition $\mathbf{e}(t) \equiv \mathbf{0}$ throughout the data fitting duration. In that case, (5-15) in combination with a dynamic transformation yields

$$\mathbf{z}_m = \mathbf{H}_m \boldsymbol{\Phi}(t_m, t_0)\mathbf{x}_0 + \boldsymbol{\epsilon}_m \tag{5-17}$$

With each $M_m \times N$ product $\mathbf{H\Phi}$ regarded as a partition of a larger matrix $\mathbf{J}$, the measurements obtained at all observation times can be collected in the form (Ex. 5-6)

$$\mathbf{z} = \mathbf{J}\mathbf{x}_0 + \boldsymbol{\epsilon} \tag{5-18}$$

in which the number of elements in $\mathbf{z}$ or $\boldsymbol{\epsilon}$ (and the number of rows of $\mathbf{J}$) is the total number of scalar measurements,

$$\mathsf{M} = \sum_m M_m \tag{5-19}$$

There are then M equations that can be used to estimate the N components of $\mathbf{x}_0$ but, in general, the unknown $\boldsymbol{\epsilon}$ prevents an exact solution of (5-18). One possible procedure is to seek an estimate $\hat{\mathbf{x}}_0$ that minimizes the quantity $|\mathbf{z} - \mathbf{J}\hat{\mathbf{x}}_0|^2$. With unequal accuracies for different measurements, however, a better procedure is to allow the more precise observations a greater influence on the estimate; suppose that

$$\mathbf{U} = \langle \boldsymbol{\epsilon}\boldsymbol{\epsilon}^{\mathrm{T}} \rangle^{-1/2} \tag{5-20}$$

A weighted sum of squares of difference elements can then be defined, in the form

$$L \triangleq \{\mathbf{U}(\mathbf{z} - \mathbf{J}\hat{\mathbf{x}}_0)\}^{\mathrm{T}}\{\mathbf{U}(\mathbf{z} - \mathbf{J}\hat{\mathbf{x}}_0)\} \tag{5-21}$$

As a positive definite quadratic function, L is minimized when its partial derivative with respect to each component of $\hat{\mathbf{x}}_0$ is set to zero (Ex. 5-7):

$$\partial L/\partial \hat{\mathbf{x}}_0 = \mathbf{0} \tag{5-22}$$

or (Ex. 5-8)

$$(\mathbf{U}\mathbf{J})^{\mathrm{T}}\{\mathbf{U}(\mathbf{z} - \mathbf{J}\hat{\mathbf{x}}_0)\} = \mathbf{0} \tag{5-23}$$

An estimate that conforms to this minimization will be written $\hat{\hat{\mathbf{x}}}_0$ to denote smoothing as defined in Section 5.0. The elements of the difference vector $(\mathbf{z} - \mathbf{J}\hat{\hat{\mathbf{x}}}_0)$ are then called *true residuals* and, with $\mathbf{S} \triangleq \mathbf{U}\mathbf{J}$, the preceding expression is rewritten

$$\hat{\hat{\mathbf{x}}}_0 = (\mathbf{S}^{\mathrm{T}}\mathbf{S})^{-1}\mathbf{S}^{\mathrm{T}}\mathbf{U}\mathbf{z}, \qquad |\mathbf{S}^{\mathrm{T}}\mathbf{S}| \neq 0 \tag{5-24}$$

This solution is called the *weighted least squares* estimate, since it minimizes the weighted sum of squares of the true residuals. The corresponding smoothed estimate for the state at another time (e.g., t_m) is

$$\hat{\hat{\mathbf{x}}}_m = \boldsymbol{\Phi}(t_m, t_0)\hat{\hat{\mathbf{x}}}_0 \tag{5-25}$$

When all components of $\boldsymbol{\epsilon}$ in (5-20) have equal variance (Ex. 5-9), $\mathbf{U}$ is the product of a scalar (e.g., $1/\sigma$) times an orthogonal matrix; (5-24) then reduces to an unweighted least squares solution,

$$\hat{\hat{\mathbf{x}}}_0 = (\mathbf{J}^{\mathrm{T}}\mathbf{J})^{-1}\mathbf{J}^{\mathrm{T}}\mathbf{z}, \qquad \mathbf{U}^{\mathrm{T}}\mathbf{U} = (1/\sigma^2)\mathbf{I}, \qquad |\mathbf{J}^{\mathrm{T}}\mathbf{J}| \neq 0 \tag{5-26}$$

Another special case follows when $\mathbf{J}$ is nonsingular, which can occur only when $\mathsf{M} = N$. In that case, estimation on the basis of (5-18) reduces to

$$\hat{\mathbf{x}}_0 = \mathbf{J}^{-1}\mathbf{z}, \qquad |\mathbf{J}| \neq 0 \tag{5-27}$$

OBSERVABILITY

A solution from any of the last four equations is contingent upon a nonvanishing determinant. To illustrate this point, the two-dimensional example introduced previously is again instructive. Consider an expression in the form of (5-8) with no forcing function, and with reference time chosen at t_0 so that

$$\begin{bmatrix} x_1 \\ x_2 \end{bmatrix} = \begin{bmatrix} 1 & t - t_0 \\ 0 & 1 \end{bmatrix} \begin{bmatrix} x_{1(t_0)} \\ x_{2(t_0)} \end{bmatrix} \tag{5-28}$$

If the only available observations were sequential measurements of x_2, every row of **J** would be

$$[0 \quad 1] \begin{bmatrix} 1 & t_m - t_0 \\ 0 & 1 \end{bmatrix} = [0 \quad 1] \tag{5-29}$$

No inversion would then be possible, since position could not be initialized. On the other hand, two or more nonsimultaneous measurements of x_1 would enable an estimate to be derived (Ex. 5-10).

The state is *observable* if and only if the necessary inversion can be performed.* For many applications, however, a sharp division between observable and unobservable cases is insufficient. A quantitative measure of accuracy is needed for those cases in which inversion is possible. From (5-18) and (5-24), the estimation error is

$$\mathbf{x}_0 - \hat{\hat{\mathbf{x}}}_0 = -(\mathbf{S}^T\mathbf{S})^{-1}\mathbf{S}^T\mathbf{U}\epsilon, \qquad |\mathbf{S}^T\mathbf{S}| \neq 0 \tag{5-30}$$

with a covariance matrix (Ex. 5-11)

$$(\mathbf{S}^T\mathbf{S})^{-1}\mathbf{S}^T\mathbf{U}\langle\epsilon\epsilon^T\rangle\mathbf{U}^T\mathbf{S}(\mathbf{S}^T\mathbf{S})^{-1} = (\mathbf{S}^T\mathbf{S})^{-1}, \qquad |\mathbf{S}^T\mathbf{S}| \neq 0 \tag{5-31}$$

This matrix provides an indication of observability, not only by the variance of uncertainty (diagonal element) corresponding to each state, but also by the ease or difficulty of obtaining an accurate inversion from inexact computation. Two situations that suggest attention to possible numerical degradation, especially with large $\bar{\mathbf{M}}$, are

(1) *High-precision measurements.* Extremely accurate observations can introduce wide spreads between elements of (5-20). In the limit, a measurement known to be exact would reduce one row of (5-18) to the form

$$\mathbf{J}_m \mathbf{x}_0 = z_m \tag{5-32}$$

* A classical definition, which formulates an *observability matrix* and tests its rank, produces a sharp division into two classes (observable or unobservable). The *information matrix* $\mathbf{S}^T\mathbf{S}$ offers a general quantitative assessment.

(2) *Highly correlated measurement errors.* Close correlation between different observation errors can produce *ill-conditioning*, in the sense of a small determinant exemplified by Ex. 5-5e. In the limit, full correlation between two measurement errors (e.g., $\epsilon_m = \epsilon_n$) can be expressed in a form

$$(\mathbf{J}_m - \mathbf{J}_n)\mathbf{x}_0 = z_m - z_n \tag{5-33}$$

As constraining relations between the state variables, (5-32) and (5-33) effectively reduce the number of independent unknowns to be estimated. Therefore, by linear operations equivalent to substitution and elimination, these relations can be used to *advantage* in a proper reformulation. This opportunity for dimensionality reduction, if unrecognized, appears as a computational *disadvantage* through violation of independence between states (Ex. 5-12).

An instance of marginal observability that differs from the two items just mentioned is a partial deficiency in observational data. Situations arise in which some state variables are accurately monitored, for example, while others are associated with only the lowest precision measurements. Even in that case, computational difficulties of (5-24) can be circumvented (Appendix 5.A), but the formulation would still leave high uncertainties in those states having little influence on the observation sequence.

Weighted least squares smoothing provides an opportunity to extract maximum information from all measurements within a data block. Its presentation has also produced much insight for the development that follows.

5.2.2 Linear Sequential Estimation

Suppose that an estimate for the state at t_{m-1} and a subsequent observation vector $\mathbf{z}_m$ are available for processing, but the earlier data vectors $\mathbf{z}_{m-1}, \mathbf{z}_{m-2}, \ldots$ are no longer accessible. Instead of repeating (5-24) with each added measurement, an update can be performed as follows: An *a priori* state estimate $\hat{\mathbf{x}}_m^-$, just before the latest observation, is obtained by extrapolating the state $\hat{\mathbf{x}}_{m-1}^+$ estimated just after the preceding observation (Ex. 5-13):

$$\hat{\mathbf{x}}_m^- = \mathbf{\Phi}(t_m, t_{m-1})\hat{\mathbf{x}}_{m-1}^+ \tag{5-34}$$

so that an anticipated value $\mathbf{y}_m^-$ for the data vector can be subtracted from $\mathbf{z}_m$, to form the *predicted residual*

$$\mathbf{z}_m - \mathbf{y}_m^- = \mathbf{z}_m - \mathbf{H}_m \hat{\mathbf{x}}_m^- \tag{5-35}$$

which, as will be demonstrated in this section, produces the desired adjustment to the estimate when linearly weighted by a matrix $\mathbf{W}_m$:

$$\hat{\mathbf{x}}_m^{+} = \hat{\mathbf{x}}_m^{-} + \mathbf{W}_m(\mathbf{z}_m - \mathbf{H}_m \hat{\mathbf{x}}_m^{-}) \tag{5-36}$$

This updating procedure is recursive, in the sense that only the most recent values are needed for all variables. Thus, in contrast to repeated application of (5-24), both computation and data storage requirements are eased.

To determine the weighting matrix $\mathbf{W}_m$ that produces a minimum variance estimate, (5-35) is combined with (5-15) and (5-16) (Ex. 5-14):

$$\mathbf{z}_m - \mathbf{y}_m^{-} = \mathbf{H}_m \tilde{\mathbf{x}}_m^{-} + \boldsymbol{\epsilon}_m \tag{5-37}$$

Both sides of (5-36) are then subtracted from $\mathbf{x}_m$ and the result is combined with (5-37):

$$\tilde{\mathbf{x}}_m^{+} = (\mathbf{I} - \mathbf{W}_m \mathbf{H}_m)\tilde{\mathbf{x}}_m^{-} - \mathbf{W}_m \boldsymbol{\epsilon}_m \tag{5-38}$$

Of immediate interest are the mean values and the variances of the estimation errors, as propagated through this transformation. In particular, when the initial uncertainty as well as the measurement error has zero mean, then every component of $\tilde{\mathbf{x}}^{+}$ is also a zero-mean random variable. The general definition of a covariance matrix for $\tilde{\mathbf{x}}$,

$$\mathbf{P} \triangleq \langle(\tilde{\mathbf{x}} - \langle\tilde{\mathbf{x}}\rangle)(\tilde{\mathbf{x}} - \langle\tilde{\mathbf{x}}\rangle)^{\mathrm{T}}\rangle \tag{5.39}$$

then reduces to

$$\mathbf{P} = \langle\tilde{\mathbf{x}}\tilde{\mathbf{x}}^{\mathrm{T}}\rangle, \qquad \langle\tilde{\mathbf{x}}\rangle = \mathbf{0} \tag{5-40}$$

By substitution of (5-9) and (5-38) into (5-40), under conditions of no correlation between $\tilde{\mathbf{x}}_m^{-}$ and $\boldsymbol{\epsilon}_m$,

$$\mathbf{P}_m^{+} = (\mathbf{I} - \mathbf{W}_m \mathbf{H}_m)\mathbf{P}_m^{-}(\mathbf{I} - \mathbf{H}_m^{\mathrm{T}}\mathbf{W}_m^{\mathrm{T}}) + \mathbf{W}_m \mathbf{R}_m \mathbf{W}_m^{\mathrm{T}} \tag{5-41}$$

All that remains to determine $\mathbf{W}_m$ is the following succession:

(1) Derive a modified covariance matrix $\boldsymbol{P}_m$ resulting from a modified weighting matrix $(\mathbf{W}_m + \boldsymbol{\delta}\mathbf{W}_m)$.
(2) Obtain the difference $\boldsymbol{\delta}\mathbf{P} = \boldsymbol{P}_m - \mathbf{P}_m^{+}$.
(3) Choose $\mathbf{W}_m$ such that $\boldsymbol{\delta}\mathbf{P}$ is nonnegative definite.

This will provide a minimum variance estimate (Ex. 5-15). After replacement of $\mathbf{W}_m$ by $\mathbf{W}_m + \boldsymbol{\delta}\mathbf{W}_m$ in (5-41), followed by subtraction of $\mathbf{P}_m^{+}$, the result can be written

$$\boldsymbol{\delta}\mathbf{P} = \mathscr{P} + \mathscr{P}^{\mathrm{T}} + \boldsymbol{\delta}\mathbf{W}_m(\mathbf{H}_m \mathbf{P}_m^{-} \mathbf{H}_m^{\mathrm{T}} + \mathbf{R}_m)\, \boldsymbol{\delta}\mathbf{W}_m^{\mathrm{T}} \tag{5-42}$$

where

$$\mathscr{P} \triangleq \boldsymbol{\delta}\mathbf{W}_m\{\mathbf{H}_m \mathbf{P}_m^{-}(\mathbf{I} - \mathbf{H}_m^{\mathrm{T}}\mathbf{W}_m^{\mathrm{T}}) - \mathbf{R}_m \mathbf{W}_m^{\mathrm{T}}\} \tag{5-43}$$

This matrix *vanishes* (Ex. 5-16) when $\mathbf{W}_m$ is chosen as (Ref. 5-1)

$$\mathbf{W}_m = \mathbf{P}_m^{-}\mathbf{H}_m^{\mathrm{T}}(\mathbf{H}_m\mathbf{P}_m^{-}\mathbf{H}_m^{\mathrm{T}} + \mathbf{R}_m)^{-1} \tag{5-44}$$

in which case (5-42) reduces to

$$\boldsymbol{\delta}\mathbf{P} = \boldsymbol{\delta}\mathbf{W}_m\mathbf{Z}_m\boldsymbol{\delta}\mathbf{W}_m^{\mathrm{T}} \tag{5-45}$$

where $\mathbf{Z}_m$ is the covariance matrix of predicted residuals from (5-37):

$$\mathbf{Z}_m = \mathbf{H}_m\mathbf{P}_m^{-}\mathbf{H}_m^{\mathrm{T}} + \mathbf{R}_m \tag{5-46}$$

Since this symmetric matrix is positive definite, $\boldsymbol{\delta}\mathbf{P}$ in (5-45) must be nonnegative definite. Thus, every component of $\tilde{\mathbf{x}}_m^{+}$ has minimum variance when the weighting matrix of (5-44) is used in (5-36). This approach to the estimator derivation appears in Ref. 5-2; for a general and rigorous derivation, the reader is referred to Ref. 5-1. Further development and applications are found in Ref. 5-3.

COVARIANCE MATRIX PROPAGATION

Dependence of the weighting matrix on $\mathbf{P}_m^{-}$ calls for a specification of

(1) an "initial" covariance matrix of uncertainty at some reference time, e.g., $\mathbf{P}_0$ at time t_0;
(2) the dynamic behavior of $\mathbf{P}$ between observations; and
(3) changes in $\mathbf{P}$ at discrete measurement instants.

The initial matrix $\mathbf{P}_0$ can start a cycle of dynamic adjustments in the estimator uncertainty, characterized as follows: Subtraction of (5-34) from (5-13), evaluated at t_m, yields

$$\tilde{\mathbf{x}}_m^{-} = \boldsymbol{\Phi}(t_m, t_{m-1})\tilde{\mathbf{x}}_{m-1}^{+} + \int_{t_{m-1}}^{t_m}\boldsymbol{\Phi}(t_m, \tau)\mathbf{e}(\tau)\,d\tau \tag{5-47}$$

and, by substitution into (5-40), the development at the end of Appendix II can be invoked directly to yield

$$\mathbf{P}_m^{-} = \boldsymbol{\Phi}(t_m, t_{m-1})\mathbf{P}_{m-1}^{+}\boldsymbol{\Phi}^{\mathrm{T}}(t_m, t_{m-1}) + \int_{t_{m-1}}^{t_m}\boldsymbol{\Phi}(t_m, \tau)\mathbf{E}(\tau)\boldsymbol{\Phi}^{\mathrm{T}}(t_m, \tau)\,d\tau \tag{5-48}$$

Either this relation or its derivative (Ex. 5-17)

$$\dot{\mathbf{P}} = \mathbf{A}\mathbf{P} + \mathbf{P}\mathbf{A}^{\mathrm{T}} + \mathbf{E} \tag{5-49}$$

can provide $\mathbf{P}_m^{-}$, given the uncertainty covariance immediately after the preceding update (or, in the case $m = 1$, given $\mathbf{P}_0$).

The uncertainty adjustment at a discrete update has already been expressed by (5-41); combination with (5-44) provides the simplified expression

$$\mathbf{P}_m^{\ +} = (\mathbf{I} - \mathbf{W}_m \mathbf{H}_m)\mathbf{P}_m^{\ -} \tag{5-50}$$

which, by substitution of (5-44) and (5-46), corresponds to a change in $\mathbf{P}$ by an amount

$$\Delta\mathbf{P}_m = -\mathbf{P}_m^{\ -}\mathbf{H}_m^{\ \mathrm{T}}\mathbf{Z}_m^{-1}\mathbf{H}_m \mathbf{P}_m^{\ -} \tag{5-51}$$

The negative quadratic nature of this symmetric matrix indicates a reduction in uncertainty, as each properly weighted observation is used to adjust $\hat{\mathbf{x}}$.

INTERPRETATION OF SEQUENTIAL WEIGHTING

Existence of an initial state estimate $\hat{\mathbf{x}}_0^{\ +}$, with an uncertainty $\tilde{\mathbf{x}}_0^{\ +}$ characterized by zero mean and a positive definite covariance matrix $\mathbf{P}_0$, ensures positive definiteness of $\mathbf{P}$ for *all* time with (5-11) satisfied (e.g., no exact measurements are taken). Although no state variable uncertainty can then be completely eliminated, there is a sudden reduction in accordance with (5-50) and (5-51) at each discrete update. The following example illustrates the general sequence.

Scalar static state with direct observations, $\dot{x} = 0$, $\mathsf{H} = 1$. With sequentially uncorrelated measurement errors of equal variance R, the minimum variance estimate at any time as obtained from (5-24) or (5-26) is the average of all available measured values. From (5-31), the variance of estimation error is then inversely proportional to the number of observations taken thus far. In particular, after $\mathsf{M} - 1$ observations,

$$\mathsf{P}_\mathsf{M}^{\ -} = \mathsf{P}_{\mathsf{M}-1}^{+} = \mathsf{R}/(\mathsf{M} - 1) \tag{5-52}$$

so that, by (5-36) and (5-44) the update at the Mth measurement is

$$\hat{x}_\mathsf{M}^{\ +} = [1/(\mathsf{M} - 1)] \sum_{m=1}^{\mathsf{M}-1} z_m + [\mathsf{P}_\mathsf{M}^{\ -}/(\mathsf{P}_\mathsf{M}^{\ -} + \mathsf{R})]\left\{z_\mathsf{M} - [1/(\mathsf{M} - 1)] \sum_{m=1}^{\mathsf{M}-1} z_m\right\} \tag{5-53}$$

which reduces to

$$\hat{x}_\mathsf{M}^{\ +} = (1/\mathsf{M}) \sum_{m=1}^{\mathsf{M}} z_m, \qquad \mathsf{P}_\mathsf{M}^{\ -}/(\mathsf{P}_\mathsf{M}^{\ -} + \mathsf{R}) = 1/\mathsf{M} \tag{5-54}$$

and, from (5-50), mean squared uncertainty after the current update is given by

$$\mathsf{P}_\mathsf{M}^{\ +} = \mathsf{R}/\mathsf{M} \tag{5-55}$$

Although (5-54) and (5-55) were obtained from sequential estimation expressions, they agree with the estimate and the covariance that would have resulted from block processing of all $\mathbf{M}$ observations.* This correspondence illustrates how the *method* of deriving a minimum variance estimate can be immaterial, as long as consistent *conditions* are applied to alternative formulations. For any system having known conformance to (5-2) with $\mathbf{e} \equiv \mathbf{0}$, estimates and covariances corresponding to (5-25) and (5-36) will agree when given the same total information to be processed. At any observation before the last one in a block, however, (5-24) and (5-25) provide an interpolation with lower probabilistic error than the corresponding filtered result.

A crucial property of the filter weighting is the relative influence of estimation uncertainty versus measurement error. When $\mathbf{P}_{\mathrm{M}}^{-} \gg \mathbf{R}$ in (5-53), the observation exerts considerable influence in adjusting the estimate. This may be the case, for example, as an initial estimate is being formed. After processing of much data, however, the current estimate may be trusted more readily than a single observation; weighting is then reduced. As applied to dynamic systems, this variable weighting resolves a dilemma inherent in fixed-parameter filters. In particular, if a constant matrix were substituted for $\mathbf{W}$ in (5-36), the estimate could have either a high vulnerability to observation error (with heavy weighting) or a slow transient response (e.g., if small weights retarded the adjustment to initial conditions). Because (5-44) is evaluated from current values (at time t_m) of $\mathbf{P}$, $\mathbf{H}$, and $\mathbf{R}$, each measurement is fully exploited to obtain the greatest reduction in estimation uncertainty.

Versatility of the linear estimation approach is clear from the general form of the equations. The scheme readily accommodates multidimensional systems with time-varying parameters and random forcing functions, observed at irregular intervals by changing combinations of sensors with variable accuracies. With these general conditions, dynamic behavior of rms uncertainty $\tilde{x}_{j\mathrm{r}}$ (square root of a diagonal element in $\mathbf{P}$) for a typical state variable x_j may resemble Fig. 5-2. Initially the estimate can have a large error, but one sufficiently accurate measurement (at t_1) that is sensitive to x_j can reduce its uncertainty to the sensor tolerance level. Accompanying uncertainties in derivatives then account for the rise between t_1 and t_2, along a curve determined by the influence of other states and $\mathbf{E}$ (Ex. 5-18). Another observation at t_2 then reduces $\tilde{x}_{j\mathrm{r}}$ *and* the uncertainty in its derivative; thus the curve from t_2 to t_3 rises more slowly. Measurement of x_j by a crude instrument at t_4 would not be very effective and, while update timing is not critical, extended observationless periods would allow

* It is well known that equivalence between block and sequential estimation extends to more general conditions; e.g., see Ref. 5-4.

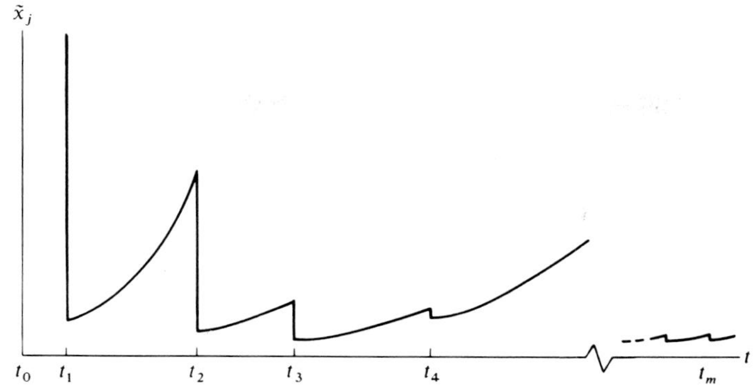

FIG. 5-2. rms uncertainty vs. time.

unattended growth of $\tilde{x}_{jr}$. Finally, subsequent observations could produce a "quasi-steady-state" uncertainty level, somewhat below the value dictated by the measurement error. Substantial reduction below that value would require many independent observations (Ex. 5-19), and the positive definiteness of $\mathbf{P}$ precludes a zero value at any time.

No particular probability distribution is assumed throughout this development. Although gaussian random variables allow further insights, they would be inapplicable to many degradations common to navigation systems.*

OBSERVABILITY AND THE DATA WINDOW

Linear filtering enables estimates to evolve from weighted combinations of past and present measurements, according to prescribed dynamic and observational models. For reasons to be more fully explained in Section 5.4, imperfect modeling calls for intentional suppression of data from the remote past. As an initial example involving an accumulating observation sequence, imagine repeated evaluations of block estimates that always include only the measurements taken within the past T sec. The advancing data block of fixed duration T can be approximated by gradually attenuating the influence that past observations can exert on $\hat{\mathbf{x}}$. The ensuing discussion demonstrates how a gradual suppression of past information is inherently produced by the "process noise" covariance matrix $\mathbf{E}$, so that sequential estimates with $\mathbf{E}$ are comparable to fixed duration block estimates without

* Gaussian variates guarantee (1) equivalence of conditional mean, maximum likelihood, and minimum variance estimates, and (2) complete statistical independence between variables that are linearly uncorrelated. These properties, and the items described in Ref. 5-5, are significant but not essential for applying estimation theory to navigation (Ex. 5-20).

E. A specific mathematical relation is also given between **E** and a nominal *data window* T for sequential estimation.

With nonzero elements in **E**, (5-12) ensures that mean squared estimation uncertainty will increase between measurements. The cumulative contribution of **E** to **P**, already exemplified by the last term of (5-48), is expressed for a general time interval as

$$\mathscr{Q}(t, \ell) \triangleq \int_{t-\ell}^{t} \mathbf{\Phi}(t, \mathrm{r})\mathbf{E}(\mathrm{r})\mathbf{\Phi}^{\mathrm{T}}(t, \mathrm{r})\, d\mathrm{r} \tag{5-56}$$

As this contribution increases, so does the relative influence of state versus measurement uncertainty upon residual covariance. For any given scalar measurement y (or any component y of a given data vector), having a $1 \times N$ proportionality matrix $\mathscr{H} \triangleq \partial y/\partial \mathbf{x}$ and a variance R, the defining relation for T is now given as*

$$\mathscr{H}\mathscr{Q}(t, T)\mathscr{H}^{\mathrm{T}} = \mathrm{R} \tag{5-57}$$

so that, over longer durations,

$$\mathscr{H}\mathscr{Q}(t, \ell)\mathscr{H}^{\mathrm{T}} \gg \mathrm{R}, \qquad \ell \gg T \tag{5-58}$$

To apply these last three relations in a navigation scheme, consider an update in (5-36) with a scalar position measurement y_m having a proportionality matrix $\mathscr{H}_m$ and a variance R_m, under the following conditions:

(1) In the duration since the last observation, an accumulation $\mathscr{Q}(t_m, t_m - t_{m-1})$ contributes appreciable variances, on the right of (5-48), to estimates of the position states affecting y_m.

(2) (5-58) is satisfied, from which it follows that $\mathscr{H}_m \mathbf{P}_m^- \mathscr{H}_m^{\mathrm{T}} \gg \mathrm{R}_m$, and the measurement at time t_m is weighted heavily.

Thus, when the time between observations exceeds T, updated estimates of states affecting y_m are dictated primarily by the latest observational data. Relative emphasis of the residual in (5-36) is synonymous with relative de-emphasis of $\hat{\mathbf{x}}_m^-$, extrapolated from the past. A more balanced influence of past and current information would have resulted, if condition (2) had been based on (5-57) rather than (5-58). Intervals of shorter

* Though not an explicit formula for T nor for elements of **E**, (5-57) can always be solved for given dynamic and observation characteristics. The solution can be obtained numerically, if necessary, when the functional form of $\mathbf{\Phi}$ is variable or highly complex. Note that (5-57) is a design equation, for which repeated solution during the filtering operation is not necessary. The immediate development is not concerned with the means of solving (5-57), nor with whether **E** or T is specified independently. In practice, **E** must often conform to an independently specified T, not to exceed the duration of model fidelity (Section 5.4). Specific examples for certain operating modes are given in subsequent chapters.

duration ($t_m - t_{m-1} < T$) would imply that recent observations have not yet receded into the remote past (Ex. 5-21).

Due to almost unlimited variety of measurement types and schedules, dynamic system characteristics, and parameter values, the extent and duration of each measurement's influence defies an exact simple general analysis. In navigation, however, a representative duration of influence for position observations is given by (5-57). The most sensitive measurement for each direction will govern overall performance, for which an approximate characterization is generally permissible (Ex. 5-22). Just as a decaying function of time is not completely attenuated after one time constant, the effect of a particular observation need not disappear after T.

As Section 5.2.1 imposed requirements on the data within a block, continuously acceptable performance of sequential estimates is based on measurements within each data window. To obtain position and velocity estimates commensurate with sensor accuracies, there should be at least one position measurement per data window for each direction. Position uncertainty is then determined by $\mathscr{H}$ and R, while velocity can be derived to within $(1/T) \times$ (position uncertainty). When the state also includes acceleration, a minimum of one nonsimultaneous position measurement per data window must be added. Redundant observations should be introduced, within practical limits, to reduce estimation error below the levels dictated by the sensors (Ex. 5-23).

5.3 SYSTEMATIC ERROR SOURCES

If an ensemble of navigation instruments were known to produce nonzero average errors, simple recalibration or computation could nullify these effects. No generality is lost, therefore, in assuming zero mean errors throughout this chapter. Any particular instrument selected for a given flight, however, may be biased. *Systematic* degradations, in the form of constant or slowly varying biases, then produce sequential correlations that necessitate generalization of the material just presented.* A reformulation is theoretically obtainable, in terms of an *augmented* state that includes the biases themselves. Extension of the theory is accomplished by re-expressing the equations, with systematic errors in dynamics and/or observations, in the same form as (5-2) and (5-15). This section provides the necessary reformulation, and explains the practical limitations. In-flight calibration, applied to the isolated vertical channel, serves as an illustrative example.

* Systematic errors can be introduced by instruments and also by incomplete knowledge of physical parameters. Uncertainty in present local barometric pressure at mean sea level, in Fig. 1-14, would produce a persistent offset in altitude as determined from barometric data.

5.3.1 Bias Effects in System Dynamics

Consider a re-expression of (5-7) with $\tilde{Z}_V$ decomposed as

$$\tilde{Z}_V = \mathsf{b}_Z + e_Z \tag{5-59}$$

where b_Z, e_Z denote constant and purely random components, respectively, of total vertical acceleration error. Immediately, (5-7) can then be replaced by

$$\begin{bmatrix} -\dot{\tilde{h}} \\ \dot{\tilde{V}}_V \\ \dot{\mathsf{b}}_Z \end{bmatrix} = \begin{bmatrix} 0 & 1 & 0 \\ 0 & 0 & 1 \\ 0 & 0 & 0 \end{bmatrix} \begin{bmatrix} -\tilde{h} \\ \tilde{V}_V \\ \mathsf{b}_Z \end{bmatrix} + \begin{bmatrix} 0 \\ e_Z \\ 0 \end{bmatrix} \tag{5-60}$$

which has *only* sequentially uncorrelated error for a forcing function. Estimation equations already presented are then applicable to an augmented vertical channel, in which sequential altitude observations determine $\hat{\mathsf{b}}_Z$ as well as $\hat{h}$ and $\hat{V}_V$ (Ex. 5-24), with [1, 0, 0] substituted for $\mathbf{H}_m$ in (5-36) and (5-44). In the corresponding time history of altitude uncertainty (e.g., Fig. 5-2), the curvature between any two updates would depend primarily on the amount of undetected acceleration bias remaining at that time. This bias, as well as the initial slope of the uncertainty rise, would be more quickly reduced by direct measurements of velocity. If initial transients are acceptable, however, position information alone can adequately determine $\hat{V}_V$ and $\hat{\mathsf{b}}_Z$ after several observations are processed.

The augmenting procedure just described is easily modified for exponentially varying bias, corresponding to a band-limited sensor error of the form

$$\dot{x}_3 = (-1/\tau_3)x_3 + e_3 \tag{5-61}$$

so that

$$\begin{bmatrix} -\dot{\tilde{h}} \\ \dot{\tilde{V}}_V \\ \dot{x}_3 \end{bmatrix} = \begin{bmatrix} 0 & 1 & 0 \\ 0 & 0 & 1 \\ 0 & 0 & -1/\tau_3 \end{bmatrix} \begin{bmatrix} -\tilde{h} \\ \tilde{V}_V \\ x_3 \end{bmatrix} + \begin{bmatrix} 0 \\ e_Z \\ e_3 \end{bmatrix} \tag{5-62}$$

with a transiton matrix (Ex. 5-25)

$$\mathbf{\Phi}_{3\times 3} = \begin{bmatrix} 1 & t - t_m & \tau_3(t - t_m - v\tau_3) \\ 0 & 1 & v\tau_3 \\ 0 & 0 & 1 - v \end{bmatrix}, \qquad v \triangleq 1 - \exp\left(\frac{t - t_m}{-\tau_3}\right) \tag{5-63}$$

which reduces to the intuitive form for fixed acceleration bias,

$$\lim_{\tau_3 \to \infty} \mathbf{\Phi}_{3\times 3} = \begin{bmatrix} 1 & t - t_m & \frac{1}{2}(t - t_m)^2 \\ 0 & 1 & t - t_m \\ 0 & 0 & 1 \end{bmatrix} \tag{5-64}$$

Application of augmented dynamics to multiple directions is straightforward. In fact, the text accompanying (4-4) foreshadows augmentation of the horizontal channels with acceleration bias and gyro drift, discussed more fully in Section 6.4. Extension is also possible to multiple bias components in any one direction, e.g., the relation

$$\begin{bmatrix} -\dot{\tilde{h}} \\ \dot{\tilde{V}}_V \\ \dot{x}_3 \\ \dot{x}_4 \end{bmatrix} = \begin{bmatrix} 0 & 1 & 0 & 0 \\ 0 & 0 & 1 & 1 \\ 0 & 0 & -1/\tau_3 & 0 \\ 0 & 0 & 0 & 0 \end{bmatrix} \begin{bmatrix} -\tilde{h} \\ \tilde{V}_V \\ x_3 \\ x_4 \end{bmatrix} + \begin{bmatrix} 0 \\ e_Z \\ e_3 \\ 0 \end{bmatrix} \tag{5-65}$$

can describe a vertical channel with both a constant and a dynamic acceleration bias. If $\tau_3 > T$, however, the "time-varying" x_3 is hardly distinguishable from x_4, and (5-60) or (5-62) would be preferable to (5-65). A quantity should not be an augmenting state unless its effects are (1) discernible from the measurements within a data window, and (2) distinguishable from the influence of all other state variables, including all other augmenting states.

5.3.2 Observation Bias

When observation errors are sequentially correlated, $\boldsymbol{\epsilon}$ can be supplemented by a bias term, in analogy with (5-59). This section describes state augmentation to detect observation bias, and identifies pitfalls to be avoided. To introduce the theory, the familiar vertical channel application is again instructive.

Consider both a fixed bias and a sequentially uncorrelated random error ϵ in the barometric observations of Fig. 1-14. With this bias chosen as a third state for augmentation of (5-7), the conventions

$$\mathbf{A}^* = \begin{bmatrix} 0 & 1 & 0 \\ 0 & 0 & 0 \\ 0 & 0 & 0 \end{bmatrix}, \qquad \mathbf{e}^* = \begin{bmatrix} 0 \\ \tilde{Z}_V \\ 0 \end{bmatrix} \tag{5-66}$$

are consistent with (5-2), while conformance to (5-9), (5-11), and (5-15) is satisfied with the 1×3 row matrix $[1 \quad 0 \quad 1]$ substituted for $\mathbf{H}$. The augmented state would be unobservable, however, since barometric bias could be mistaken for a deviation in altitude. For a mathematical demonstration of unobservability in this case, apply Section 5.2.1 to the system of (5-66) with $\mathbf{e}^*$ nulled and

$$\mathbf{\Phi}_m{}^* = \begin{bmatrix} 1 & t_m - t_0 & 0 \\ 0 & 1 & 0 \\ 0 & 0 & 1 \end{bmatrix} \tag{5-67}$$

so that every row of $\mathbf{J}$ in (5-18) would have the same element (unity) in the first and third column. The determinant $|\mathbf{J}^T\mathbf{J}|$ would therefore vanish (Exs. 5-26 and 5-27).

By adding more information to the system just described, the bias could be made observable. For example, an unbiased radar altimeter measurement over the sea would be characterized here by the $\mathscr{H}$-matrix [1, 0, 0] and a corresponding row $[1, t_m - t_0, 0]$ in $\mathbf{J}$. Mathematically, $\mathbf{J}^T\mathbf{J}$ would no longer be singular; physically, the additional information would provide a basis for comparison with the biased data. The comparison becomes less effective, however, with increasing rms error in the unbiased information (Ex. 5-28). These considerations now enable a broader analysis.

GENERALIZED MEASUREMENT FORMULATION

Suppose that observations are corrupted by the sequentially uncorrelated error vector $\boldsymbol{\epsilon}$, *and* by elements of a bias vector $\boldsymbol{\beta}$ having dynamics similar to (5-61),

$$\dot{\boldsymbol{\beta}} = -\boldsymbol{\tau}^{-1}\boldsymbol{\beta} + \mathbf{e}_\beta \tag{5-68}$$

where $\boldsymbol{\tau}$ and $\mathbf{e}_\beta$ contain the time constants and white driving noise, respectively, for the biases. For this general linear bias representation (Ex. 5-29), let $\mathbf{D}$ express the proportionality between each bias and each measured quantity; (5-15) is then replaced by

$$\mathbf{z}_m{}^* = \mathbf{H}_m \mathbf{x}_m + \mathbf{D}_m \boldsymbol{\beta}_m + \boldsymbol{\epsilon}_m \tag{5-69}$$

or

$$\mathbf{z}_m{}^* = \mathbf{H}_m{}^*\mathbf{x}_m{}^* + \boldsymbol{\epsilon}_m \tag{5-70}$$

where

$$\mathbf{H}_m{}^* = [\mathbf{H}_m \,\vdots\, \mathbf{D}_m], \qquad \mathbf{x}_m{}^* = \begin{bmatrix} \mathbf{x}_m \\ \hdashline \boldsymbol{\beta}_m \end{bmatrix} \tag{5-71}$$

Also, (5-68) can be combined with (5-2) to yield

$$\begin{bmatrix} \dot{\mathbf{x}} \\ \hdashline \dot{\boldsymbol{\beta}} \end{bmatrix} = \begin{bmatrix} \mathbf{A} & \mathbf{O} \\ \mathbf{O} & -\boldsymbol{\tau}^{-1} \end{bmatrix} \begin{bmatrix} \mathbf{x} \\ \hdashline \boldsymbol{\beta} \end{bmatrix} + \begin{bmatrix} \mathbf{e} \\ \hdashline \mathbf{e}_\beta \end{bmatrix} \tag{5-72}$$

Since (5-70) and (5-72) have the same form as the original equations for observations and dynamics, respectively, the development of Section 5.2 and related interpretive discussions are now applicable to the augmented state.

The approach just formulated retains $\boldsymbol{\epsilon}$, with positive definite covariance matrix $\mathbf{R}$, in addition to any bias that may be present (Ref. 5-6). Reasons can be explained by conditions that could arise if ϵ were *not* retained throughout, as follows:

(1) $\mathbf{R}_m$ would be allowed to become singular in augmented formulations.

(2) With any component of ϵ_m set to zero, (5-70) would constrain the augmented state, and $\langle \tilde{\mathbf{x}}_m{}^* \tilde{\mathbf{x}}_m^{*\mathrm{T}} \rangle$ would be singular immediately after the mth measurement.

Singularity in covariance matrices of both state and measurement uncertainty would invite possible computational breakdown in subsequent evaluations of (5-44). In general, the inversion could not be performed unless $\mathbf{E}$ restored positive definiteness to $\mathbf{P}$ between observations.*

When restoration by $\mathbf{E}$ does not occur, Refs. 5-7 and 5-8 offer an alternate approach; the estimate in the presence of bias is reformulated with dimensionality equal to that of the unaugmented state. The natural counterpart in block processing is the addition of off-diagonal elements in (5-20). In fact, (5-9) could be generalized to allow correlations between any pair of measurements, simultaneous or sequential, and (5-24) would remain applicable with no change in dimension. For further interpretation, Appendix 5.B demonstrates a bias estimate implicitly accounted for without augmentation, *if* correlations between observation errors are known. It follows that augmentation is an expedient to be used judiciously, and the scope of this subject must now expand to include estimates based on inexact characterizations.

5.4 PRACTICAL REALIZATION OF CONTINUOUS–DISCRETE ESTIMATION

The development presented thus far has concentrated primarily on estimation for linear systems with known characteristics. A practical estimator, however, must reconstruct the state dynamics from observations without full adherence to the conditions concluding Section 5.1. Possible departures from ideal conditions include

(1) nonlinearities in dynamic equations [e.g., the geographic velocity dynamics in (3-32) do not conform to (5-2)];

* Restoration of positive definiteness by $\mathbf{E}$ is not a general condition. Therefore, (5-11) cannot be relaxed in a general augmented formulation; $\mathbf{E}$ can satisfy (5-12) without affecting any specified element of $\mathbf{P}$. In fact, even a null matrix satisfies (5-12). Because (5-12) imposes a weaker condition than (5-11), the development in Section 5.3.1 is less complicated than augmentation for observation bias. Note in any case that (5-11) can be a conservative requirement when positive definiteness is restored to $\mathbf{P}$ by appropriate elements of $\mathbf{E}$. This conservatism enhances computation of the inverse required in (5-44) (Ex. 5-30).

(2) nonlinear relations between states and observables (e.g., range is the root sum square of Cartesian displacement components);

(3) terms missing from equations involving observables and/or state dynamics (e.g., due to unknown forcing functions or omission of horizontal/vertical channel interaction);

(4) imperfect parameter values in applicable equations;

(5) timing and spectral approximations (e.g., discret*ized* rather than discrete observations, simplified bias error models, etc.);

(6) incomplete knowledge of covariances (e.g., approximation of **R** and very rough characterization of initial **P**-matrix);

(7) unforeseen measurement degradations—"wild" data points;

(8) computational discrepancies (e.g., roundoff).

These effects, unless counteracted, could cause an estimate to diverge as follows: Shortly after initialization, observations are processed by imprecisely determined weighting elements. Equation (5-50) nevertheless computes the maximum reduction in uncertainty that is theoretically obtainable from each measurement. As a result, optimistically low uncertainty covariances in subsequent applications of (5-44) will provide inadequate weighting of new observational data. This effect is compounded as more measurements are processed, while (5-48)–(5-50) are repeatedly applied with blissful unawareness of the increasingly unrealistic nature of **P**. The estimated state is eventually corrupted by a useless accumulation of improperly weighted residuals.

The catastrophic sequence just described can be avoided by methods to be presented here. Instead of *optimizing* the estimate in a minimum variance sense, computations involving (5-44) are stabilized by driving **P** with **E**. The "minimum" variance property is thus compromised in favor of positive definiteness in **P**, and data spanning a duration T can be fitted to an *imperfect* model with reasonable fidelity over T. For instance, the points in Fig. 5-3 are acceptably close to the curve shown with a 1% frequency

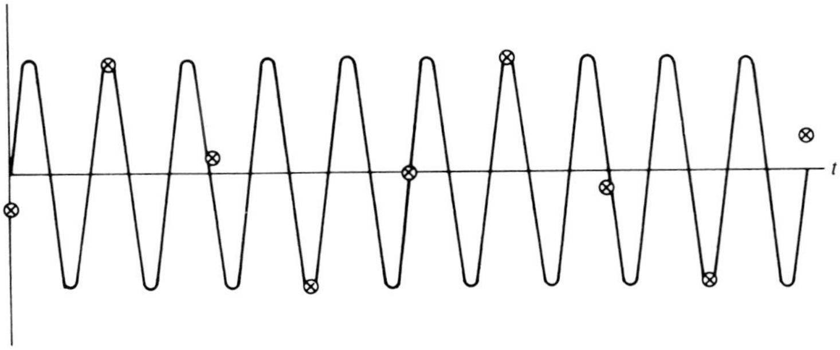

FIG. 5-3. Ten-cycle fit with 1% frequency error.

error in (5-4). Practically the same fitting quality would be obtained over a duration T corresponding to nine or eleven cycles. The example could also be extended, by adding a small nonlinearity or by superimposing another sinusoid of low amplitude (Ex. 5-31).

From the above discussion it follows that there can be some flexibility in the model used for estimation, and in the data window T. With suppression of information from the remote past via (5-58), a modified Kalman estimator acquires the following advantages:

(1) insensitivity to initial conditions and to minor imperfections in the model;
(2) tolerance for limited duration of model fidelity;
(3) an advancing data window, always terminating with the current observation;
(4) increased computational stabilization (recall Ex. 5-30); and
(5) applicability of estimation methods already presented, even when $\mathbf{e}$ and $\boldsymbol{\epsilon}$ deviate significantly from white noise generating processes.

In exchange for these crucial properties, the only item compromised is theoretical optimality. The estimate is somewhat suboptimal, rather than optimum in a minimum variance sense, because influence of old information is sacrificed. The means of maintaining practical estimator performance are described in this section.

5.4.1 Nonlinearity and the Extended Kalman Filter

Although many systems do not conform to (5-2), an estimator can capitalize on the existence of a general functional relation (5-1). In particular, variables representing unknown position, velocity, etc., can be regarded as components of a *true* state $\mathbf{X}$ (Ex. 5-32), with a unique dynamic history given by

$$\dot{\mathbf{X}} = \boldsymbol{f}(\mathbf{X}, t), \qquad \mathbf{X}(t_0) = \mathbf{X}_0 \tag{5-73}$$

The solution to this differential equation can be represented by a vector function $\boldsymbol{\Phi}$,

$$\mathbf{X} = \boldsymbol{\Phi}(\mathbf{X}_0) \tag{5-74}$$

which is always identifiable numerically, irrespective of whether a corresponding closed form expression may exist. Now let an estimate $\hat{\mathbf{X}}$ deviate from $\mathbf{X}$ by an *error state* $\mathbf{x}$,

$$\mathbf{x} = \mathbf{X} - \hat{\mathbf{X}} \tag{5-75}$$

with a dynamic representation obtained from *linearization* of (5-73),

$$\dot{\mathbf{x}} = \boldsymbol{f}(\mathbf{X}, t) - \hat{\boldsymbol{f}}(\hat{\mathbf{X}}, t) \doteq \mathbf{A}\mathbf{x} + \mathbf{e} \tag{5-76}$$

where (Ex. 5-33)

$$\mathbf{A} = \partial \boldsymbol{f}/\partial \mathbf{X} \tag{5-77}$$

and $\mathbf{e}$ now contains dynamic modeling errors. Linearization errors are generally minimized by evaluating the above partial derivative function as close to $\mathbf{X}$ as possible, i.e., at $\hat{\mathbf{X}}$. Also, the estimated value $\hat{\mathbf{x}}$ of the error state can be zero, since any known departure from $\mathbf{X}$ could immediately be used to refine $\hat{\mathbf{X}}$; therefore the standard condition assumed here is*

$$\tilde{\mathbf{x}} = \mathbf{x} \tag{5-78}$$

Linearization of measurement relationships is now defined in terms of a general vector function $\boldsymbol{H}$, which may be nonlinear,

$$\mathbf{Y}_m = \boldsymbol{H}_m(\mathbf{X}_m) \tag{5-79}$$

so that (Ex. 5-34)

$$\boldsymbol{H}_m(\mathbf{X}_m) - \boldsymbol{H}_m(\hat{\mathbf{X}}_m) \doteq \mathbf{H}_m \mathbf{x}_m \tag{5-80}$$

where

$$\mathbf{H}_m = \partial \boldsymbol{H}_m/\partial \mathbf{X}_m \tag{5-81}$$

and error in the approximation (5-80) can be included in ϵ_m.

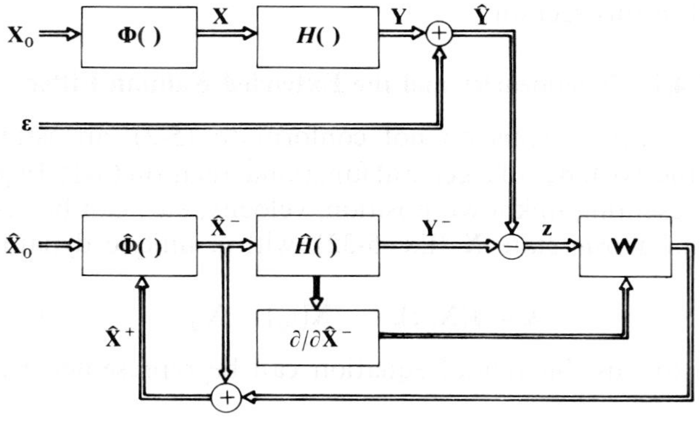

FIG. 5-4. Block diagram of estimation process.

With the foregoing conventions it is possible to describe an estimated dynamic history for the state in Fig. 5-4. An initial state $\mathbf{X}_0$ is fed to a dynamic transformation that produces the current state $\mathbf{X}_m$ via (5-74). An error-free data vector $\mathbf{Y}_m$ would then have a value given by (5-79). The observed data vector $\hat{\mathbf{Y}}_m$, corrupted by ϵ_m, is fed to an estimator assigned

* One departure from (5-78) involving INS update mechanization is discussed near the end of Section 6.3.

the task of constructing state estimates. The updated value $\hat{\mathbf{X}}_m^+$ after each measurement (or, at the outset, the initial estimate $\hat{\mathbf{X}}_0$) is then substituted into a modeled vector function $\hat{\mathbf{\Phi}}$ used to represent the dynamics,

$$\hat{\mathbf{X}}_m^- = \hat{\mathbf{\Phi}}_m(\hat{\mathbf{X}}_{m-1}^+) \tag{5-82}$$

and the result is used to predict the data vector from a modeled measurement/state relation $\hat{\boldsymbol{H}}_m$:

$$\mathbf{Y}_m^- = \hat{\boldsymbol{H}}_m(\hat{\mathbf{X}}_m^-) \tag{5-83}$$

A predicted residual, now obtained from

$$\mathbf{z}_m = \hat{\mathbf{Y}}_m - \mathbf{Y}_m^- \tag{5-84}$$

is used to update the state by a modified form of (5-36),

$$\hat{\mathbf{X}}_m^+ = \hat{\mathbf{X}}_m^- + \mathbf{W}_m \mathbf{z}_m \tag{5-85}$$

in which $\mathbf{W}_m$ is obtained from (5-44) *with previously noted imperfections.* For example, (5-81) is evaluated from $\hat{\boldsymbol{H}}_m$, which may differ from $\boldsymbol{H}_m$; $\mathbf{R}_m$ may not represent the true mean squared values for components of ϵ_m; $\mathbf{P}$ may be computed from (5-49) with an approximate dynamics matrix,

$$\hat{\mathbf{A}} = \partial \hat{\boldsymbol{f}} / \partial \hat{\mathbf{X}} \tag{5-86}$$

instead of the matrix in (5-77). The level of imperfection allowed in any application is generally dictated by heuristic analysis and experience, rather than mathematical rigor. Questions involving nonlinear transformations of random variables are thus avoided; (5-85) and the $\mathbf{P}$-matrix equations from Section 5.2.2 are linear transformations, while other operations represented in Fig. 5-4 may be nonlinear.

5.4.2 System Modeling

Since estimation can be achieved by fitting data to an imperfect model, equations chosen to represent a system may be simplified intentionally. If uncorrectable model discrepancies exceed the effects of conscious omissions, for example, little is sacrificed in attainable performance. At the same time, undue complexity is avoided by observing the procedures that follow.

MINIMUM STATE SELECTION

Whereas a true linear system model has unalterable dimensionality, approximate models for linearized systems can and *should* be represented by the fewest possible states. For a concrete illustration, (3-53) expresses an approximate error state for east position as a power series in time.

Consider a data window extending over a 5-min course leg (Ex. 5-35), so that

$$\begin{bmatrix} \tilde{\mathbf{x}}_\mathrm{E} \\ \tilde{V}_\mathrm{E} \end{bmatrix} = \begin{bmatrix} 1 & t \\ 0 & 1 \end{bmatrix} \begin{bmatrix} \tilde{\mathbf{x}}_{\mathrm{E}(0)} \\ \tilde{V}_{\mathrm{E}(0)} \end{bmatrix} + \begin{bmatrix} \frac{1}{2}(g\psi_{\mathrm{N}(0)} + n_{\mathrm{a}2})t \\ g\psi_{\mathrm{N}(0)} + n_{\mathrm{a}2} \end{bmatrix} t \tag{5-87}$$

in which the drift effect in Fig. 3-8 is ignored over a 5-min interval. In fact, over this duration, the effect of acceleration error on $\tilde{x}_\mathrm{E}$ is only marginally significant. Any attempt to estimate the coefficients of t^2 and t^3 in (3-53) would therefore be wasted effort. For practical purposes only the position and velocity error states would be observable; effects of INS misorientation and drift in this case could be absorbed in **E**.

Prevention of wasted computation is not the only motive for minimizing state dimension. In block estimation, particularly with an imperfect model, (5-24) can produce large errors in marginally observable states. The reason is explained by comparing terms in the following expansion for (3-51):

$$\begin{aligned} \tilde{x}_\mathrm{E} = \tilde{x}_{\mathrm{E}(0)} + \tilde{V}_{\mathrm{E}(0)}(t) + (g\psi_{\mathrm{N}(0)} + n_{\mathrm{a}2})(t^2/2!) + gd(t^3/3!) \\ + W^2 V_\mathrm{N} n_{\omega 3}(t^4/4!), \qquad t < 0.1 \times 2\pi/W \end{aligned} \tag{5-88}$$

After five minutes at $V_\mathrm{N} = 1500$ fps, a 0.1-fps initial velocity discrepancy produces more final position uncertainty than 5°/hr azimuth gyro drift. It is therefore unwise to estimate all coefficients in the above expansion by minimizing a scalar as in (5-22); enormous drifts in higher order terms could counterbalance each other or slight adjustments in velocity, to produce a trivial reduction in the scalar to be minimized. Unrealistic, if less extreme, estimates could also be obtained for INS misorientation under the same conditions.

Degraded estimation with excess dimensionality above is *not* traceable to numerical difficulty. Exact inversion of (5-24) or replacement by the algorithm in Appendix 5.A would not remove model errors nor make drifts observable within 5 min. For short intervals, the contrasts in Fig. 3-8 stay hidden from any estimation algorithm, block or sequential.

AUGMENTATION PRINCIPLES

In applications of estimation to navigation systems, gyro drift rates are often considered as candidate augmenting states. The preceding example suggests that only a poor quality INS, or an INS in warmup, would have a discernible drift effect within a short data window. A partial demonstration of the principle that closes Section 5.3.1 has thus been presented. Distinguishability of augmenting states will now be discussed on a spectral basis.

Dynamic behavior of an error source can be described in terms of a spectral density function, which specifies an average noise energy within all frequency bands of interest. At one extreme, full concentration at zero frequency produces a fixed bias; the other extreme is white noise, with equal distribution of noise energy over all frequencies. Between these extremes, the spectral density function for an error source could in principle be any arbitrary single-valued function of frequency. For practical estimation, however, only an approximate characterization of spectral distribution is necessary. Consider, for example, the error in a scalar observation repeated periodically within a data window T. The spectrum for this error could be separated into three regions, as follows:

(1) *Low frequency.* A band between zero and $1/(2\pi T)$ Hz includes a total bias, essentially fixed throughout T. This total should be disregarded if overshadowed by random error (Ex. 5-36).

(2) *Midfrequency.* The low frequency edge of this band, at $1/(2\pi T)$ Hz, contains systematic errors that are hardly distinguishable from fixed biases. The nominal upper edge of this band can be the reciprocal of the inter-observation interval. It is only between these edges that band limited noise in (5-61) could be worth considering. Moreover, multiple biases in any one source should be treated as a combined total, unless components can be distinguished from each other (Ex. 5-37).

(3) *High frequency.* Above the upper edge of the mid frequency band, spectral components contribute toward a resultant ϵ, essentially uncorrelated from one observation to the next.

Applying the principles just described, to biases in dynamics and observations, will guard against addition of unobservable states. The estimator will then fit the data to the chosen model, without wasting information or computation.

MODE MANAGEMENT

A mode of navigation can be identified by specifying all states and data sources employed. Expressed or implied in such a definition are

(1) coordinates used (e.g., Cartesian or spherical),

(2) number of derivatives for each position coordinate [e.g., order of the series truncated from (5-88)], and

(3) a reasonable data window length.

Mode parameters thus determine the required model complexity. For example, earth curvature can be ignored over short distances and durations. Another modal feature is the extent of interaction across directional channels. For reasons traceable to latitude/longitude/altitude conventions and gravity effects, the vertical channel already described is directly usable, decoupled from horizontal navigation in many modes. Separation of channels replaces large **H**, **P**, and **W** matrices by groups of smaller matrices, thus reducing computational requirements.

A typical flight includes several mode changes, with variations in aircraft operation, flight path, or switched sensor configuration. Within each mode, the factors just described form the basis of dynamic characterization given in Chapter 6.

EQUATIONS REPRESENTING TRANSFORMATIONS

Acceptability of minor model imperfections allows simplified representation of some dynamic and observable relations. In general, state *vector* transformations (5-82) and (5-83) are faithfully modeled to minimize discrepancies in (5-84). Less modeling fidelity is permissible in partial derivative relationships such as (5-81) and (5-86). For example, even when a predicted VOR bearing from Fig. 1-10 includes earth curvature effects, **H** may be computed from a flat earth model.

Linearized covariance matrix extrapolation cannot, and need not, correspond exactly to a nonlinear dynamic relation (5-82). If a transition matrix can be adequately represented in closed form, (5-48) should be applied with an *analytical* solution for the integral. If not, Ref. 5-9 explains that numerical solution of (5-49) is more efficient than integrating both (II-55) and (5-48).

Since airborne computers are designed for economy, chosen navigation models and algorithms must not be sensitive to roundoff. Units can be selected so that typical values for all states have similar orders of magnitude.* Also, for some applications, Ref. 5-9 explains that replacement of (5-50) by (5-41) is advisable. In either case, symmetry of **P** is often exploited by computing only half the off-diagonal elements in a transformation; the same is true for (5-48) and (5-49).

5.4.3 Control of Matrix Elements

Most departures from theory considered thus far arise from incomplete knowledge, marginal observability, or finite computing resources. Additional departures, some of which may be large, are often introduced to enhance convergence of an estimate. Applicable procedures include overriding of pertinent matrix equations, with independently chosen element values or coefficients. Elements of **W** are sometimes multiplied by scalars, for example, to alter the influence of updates in (5-85). This section describes means of controlling these weights indirectly, via **R** and **P**.

When **R** in (5-44) is formed from pessimistic sensor tolerances, an

* Since unit conversion is straightforward, no further illustration is warranted here. A consistent set of units (Appendix I) enables brevity of this presentation.

update is obtained that *could* have had minimum variance *if* observation errors had been as large as the values assigned. Both $\hat{\mathbf{X}}^+$ and $\mathbf{P}^+$ would be realistic, since components of ϵ would tend toward reasonable probabilistic levels (Ex. 5-38). Moderation of the decrement in $\mathbf{P}$ enhances positive definiteness on subsequent updates, thus stabilizing the computation. This is particularly important for gradualizing initial transients, such as the first update of Fig. 5-2, with measurement nonlinearities and/or other model imperfections. Note that weights in (5-44) can be reduced by conservatism in either $\mathbf{R}$ or the positive definite quadratic form $\mathbf{HPH}^{\mathrm{T}}$. By amplifying the quadratic form, Ref. 5-6 ensures that underweighting will be most effective when elements of $\mathbf{W}$ would have been abnormally large.

Gradualization of the initial transient can be combined with three techniques for controlling the elements of $\mathbf{P}$:

(1) An initial setting $\mathbf{P}_0$ with large diagonal elements. A diagonal $\mathbf{P}$-matrix is conservative (Refs. 5-10 and 5-11).

(2) Sudden incrementing of diagonal elements upon changing conditions (e.g., Ex. 4-15c), or recognition of abnormal residual patterns (e.g., large values, consistent algebraic signs).

(3) Reinitialization to $\mathbf{P}_0$ at extreme changes in conditions (e.g., mode switch or evidence of imminent divergence).

These techniques can be applied, subject to ceilings on allowable element values, with little sacrifice in obtainable performance (Ex. 5-39).

In contrast to underweighting for large elements of $\mathbf{W}$, residuals should continue to influence the estimate after many updates, when theoretical weights could be small (see pp. 307–311 of Ref. 5-3). Diagonal elements of $\mathbf{P}$, for example, could be inhibited from falling below acceptable minima. Positive definiteness would then be enhanced and, for any given $\mathbf{R}$, weights in (5-44) would be increased. This same objective, of course, motivates usage of $\mathbf{E}$ in (5-48) and (5-49). An approach for choosing elements of $\mathbf{E}$, on the basis of (5-56) and (5-57), has already been given. Alternatively, $\mathbf{E}$ can be determined (Ref. 5-12) from the dynamic environment and instrument characteristics described in the preceding chapter (Ex. 5-40).

The matrix adjustments described above oppose the divergence sequence previously described (i.e., unduly large decrementing of initial uncertainty covariances and inadequate weighting of subsequent observations). It follows that *less* variation in weighting, while theoretically suboptimal, has a stabilizing effect in practical estimation. To carry this trend further, elements of $\mathbf{P}$ can be held *constant* until severe transients due to $\tilde{\mathbf{x}}_0$ are removed; suboptimal response can then be chosen by conventional means (Ex. 5-41).

5.4.4 Observation Data Handling

Sections 4.3.1 and 4.3.2 mentioned lag equilization for inertial sensor data. The time of an observation must also be identified in relation to the dynamic state estimate. For continuous–discrete estimation, a prefiltering operation may reference an update at the center of a measurement averaging interval. When the measured quantity varies appreciably and nonlinearly within this interval, it is appropriate to average residuals instead of measurements (Ref. 5-13, p. 242). In either case, an opportunity is thus presented for convenient spacing of discrete updates. Multiple observations can be staggered in time to minimize the number of residuals simultaneously processed; inversion of large matrices in (5-44) is then avoided. Undue processing of observations with sequentially correlated errors can also be avoided through separation of updates by intervals exceeding the reciprocal bandwidth of each sensor.

Editing should be employed, to remove unexplained spurious observational data. A scalar measurement can be omitted, for example, when the square of the residual from (5-84) far exceeds the covariance from (5-46). Observations may be rejected also for geometries that introduce ambiguity or nonlinearity effects exceeding the random errors. Such a selective omission of data improves the extended Kalman filter (e.g., see the simulated space applications of Refs. 5-14 and 5-15).

Nonlinear degradations can be minimized by smoothing with recycled data. The state at any time t_m will benefit, as already explained, from all observations before and after t_m in a data block. With partial derivatives computed from best available estimates, recycling of data therefore improves the fit.

5.4.5 Further Concepts in Estimation

While not essential for many navigation applications, certain additional topics in estimation theory are worth noting here.

Some items already discussed can call for straightforward modification of the matrix equations presented. Prefiltering over long intervals, for example, can produce appreciable correlation between $\mathbf{e}$ and $\boldsymbol{\epsilon}$ (Ref. 5-16). The same type of correlation is involved in the Bryson–Johansen/Bryson–Henrikson dimensionality reduction mentioned in Section 5.3.2.

In Refs. 5-7 and 5-8, the augmented state dimension was reduced by the number of observations free of white noise. A related concept is reduction of dimension to $N - M$, for a linear observable system with N states and M sources of observational data, with or without noise. This is the basis of *minimal order observer* theory; more generally, an observer is a linear estimator driven by measured data (Refs. 5-17–5-19). As a special case, the Kalman filter is an N-dimensional observer, optimum in a minimum variance sense. The references just noted, plus the sources cited therein, introduce a body of literature on reduced-state estimation.

Additional means of improving estimation performance for a given amount of computation include the following:

(1) Replacement of (5-24) by the algorithm in Appendix 5.A for smoothing, with recycling of data and recomputation of (5-81)–(5-83) and (5-86) on each iteration, to reduce nonlinearity effects. Recycling can continue until satisfactory results are obtained and/or no further reduction in residuals occurs.

(2) A square root filtering algorithm (e.g., Refs. 5-20 and 5-21), which increases computational precision, especially for systems of large dimension.

(3) Methods of accounting for biases without the full computational burden of augmentation. This includes the *consider* option (Ref. 5-22) and a decoupled adjustment of the bias-free estimate, based on the residuals (Ref. 5-23).

There is also much informative literature on nonlinear estimation and continuous observations (Ref. 5-3). Most aircraft navigation problems, however, can be solved without those techniques.

5.4.6 Verification of Practical Estimation Approach

A navigation system is ultimately judged by its performance in flight. Before a scheduled test date arrives, however, much preparatory validation should be made. A key to validation is a *two-tier* simulation depicted by Fig. 5-4, in which corrupted observations are the only direct connection between the upper (inaccessible) tier and the airborne system. Although the "real world" defies exact representation, realistic differences between upper and lower tier models can be simulated. The items enumerated at the start of Section 5.4, and all corrective measures subsequently prescribed, can then be evaluated. As pointed out on p. 22 of Ref. 5-24, this simulation approach goes far beyond linearized covariance analysis in the investigation of errors that are potentially the most damaging. Insight gained from this approach can exert a decisive influence on successful navigation system design.

5.5 CARTESIAN POSITION AND VELOCITY APPLICATION

For further insight on usage of estimation in navigation, the single-channel analysis presented earlier will now be extended to three-dimensional space. Let an error state vector

$$\begin{bmatrix} \tilde{\boldsymbol{R}} \\ \hdashline \dot{\tilde{\boldsymbol{R}}} \end{bmatrix}$$

consist of position uncertainty components in three orthogonal directions, plus velocity uncertainties in the same directional sequence.* The dynamics matrix in (5-7) then generalizes to

$$\mathbf{A}_{6\times 6} = \begin{bmatrix} \mathbf{O}_{3\times 3} & \mathbf{I}_{3\times 3} \\ \mathbf{O}_{3\times 3} & \mathbf{O}_{3\times 3} \end{bmatrix} \tag{5-89}$$

which for any time interval Δt produces the transition matrix

$$\mathbf{\Phi}_{6\times 6} = \begin{bmatrix} \mathbf{I}_{3\times 3} & (\Delta t)\mathbf{I}_{3\times 3} \\ \mathbf{O}_{3\times 3} & \mathbf{I}_{3\times 3} \end{bmatrix} \tag{5-90}$$

Off-diagonal elements in these matrices couple each position state to the corresponding velocity state only; there is no cross-axis coupling in these equations. Much of the single-channel analysis already presented can therefore be applied for each direction here, but channel interaction can be introduced through measurement geometry (Ex. 5-42). The plurality of possible measurement accuracies and schedules, initial conditions, etc., allows a vast array of error time histories, but general insights are obtainable from preceding analyses. Repeated substitution of (5-89) into (5-49) or (5-90) into (5-48), with occasional decrementing via (5-50), produces position error ellipsoids depicted in Fig. 5-5. Ellipsoid size at any time is

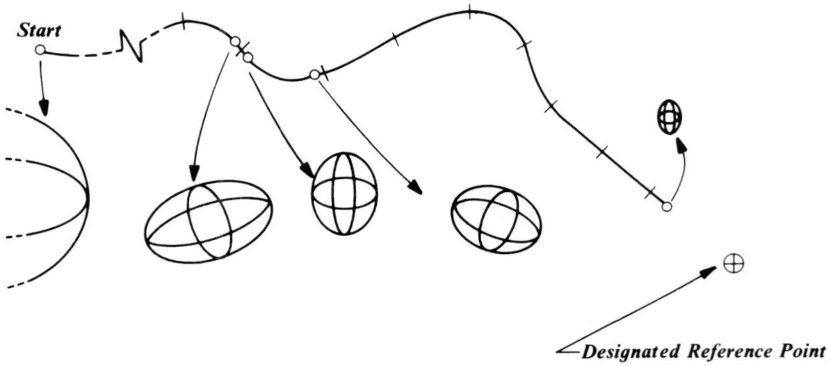

FIG. 5-5. Three-dimensional uncertainty ellipsoid dynamics.

related to mean squared error in estimated position relative to the origin shown. Initially all errors may be large, but corrective observations are made at the indicated points. The volume of uncertainty shrinks at the measurement points and expands between these measurements.

This three-dimensional example illustrates the usage of intermittent, indirect, incomplete, inexact observations described at the beginning of this

* Implementation of this navigation mode is described in Section 6.5.

chapter. Position uncertainty in each direction grows between discrete reductions as in Fig. 5-2, with decrements governed by observation type and precision. Distance measurements in a plane, for example, can correct uncertainties along directions in that plane. Within each data window, there must be enough observational information to maintain acceptable accuracies of position and velocity estimates in all directions.

The size, shape, and orientation of ellipsoids in Fig. 5-5 vary along the flight path. The position uncertainty covariance matrix (denoted $\mathbf{p}$, representing an upper left 3×3 partition of $\mathbf{P}$ in this example) at any time can be transformed to another directional reference. In particular, the eigenvectors of $\mathbf{p}$ are of special interest here. By the methods of Appendix II, Ex. 5-43 identifies mean squared errors along these principal directions as eigenvalues, represented by ellipsoid semimajor axes in Fig. 5-5. Similar geometric interpretations benefit other navigation modes as well.

Principal axis transformations are useful also in assessing the need for various observations. For example, when the eigenvector $\boldsymbol{T}$ corresponding to the largest eigenvalue $\Lambda_{\mathbf{max}}$ of $\mathbf{p}$ coincides with the vertical, an altitude measurement is most desirable. More generally, as pointed out in Ref. 5-2, eigenvectors of $\mathbf{p}$ indicate which direction affords the greatest possible reduction in uncertainty. Consider a scalar position measurement with the proportionality matrix $\mathscr{H}$. When $\mathscr{H}^{\mathrm{T}}$ is parallel to $\boldsymbol{T}$, application of (II-20) to (5-51) can produce an uncertainty decrement (trace $\mathbf{\Delta p}$) close to $\Lambda_{\mathbf{max}}$. Substitution of (II-20) into the denominator also shows that, with measurement imperfections, the reduction in total mean squared uncertainty will always be less than $\Lambda_{\mathbf{max}}$.

INTERPRETATION FROM CLOSED-FORM EXPRESSION

For a constant diagonal process noise matrix, with velocity and acceleration spectral densities η_{v} and η_{a}, respectively,

$$\mathbf{E}_{6 \times 6} = \begin{bmatrix} \eta_{\mathrm{v}} \mathbf{I}_{3 \times 3} & \mathbf{O}_{3 \times 3} \\ \mathbf{O}_{3 \times 3} & \eta_{\mathrm{a}} \mathbf{I}_{3 \times 3} \end{bmatrix} \tag{5-91}$$

Combination with (5-48) and (5-90) provides a solution of the form

$$\mathbf{P}_m^{-} = \mathbf{\Phi} \mathbf{P}_{m-1}^{+} \mathbf{\Phi}^{\mathrm{T}} + \left[\begin{array}{c:c} [\eta_{\mathrm{v}}(\Delta t) + \frac{1}{3}\eta_{\mathrm{a}}(\Delta t)^3]\mathbf{I}_{3 \times 3} & \frac{1}{2}\eta_{\mathrm{a}}(\Delta t)^2 \mathbf{I}_{3 \times 3} \\ \hdashline \frac{1}{2}\eta_{\mathrm{a}}(\Delta t)^2 \mathbf{I}_{3 \times 3} & \eta_{\mathrm{a}}(\Delta t)\mathbf{I}_{3 \times 3} \end{array}\right] \tag{5-92}$$

The uncertainty contributed by process noise has a straightforward interpretation here. For example, a velocity error spectral density of η_{v} $(\text{fps})^2/\text{Hz}$, flat to a cutoff frequency of $\mathscr{B}$ Hz, produces a mean squared velocity error of $\eta_{\mathrm{v}} \mathscr{B}$ $(\text{fps})^2$, with an autocorrelation time of $(1/\mathscr{B})$ sec (Ex. 5-44). The accumulated effect over an inverval $(1/\mathscr{B})$ is then a mean squared position error of $(\eta_{\mathrm{v}}/\mathscr{B})$ ft^2. Repetition throughout a duration

Δt constitutes a random walk with $(\Delta t)/(1/\mathscr{B})$ independent steps of $(\eta_v/\mathscr{B})$ ft^2 mean squared error per step. This results in a mean squared integrated position error of $\eta_v \, \Delta t$ ft^2; note that the process noise need not be white, so long as its autocorrelation time is short compared with the interval between measurements ($\mathscr{B}\,\Delta t \gg 1$).

APPENDIX 5.A: GENERALIZED BLOCK ESTIMATION

Section 5.2.1 mentioned a case of marginal observability for which numerical inversion in (5-24) presents inescapable computational difficulties. An attempt to use (5-24) in that case can often produce large estimation errors, even in those states that could be adequately determined from the data. A computationally efficient procedure, particularly for large data blocks, is replacement of (5-24) by

$$\hat{\hat{\mathbf{x}}}_0 = \mathbf{S}^{\#}\mathbf{U}\mathbf{z} \tag{5-93}$$

where the operator ( )$^{\#}$ denotes the *Penrose generalized inverse*, defined in Ref. 5-25 as

$$\mathbf{S}^{\#} = \lim_{\delta \to 0} \{(\mathbf{S}^{\mathrm{T}}\mathbf{S} + \delta\mathbf{I})^{-1}\mathbf{S}^{\mathrm{T}}\} \tag{5-94}$$

The limit always exists, is unique, and can be computed from the four Penrose conditions, which, for real matrices, are

$$\mathbf{S}\mathbf{S}^{\#}\mathbf{S} = \mathbf{S}, \qquad \mathbf{S}^{\#}\mathbf{S}\mathbf{S}^{\#} = \mathbf{S}^{\#}, \qquad \mathbf{S}\mathbf{S}^{\#} = (\mathbf{S}\mathbf{S}^{\#})^{\mathrm{T}}, \qquad \mathbf{S}^{\#}\mathbf{S} = (\mathbf{S}^{\#}\mathbf{S})^{\mathrm{T}} \tag{5-95}$$

A powerful algorithm for computing $\mathbf{S}^{\#}$ appears in Ref. 5-26; the formulation is valid even when $|\mathbf{J}^{\mathrm{T}}\mathbf{J}|$ vanishes, so long as $\mathbf{U}$ is nonsingular.

The case of a vanishing determinant $|\mathbf{J}^{\mathrm{T}}\mathbf{J}|$ is exemplified by a column of zeros in $\mathbf{S}$ (due to a lack of measured data related to one of the states; Ex. 5-45). There will then be a row of zeros in $\mathbf{S}^{\#}$, and (5-24) will be inapplicable. A weighted least squares estimate for the observable states is obtainable in any case from (5-93), which would also null the estimate of the unobservable state in this example (Ex. 5-46). More generally, in the presence of unobservable states, (5-93) simultaneously minimizes L of (5-21) and $|\hat{\hat{\mathbf{x}}}_0|$ itself.

APPENDIX 5.B: ESTIMATION WITH KNOWN MEASUREMENT ERROR COVARIANCE: FURTHER CONSIDERATIONS

The purpose of this appendix is to illustrate how knowledge of observation error correlations inherently implies a bias estimate, without state augmentation. An example with a closed-form solution is used for clarity. Suppose that a static two-state system is observed directly by two independent instruments A and B, such that A measures x_1 only, while B measures each state once. It follows that

$$y_1 = y_2 = x_1, \qquad y_3 = x_2 \tag{5-96}$$

or, for (5-18),

$$\mathbf{J} = \begin{bmatrix} 1 & 0 \\ 1 & 0 \\ 0 & 1 \end{bmatrix} \tag{5-97}$$

and, denoting rms errors in A and B and the correlation between the two B measurement errors as a, b, and ρ, respectively (Ex. 5-47),

$$\langle \boldsymbol{\epsilon}\boldsymbol{\epsilon}^{\mathrm{T}} \rangle = \begin{bmatrix} a^2 & 0 & 0 \\ 0 & b^2 & \rho b^2 \\ 0 & \rho b^2 & b^2 \end{bmatrix} \tag{5-98}$$

For $|\rho| < 1$ this matrix is nonsingular and (5-20) provides a weighting matrix,

$$\mathbf{U} = \begin{bmatrix} \dfrac{1}{a} & 0 & 0 \\ 0 & \dfrac{(1+u)^{1/2}}{\sqrt{2}\,bu} & \dfrac{-\rho}{\sqrt{2}\,bu(1+u)^{1/2}} \\ 0 & \dfrac{-\rho}{\sqrt{2}\,bu(1+u)^{1/2}} & \dfrac{(1+u)^{1/2}}{\sqrt{2}\,bu} \end{bmatrix}, \qquad u \triangleq (1-\rho^2)^{1/2} \tag{5-99}$$

from which it follows that

$$\mathbf{S} = \mathbf{U}\begin{bmatrix} 1 & 0 \\ 1 & 0 \\ 0 & 1 \end{bmatrix} = \begin{bmatrix} \dfrac{1}{a} & 0 \\ \dfrac{(1+u)^{1/2}}{\sqrt{2}\,bu} & \dfrac{-\rho}{\sqrt{2}\,bu(1+u)^{1/2}} \\ \dfrac{-\rho}{\sqrt{2}\,bu(1+u)^{1/2}} & \dfrac{(1+u)^{1/2}}{\sqrt{2}\,bu} \end{bmatrix} \tag{5-100}$$

and therefore

$$(\mathbf{S}^{\mathrm{T}}\mathbf{S})^{-1}\mathbf{S}^{\mathrm{T}} = \frac{1}{a^2+b^2}\begin{bmatrix} ab^2 & \left(\dfrac{1+u}{2}\right)^{1/2} a^2 b & \dfrac{\rho a^2 b}{[2(1+u)]^{1/2}} \\ \rho a b^2 & \dfrac{\rho b(a^2-b^2u)}{[2(1+u)]^{1/2}} & b(a^2+b^2u)\left(\dfrac{1+u}{2}\right)^{1/2} \end{bmatrix} \tag{5-101}$$

Combining this with (5-24) and (5-99),

$$\hat{\hat{\mathbf{x}}} = \frac{1}{a^2+b^2} \times \begin{bmatrix} b^2 z_1 + \dfrac{a^2}{2u}\{(1+u)z_2 - \rho z_3\} + \dfrac{a^2}{2u}\left\{\dfrac{-\rho^2}{1+u} z_2 + \rho z_3\right\} \\ \rho b^2 z_1 + \dfrac{\rho(a^2-b^2u)}{2u}\left\{z_2 - \dfrac{\rho}{1+u} z_3\right\} + \dfrac{a^2+b^2u}{2u}\{-\rho z_2 + (1+u) z_3\} \end{bmatrix} \tag{5-102}$$

which reduces to

$$\hat{\hat{x}}_1 = \frac{1}{a^2+b^2}\left\{b^2 z_1 + \frac{a^2}{2u}[1+u+(u-1)]z_2\right\} = \boxed{\frac{b^2 z_1 + a^2 z_2}{a^2+b^2}} \tag{5-103}$$

and

$$\begin{aligned} \hat{\hat{x}}_2 &= \frac{1}{a^2+b^2}\left\{\rho b^2 z_1 + \rho\left[\frac{a^2-b^2u}{2u} - \frac{a^2+b^2u}{2u}\right] z_2 \right. \\ &\quad \left. + \left[\frac{(u+1)(u-1)(a^2-b^2u)}{2u(1+u)} + \frac{(a^2+b^2u)}{2u}(1+u)\right] z_3\right\} \\ &= \frac{1}{a^2+b^2}\left\{\rho b^2(z_1-z_2) + \frac{1}{u}(ua^2+ub^2)z_3\right\} \\ &= \boxed{z_3 - \frac{\rho b^2}{a^2+b^2}(z_2-z_1)} \end{aligned} \tag{5-104}$$

Although intermediate quantities in this derivation become indeterminate for $\rho = \pm 1$, no such restriction is placed on the final results on the right of these last two expressions. Note that (5-103) conforms to (5-53) (Ex. 5-48), and the latter term in (5-104) represents estimated bias in instrument B.

The results just derived are not at all surprising; minimum variance estimation provides appropriate adjustments based on prescribed correlation properties. The adjustments are correct to the extent that prescribed correlations are realistic. Biases are then implicitly taken into account, but do not appear explicitly unless the estimation algorithm is supplemented by further computation.

The means of accounting for biases must not introduce enough numerical degradation to cause ill-conditioning, which would hamper matrix inversions required for estimation. For "judicious" use of augmentation mentioned at the end of Section 5.3, it is not enough for an inverse to exist in an abstract sense; consistent positive definiteness is required, in the presence of computational degradations. Methods of accounting for correlations may involve regression analysis, estimation with constraints, augmentation with retention of additional white noise (Ref. 5-6), reformulation with reduced dimensionality (Refs. 5-7 and 5-8), "consider" estimation with its various subdivisions and options (Ref. 5-22), separable supplemental computations (Ref. 5-23), combinations of data having infinitesimal, finite, and infinite error variances, or applications with unknown bias statistics. T. S. Englar, in some unpublished work, describes relationships between many of these techniques, while identifying permissible and prohibitive roles of pseudoinverses.

EXERCISES

5-1 (a) Show that (5-4) conforms to (5-2) with

$$x_1 = x, \qquad x_2 = \dot{x}_1, \qquad \mathbf{x} = \begin{bmatrix} x_1 \\ x_2 \end{bmatrix}, \qquad \mathbf{A} = \begin{bmatrix} 0 & 1 \\ -\omega^2 & 0 \end{bmatrix}, \qquad \mathbf{e} = \mathbf{0}$$

Find the eigenvalues of $\mathbf{A}$ and analyze the dynamics as in Section II.3. How does this compare with a similar development of (4-1), for the special case in which $\boldsymbol{\varpi}$ and the forcing function are nulled?

(b) Write a 1×2 matrix of proportionality constants for direct observations of $\mathbf{x}$, and apply (5-3) to Fig. 5-1b. Using

$$y_1 = C \sin(\omega t_1 + \theta), \qquad y_2 = C \sin(\omega t_2 + \theta),$$

relate the two points shown to an amplitude C and an initial phase angle θ. Describe a complete dynamic solution in terms of these quantities, when they are known exactly. Compare versus a solution based on initial conditions x, $\dot{x}$.

5-2 (a) "Given a linear dynamic system (5-2) with all parameters and forcing functions fully specified, the state is the minimum amount of information needed to complete the solution." How is this statement related to the two conditions itemized in Section 5.1? Could any system, properly characterized by an N-dimensional state, be re-expressed with more or fewer states? Why is (II-46) a valid re-expression of (II-41)? Are state variables functionally independent, or can any of them be determined from the others?

(b) Express the initial values of the state variables as "constants of integration." Explain that a new set of constants is required when the reference time is shifted.

5-3 Show that (5-5) conforms to (5-2), and compare with Ex. 5-1a. Express Z_{V} in terms of A_{V} and g from Chapter 3.

5-4 (a) Show by differentiation that (5-8) agrees with (5-7). Explain why, with inexact derivatives, continued integration of (5-6) would eventually produce a useless solution.

(b) Recall the footnote regarding stability in Section II.3. Apply this to (5-7) with $\tilde{Z}_{\mathrm{V}}$ set to zero.

(c) Recall the discussion after (5-2). With altitude divergence (Ex. 3-21) taken into account, is (5-7) appropriately expressed in the form of (5-2) with all system parameters and forcing functions independent of the state?

5-5 (a) Explain why the formalism in Section 5.1 allows partial correlations between different components of simultaneous measurement errors and between different components of the instantaneous white noise **e**.

(b) Isolate the expression for $\tilde{V}_{\mathrm{V}}$ from (5-8). Explain why it is possible to do so. Assume an initial value of zero and square the result, writing the squared integral as a double integral:

$$\tilde{V}_{\mathrm{V}}{}^2 = \int_{t_m}^{t} \int_{t_m}^{t} \tilde{Z}_{\mathrm{V}}(\mathfrak{r})\tilde{Z}_{\mathrm{V}}(\ell)\, d\mathfrak{r}\, d\ell$$

Form averages, noting that the linear operations of integration and averaging commute,

$$\langle \tilde{V}_{\mathrm{V}}{}^2 \rangle = \int_{t_m}^{t} \int_{t_m}^{t} \langle \tilde{Z}_{\mathrm{V}}(\mathfrak{r})\tilde{Z}_{\mathrm{V}}(\ell) \rangle\, d\mathfrak{r}\, d\ell$$

Apply (5-10) in scalar form, assuming a random driving acceleration error with fixed spectral density η_{a} (note units of η_{a} and $\boldsymbol{\delta}$ from Appendix I):

$$\langle \tilde{V}_{\mathrm{V}}{}^2 \rangle = \int_{t_m}^{t} \int_{t_m}^{t} \eta_{\mathrm{a}}\, \boldsymbol{\delta}(\mathfrak{r} - \ell)\, d\mathfrak{r}\, d\ell$$

and integrate to obtain the result

$$\langle \tilde{V}_{\mathrm{V}}^{2} \rangle = \eta_{\mathrm{a}}(t - t_m)$$

Draw an analogy between this exercise and the development of (4-38). If a time-varying spectral density had been assumed here, the result could have been left in what form?

(c) For an *erogodic* white noise $e(t)$, a probabilistic average of the form

$$\langle e(t)e(t + \mathrm{r}) \rangle = \eta \, \delta(\mathrm{r})$$

is equivalent to an average over all time,

$$\lim_{T \to \infty} \left\{ [1/(2T)] \int_{-T}^{T} e(t)e(t + \mathrm{r}) \, dt \right\}$$

which is the autocorrelation function for e. Since this function must equal the Dirac delta weighted by a constant η, its Fourier transform must be

$$\eta \int_{-\infty}^{\infty} \exp(-j\omega \mathrm{r}) \, \delta(\mathrm{r}) \, d\mathrm{r} = \eta$$

The Fourier transform of the autocorrelation function is the power spectrum for e. Sketch this power spectral density as a function of frequency. Describe the power spectrum of white noise.

(d) Relate the inequality included in (II-30) to (5-11). Does the formalism in Section 5.1 allow full correlation between different components of simultaneous measurement error?

(e) Use (II-30) to form a 2×2 covariance matrix, and find the determinant corresponding to 99% correlation. Recall the discussion regarding eigenvalues and determinants following (II-30) and (II-36). Can a covariance matrix with full correlation be positive definite? Relate your answer to the principle that if x_i and x_j were completely correlated, it would be redundant to specify both.

(f) Can any of the diagonal elements of $\mathbf{R}$ be zero? Of $\mathbf{E}$? Use (5-11) and (5-12) to give reasons.

5-6 (a) Show consistency between conventions used in (3-34), (3-37), (4-39), (4-47), and (5-16). Cite another example from Appendix 3.B.

(b) Illustrate how (5-18) can be assembled from nonsimultaneous data vectors.

5-7 Expand (5-21) for the case of a diagonal weighting matrix, a three-dimensional data vector, and a two-dimensional state. Show that the expression is positive definite, i.e., $L \geq 0$, with equality holding only at the null state.

5-8 (a) Let $\ell_1 = \mathbf{x}^T\mathbf{z}$ and $\ell_2 = \mathbf{x}^T\mathbf{x}$. Write column vectors representing partial derivatives w.r.t. **x**, as **z** and 2**x**, respectively.

(b) Expand (5-21). Apply part (a) to link the left of (5-22) with (5-23).

5-9 (a) Apply (II-35) to (5-9), and use the equivalence between transpose and inverse for orthogonal matrices (II-23), to obtain (5-26) in the case of uniform variance.

(b) Show that (5-27) solves the estimation problem when $\mathbf{J}$ is invertible. What would happen if weighting were attempted in that case?

5-10 (a) Can velocity be inferred from position measurements separated by known time intervals? Relate this question to reversal of the row matrix elements on the left of (5-29). What would then appear on the right of that equation? Interpret the result in terms of a maximum rank matrix $\mathbf{J}$, according to the definition preceding (II-8).

(b) In the system represented by (5-28), would a single position observation allow initialization of position? Of both position and velocity? Is Ex. 5-2 pertinent to these questions?

(c) Would one position measurement, plus a nonsimultaneous velocity measurement, enable a solution for (5-28)? How would this compare versus the approaches described in Ex. 5-1b?

5-11 (a) Substitute (5-18) into (5-24) to obtain (5-30). Postmultiply (5-30) by its transpose to obtain (5-31).

(b) If one diagonal element in (5-20) were over a billion times greater than all others, could imprecise computation lead to unduly inaccurate estimation from (5-30)? Could observations with extremely large variances be discarded?

(c) Consider a scalar static state, such that the transition matrix in (5-17) is replaced by unity. Consider also *direct* measurement of the state, by M repeated independent bias-free observations with uniform variance σ^2. Show from (5-24) that the minimum variance estimate would then be the average of all observations. Use (5-31) to show that rms uncertainty in this average is $\sigma/\mathsf{M}^{1/2}$.

(d) Let $L = \boldsymbol{v}^T\boldsymbol{v}$, where

$$\boldsymbol{v} = \mathbf{U}(\mathbf{z} - \mathbf{J}\hat{\mathbf{x}}_0) = \mathbf{S}\tilde{\mathbf{x}}_0 + \mathbf{U}\boldsymbol{\epsilon}$$

from which it follows that

$$L = \tilde{\mathbf{x}}_0^T(\mathbf{S}^T\mathbf{S})\tilde{\mathbf{x}}_0 + 2\tilde{\mathbf{x}}_0^T(\mathbf{S}^T\mathbf{U})\boldsymbol{\epsilon} + \boldsymbol{\epsilon}^T(\mathbf{U}^T\mathbf{U})\boldsymbol{\epsilon}$$

Use (5-30) to show that

$$L = \boldsymbol{\epsilon}^T\mathbf{U}^T[\mathbf{I} - \mathbf{S}(\mathbf{S}^T\mathbf{S})^{-1}\mathbf{S}^T]\mathbf{U}\boldsymbol{\epsilon}$$

Rewrite this and additional expressions in Section 5.2.1, accounting for symmetry in $\mathbf{U}$. Using the form of (II-33), show that the same orthogonal matrix diagonalizes both $\mathbf{U}$ and $\mathbf{U}^2$.

5-12 Suppose that (5-32) or (5-33) were nearly, but not quite, true. Would the corresponding state formulation be theoretically violated? What type of difficulty might be expected?

5-13 (a) Would (5-34) follow logically from (5-13) if $\mathbf{e}$ had a nonzero mean?

(b) Does (5-36) represent a specialized updating approach?

5-14 (a) A *statistic* is a quantity accessible in operation, and not merely an abstract mathematical entity. Which side of (5-37) makes it immediately clear that a residual is a statistic?

(b) If the superscripts $(\ )^-$ in (5-35) and (5-37) were replaced by $(\ )^+$ *current* residuals would be obtained. If minimum variance filtering will minimize the weighted sum of squares of current residuals, how does filtering compare with smoothing as described in the text accompanying (5-24)?

5-15 (a) A nonnegative definite symmetric matrix has no eigenvalues less than zero. How does this property compare with the description of a positive definite symmetric matrix, given prior to (II-31)? Write an expression comparable to (II-31) for a nonnegative definite symmetric matrix.

(b) Show that a quadratic form involving a symmetric matrix is a weighted sum of eigenvalues. Is this a real quantity? [Note the discussion following (II-30)].

(c) Recall the properties of a trace, given after (II-36). When $\boldsymbol{\delta}\mathbf{P}$, defined by condition (2) following (5-41), is nonnegative definite, could any modified weighting matrix $\mathbf{W}_m + \boldsymbol{\delta}\mathbf{W}_m$ provide a trace for $\mathbf{P}_m$ that is *less* than the trace of $\mathbf{P}_m{}^+$?

(d) With independence between diagonal elements, could the trace of a matrix be minimized if each diagonal element were not simultaneously minimized?

5-16 (a) What happens when (5-44) is substituted into (5-43)?

(b) Let a scalar P be a quadratic function of an independent variable W with derivative $d\mathsf{P}/d\mathsf{W}$ vanishing at some point. What is the significance of the *second*-order variation in P at that same point? How are these questions related to (5-43) and (5-45)?

(c) The qualification preceding (5-41) stipulates no correlation between current observation error and *a priori* state uncertainty. Without this stipulation, would (5-46) follow from (5-37)?

(d) Does (5-11) ensure positive definiteness of (5-46)? How does this affect (5-45)? Show symmetry by forming the transpose of (5-46).

5-17 Replace t_m by t in (5-48). Obtain (5-49) by differentiation.

5-18 (a) Apply (5-48) to (5-8) with no acceleration error. Show that rms position uncertainty would increase approximately as a linear function of time, if velocity uncertainty effects were dominant. Show that the curvature of this function would depend on the relative influence of initial velocity uncertainty variance and the off-diagonal element of the initial covariance matrix.

(b) Illustrate how an acceleration noise with fixed spectral density would introduce another term into the above function. Express the resulting rms position uncertainty as the square root of a cubic.

(c) Suppose that the present example were replaced by a system with the transition matrix of (5-64). How would acceleration bias affect concavity of the functional dependence for rms position uncertainty versus time?

5-19 (a) In the example involving (5-52)–(5-55), how many independent observations are needed for a threefold reduction in effective measurement error? For a hundredfold reduction? Could small sequential correlations or other practical considerations preclude the latter?

(b) Instead of this static example, consider the system of (5-8), with direct measurements of position only. How would the dynamics affect an attempt to estimate position with extreme accuracy?

5-20 (a) Some of the navigation modes to be described in the next chapter are generalizations of (4-1), having forcing functions that vary with the quantities in (4-31)–(4-34). Will components of $\mathbf{e}$ always be gaussian?

(b) Some observations involve resolution errors, with uniform probability distributions. Will components of ϵ always be gaussian?

5-21 (a) Use (II-30) to identify mean squared estimation errors under conditions of zero mean.

(b) Show that, for a diagonal $\mathbf{E}$ with large elements, the estimate from (5-36) will rely heavily on the most recent data. Discuss the opposite situation, with large rms observation error.

5-22 (a) Is the matrix in (4-1) constant? Suppose that a navigation mode, based on (5-2), used an extension of (4-1). Could the dynamics matrix be approximated, to provide an approximate evaluation of (5-56)?

(b) Suppose that the navigation mode in part (a) included position states, and that updates were obtained from VOR stations as in Fig. 1-10.

As the aircraft proceeds along its flight path, could changing geometry be expected to introduce variations into $\mathscr{H}$?

5-23 (a) With N states, what is the smallest number of scalar observations required for observability? Should any requirements be added, regarding measurement types? Relate to Ex. 5-10.

(b) Consider (5-56) and (5-57) applied to Fig. 1-14, with dynamics governed by (5-5)–(5-8). With only one altitude update per data window, would the system be unobservable? Nearly unobservable?

(c) If part (b) were modified to allow several updates per data window, would a nearly uniform interobservation interval be preferable to a high concentration of updates?

(d) Can any principles of Ex. 5-19 be applied to parts (b) and (c)?

5-24 (a) Extend the principles of Ex. 5-10 to show how repeated altitude updates would suffice with (5-60).

(b) In a vertical channel with substantial bias effects, should the connotation of an "augmented" formulation (5-60) imply a dimensional abnormality? Or would an "unaugmented" formulation (two states with white driving noise) be deficient in the presence of substantial bias?

(c) Apply portions of Section 3.4.2 and Appendix 3.A to show that biases in Section 5.3.1 could include short-term INS misorientation effects and gravity anomalies, as well as a vertical accelerometer discrepancy.

5-25 (a) Show that the matrices in (5-62) and (5-63) conform to (II-55).

(b) Obtain (5-64) from (5-63), and show that each conforms to (II-56).

5-26 (a) Show that the matrices in (5-66) and (5-67) conform to (II-55).

(b) Construct a matrix for (5-18) as described after (5-67), for three sequential observations. Could (5-24) or (5-26) provide a solution? Would more observations of the same type alter this conclusion?

5-27 Suppose that, instead of a bias, the bottom row in (5-67) corresponded to an extraneous parameter inadvertently included in a vertical channel formulation. With [1, 0, 0] used to compute each row of the matrix in (5-18), show that the system as formulated would be unobservable. Generalize this conclusion.

5-28 Reconsider Ex. 5-26b with one of the barometric observations replaced by an unbiased radar altimeter measurement over a calm sea. Demonstrate applicability of (5-24). For extremely crude radar data, show that observability would be marginal. Relate to the discussion near the end of Section 5.2.1.

5-29 (a) Consider a constant value for one of the bias components in

(5-68). What would appear in the corresponding matrix row and forcing function? Is there any serious singularity problem?

(b) Can the matrix in (5-68) have off-diagonal elements?

(c) Suppose that some systematic errors have dynamic representations defined by second-order differential equations. Show that these can be re-expressed for conformance to (5-68). Generalize this principle.

5-30 Consider a 2-state system with highly correlated uncertainties, as in Ex. 5-5e. Show that, with extreme accuracy in observations, computation of the inverse for (5-44) could be highly sensitive to small variations of off-diagonal elements. Demonstrate how this situation would change if **E** in (5-48) produced dominant diagonal elements in $\mathbf{P}^-$ at each observation time.

5-31 (a) Suppose that the points in Fig. 5-3 had been generated from an equation of the form

$$x = \mathscr{C}_1(\cos \omega t + 0.0001 \cos 1000\omega t) + \mathscr{C}_2(1 + 0.0001\omega t)^{1/2} \sin \omega t$$

Would the points be in near conformance to (5-4) for a ten-cycle duration? Does this example suggest qualifying a strict categorization of known and unknown dynamic models?

(b) Graph the equation $x_1 = x_2 t + 0.1t^2$ for $x_2 = 1, 0 \leq t \leq 10$ sec. Can this function be approximated by a sequence of straight line segments, each with 1-sec duration? Can a sequence of dynamic model segments, in the form of (5-28), be used for vertical channel estimation in the presence of unknown acceleration bias? Must the segments be non-overlapping for a reasonable data fit?

(c) Could (5-8) have been successfully adapted in part (b), with a 30-sec data window?

(d) Should a data window be allowed to straddle the event described in Ex. 4-15c?

(e) Consider a sign reversal for the second term on the right of (5-4). Relate to (5-7) with the gravity uncertainty of Ex. 3-21a substituted for acceleration error. Could a dynamic formulation, which accounts for gravity uncertainty as in Ex. 3-21a, largely remedy the deficiency pointed out in Ex. 5-4c?

5-32 Substitute X_1, X_2, X_3, X_4 for λ, φ, V_N, V_E, respectively, in (3-23) and (3-32). Express these equations in the form of (5-73), with $\mathscr{R}$ assumed constant.

5-33 (a) Apply (5-75)–(5-77) to the preceding exercise, to obtain

$$\dot{x}_1 \doteq (1/\mathscr{R})x_3, \qquad \dot{x}_2 \doteq (1/\mathscr{R}) \sec X_1[(X_4 \tan X_1)x_1 + x_4]$$

(b) Note the lower right element of the matrix in (5-7). Use Ex. 3-21a to replace this element by $2g/\mathscr{R}$. Relate to the last item raised in Ex. 5-31e.

(c) Replace $\mathbf{A}$ and $\mathbf{e}$ in (5-2) by $\mathfrak{a}$ and ε, respectively, to signify augmentation of dynamic and/or observation error models. Rewrite (5-76) and (5-77) accordingly.

(d) Consider a set of linear dynamic equations with fixed unknown parameters in the forcing function. Show that the state could be augmented, for identification of these parameters, without any extension of linear estimation methods. What modification would be required for identification of unknown *matrix* elements in (5-2)?

(e) Recall the restriction of functional dependence noted after (5-2). Discuss the notion that $\mathbf{A}$ can vary with $\hat{\mathbf{X}}$, but its dependence on $\mathbf{x}$ need not be precisely characterized.

5-34 Consider a formulation in which the first three states are Cartesian components of position with respect to a designated point. With scalar range substituted into (5-79), show that the first three elements of (5-81) constitute a unit vector parallel to the instantaneous position vector.

5-35 (a) What is the duration of a 100-nm eastward course leg at Mach 2 ($\doteq$1200 kt)? What data window should characterize this course leg, when it is preceded and followed by northbound segments under conditions similar to Ex. 4-15c?

(b) Obtain (5-88) as suggested in Ex. 3-25b.

(c) Differentiate (3-53). Explain any departures from the bottom row of (5-87).

(d) Draw an analogy between (5-8) and (5-87). Is there any reason to anticipate some commonality between dynamics of different Cartesian position components?

(e) A ship INS is degraded by a "Z-ramp" phenomenon, involving a growing drift rate with a coefficient expressed in $°/\text{hr}^2$. If this phenomenon cannot be detected until after the steady component of azimuth drift takes effect, can most airborne systems be formulated without regard to growth of drift rates?

5-36 Suppose that M observations are effectively averaged in a data window, with an anticipated bias level below $\sigma/\mathsf{M}^{1/2}$. Would the bias be "substantial" in the sense of Ex. 5-24b? Does Ex. 5-11c have any influence here?

5-37 Recall the closing item of Section 5.3.1. Show how marginal observability could result from error sources described in Ex. 5-29c.

5-38 (a) Recall the definitions preceding (5-9). What is the most probable value for each component of measurement error? With $\mathbf{R}$ in (5-44) replaced by $2\mathbf{R}$, show that a two-sigma observation error could be a plausible value within the prescribed distribution.

(b) If the prescribed distribution in part (a) were characterized by *smaller*-than-actual covariances (e.g., $\mathbf{R}/2$) show how the principle just established would be reversed.

(c) Explain why an estimate derived from the prescribed measurement error distribution in part (a) would not be optimal, in a minimum variance sense. Is the effect catastrophic?

5-39 (a) In (5-53), suppose that $\mathbf{P} = 10^4$ while $\mathbf{R} = 10$. Would resetting $\mathbf{P}$ to 10^3 greatly change the weighting? Could conservatism in $\mathbf{R}$ offset any threat to computational stability, incurred by disallowing very large values of $\mathbf{P}$? What is the role of $\mathbf{E}$?

(b) If estimates eventually settle to values substantially unaffected by initial conditions, will the exact levels chosen in part (a) critically affect estimator performance after large transients subside?

5-40 (a) For nav modes based on extensions of (4-1), give an example showing how Chapter 4 material enables characterization of spectral densities as well as biases. Must the characterization be exact? Can nongaussian degradations, noted in Ex. 5-20, be tolerated? Show that $\mathbf{e}_{\mathrm{a}}$ and $\mathbf{e}_{\omega}$, as functions of the dynamic environment, are correlated. Can $\mathbf{E}$ have off-diagonal elements?

(b) Can $\mathbf{E}$ be conservatively prescribed? When this is overdone, will estimates become unduly sensitive to instantaneous measurement errors?

5-41 Consider (5-38) combined with dynamics in the form of (5-28), for direct measurements of position only $\{\mathbf{H} = [1, 0]\}$. Let (5-44) be computed from a prescribed *constant* matrix substituted for $\mathbf{P}$. Show that estimation error following each update is governed by a linear difference equation of the form

$$\tilde{\mathbf{x}}_m^{\ +} = \begin{bmatrix} 1 - \not{f} & (1 - \not{f})\ell \\ -a & 1 - a\ell \end{bmatrix} \tilde{\mathbf{x}}_{m-1}^{+} - \begin{bmatrix} \not{f} \\ a \end{bmatrix} \epsilon_m$$

and show that equal eigenvalues are obtained with $a\ell = 2 - \not{f} - 2(1 - \not{f})^{1/2}$.

5-42 (a) Consider range measurements in skewed directions, with (5-50) evaluated under conditions of Ex. 5-34. Can cross-axis coupling be introduced?

(b) Substitute (5-89) into (II-43) to obtain (5-90).

(c) If all terms beyond any given order in (II-43) vanish, $\mathbf{A}$ is said to be *nilpotent*. What can be said of the matrix in (5-89)?

5-43 (a) Let $\mathbf{p}$ denote a covariance matrix partition for position uncertainty only. Define $\mathbf{T}$ as an orthogonal transformation from the reference coordinate frame (in which the position states are expressed) to another frame of interest. Show that the covariance matrix for position uncertainty, as resolved along the axes of the second frame, is $\mathbf{T}\mathbf{p}\mathbf{T}^{\mathrm{T}}$.

(b) When the matrix just formed in part (a) is diagonal, what can be said in regard to axis directions? Use (II-25), (II-27), and (II-40) to express the principal directions in terms of the elements and the eigenvalues of $\mathbf{p}$.

(c) Is the transformation in part (a) a similarity transformation? Is the trace affected by the transformation?

(d) Does total mean squared position error depend on the coordinate frame in which position is expressed? Is this question related to part (c)?

(e) Are the foregoing considerations applicable to velocity uncertainties?

5-44 (a) Suppose that the rectangular spectral density function described after (5-92), by reason of the Fourier transform relation mentioned in Ex. 5-5c, produces a ($\sin x/x$) autocorrelation curve with a null at $1/\mathscr{B}$ sec. When the nominal width of the autocorrelation interval is taken into account in a general model characterization, must exact conformance to the autocorrelation curve also be required for data fitting? Use the opening discussion of Section 5.4.2 to determine an answer.

(b) When the noise spectra for (5-92) are unknown, can conservative spectral density levels be assumed for the model?

(c) Let the autocorrelation time of an nth-order Butterworth filter be the reciprocal of a noise bandwidth given by

$$\mathscr{B} = 2\pi \mathsf{B}/[4n \sin(\pi/2n)] \text{ Hz}$$

where B denotes 3-dB cutoff frequency expressed in hertz. It can be shown that the rectangular function mentioned in part (a) encloses the same area as an nth-order Butterworth function with an attenuation characteristic shaped as $1/[1 + (f/\mathsf{B})^{2n}]$, as a function of frequency f. In view of the approximation of autocorrelation properties already present in part (a), should usage of noise bandwidth $\mathscr{B}$ critically affect estimation performance in the presence of Butterworth shaped noise spectra?

(d) Relate the last term of (5-92) to Ex. 5-18b.

5-45 (a) Consider an estimation problem in which the state dimensionality exceeds the rank of $\mathbf{J}$ by one, and $\mathbf{S}$ is of the form

$$\mathbf{S} = \mathbf{U}[\boldsymbol{J} \,\vdots\, \mathbf{0}] = [\boldsymbol{S} \,\vdots\, \mathbf{0}], \qquad |\mathbf{U}| \neq 0$$

where $\boldsymbol{J}$ and $\boldsymbol{S}$ have dimensions $M \times (N-1)$ and rank $(N-1)$, while $\mathbf{0}$

here represents the $M \times 1$ null vector. Show that

$$\begin{bmatrix} \mathbf{S}^{\#} \\ \hdashline \mathbf{0}^{\mathrm{T}} \end{bmatrix}$$

satisfies (5-95) and is thus the unique generalized inverse of $\mathbf{S}$.

(b) Show that the generalized inverse of any null matrix is another null matrix, with rows and columns interchanged. Reflect on the zero-to-zero transforming property of generalized inverses; should the unobservable state be estimated at zero in part (a)?

(c) In the example just considered, (5-31) is inapplicable, and the covariance matrix of estimation error is undefined. Use the reasoning from part (b) to explain why $\langle\tilde{\mathbf{x}}\tilde{\mathbf{x}}^{\mathrm{T}}\rangle$ cannot be determined on the basis of (5-93).

5-46 (a) Recalling Exs. 5-45a and 5-27, describe how (5-93) provides weighted least squares estimates for observable states, even when (5-24) is inapplicable.

(b) The discussion accompanying (5-88) exemplifies the nearly ambiguous effects of marginally observable states. How would the usage of (5-24) compare versus (5-93) in that case? Explain how the independence between states is similarly compromised by ignoring the principles stated at the end of Section 5.3.1. Briefly reiterate the theoretical and computational implications of augmenting states for sequentially correlated observation errors.

(c) When $|\mathbf{J}^{\mathrm{T}}\mathbf{J}|$ is zero, there is an infinity of possible initial state estimates that can minimize L in (5-21). From this infinity of "solutions," the one that also minimizes $|\hat{\mathbf{x}}_0|$ is unique. How does this situation compare with solutions to the equation $\mathbf{z} = \mathbf{J}\hat{\mathbf{x}}_0$ when the number of unknowns is greater than or equal to the number of independent equations?

5-47 Insert (II-55), (5-48), and (5-61) into the context of Appendix 5.B, to obtain a correlation coefficient in the form $\exp\{-(t_2 - t_1)/\tau\}$. Show that, for multiple systematic error components, elements of $\mathbf{R}$ could be expressed as sums of terms with factors in this form.

5-48 (a) In the context of Appendix 5.B, the a priori estimate for (5-53) is identical to the first measurement. Make this substitution in (5-53) to obtain the right of (5-103) directly.

(b) Verify (5-101).

(c) Show that, in the example considered for Appendix 5.B, (5-101) conforms to a generalized inverse. Is this to be expected from (5-94)?

REFERENCES

5-1. Kalman, R. E., "A New Approach to Linear Filtering and Prediction Problems," *ASME Trans. J. Basic Engr.* **82** (1), March 1960, pp. 35–45.
5-2. Battin, R. H., "A Statistical Optimizing Navigation Procedure for Space Flight," *ARSJ* **32** (11), Nov. 1962, pp. 1681–1696.
5-3. Jazwinski, A. H., *Stochastic Processes and Filtering Theory*, New York: Academic Press, 1970.
5-4. Friedland, B., "A Review of Recursive Filtering Algorithms," *Proc. 1972 Spring Joint Computer Conference*, Atlantic City, New Jersey, May 1972.
5-5. Gura, I. A., and Gersten, R. H., "Interpretation of *n*-Dimensional Convariance Matrices," *AIAAJ* **9** (4), April 1971, pp. 740–742.
5-6. Kriegsman, B. A., and Tao, Y. C., "Shuttle Navigation System for Entry and Landing Mission Phases," AIAA paper No. 74-866, Aug. 1974.
5-7. Bryson, A. E., and Johansen, D. E., "Linear Filtering for Time-Varying Systems Using Measurements Containing Colored Noise," *IEEE Trans* **AC-10** (1), Jan. 1965, pp. 4–10.
5-8. Bryson, A. E., and Henrikson, L. J., "Estimation Using Sampled Data Containing Sequentially Correlated Noise," *AIAA JSR* **5** (6), June 1968, pp. 662–665.
5-9. Widnall, W. S., *Applications of Optimal Control Theory to Computer Controller Design*, Research Monograph No. 48. Cambridge, Massachusetts: MIT Press, 1968, pp. 45–46.
5-10. Schlegel, L. B., "Covariance Matrix Approximation," *AIAAJ* **1** (11), Nov. 1963, pp. 2672–2673.
5-11. Hitzl, D. L., "Comments on 'Covariance Matrix Approximation'," *AIAAJ* **3** (10), Oct. 1965, pp. 1977–1978.
5-12. Widnall, W. S., and Carlson, N. A., "Post-Flight Filtering and Smoothing of CIRIS Inertial and Precision Ranging Data," *Proc. Guidance Test Symp. 6th*, AFSWC-TR-72-34, Holloman Air Force Base, New Mexico, Oct. 1972.
5-13. Huddle, J. R., "Applications of Kalman Filtering Theory to Augmented Inertial Navigation Systems," *in Theory and Applications of Kalman Filtering* (C. T. Leondes, ed.), Ch. 11. AGARDograph No. 139, Feb. 1970 (AD-704 306), pp. 231–268.
5-14. Farrell, J. L., "Simulation of a Minimum Variance Orbital Navigation System," *AIAA JSR* **3** (1), Jan. 1966, pp. 91–98.
5-15. Farrell, J. L., "Attitude Determination by Kalman Filtering," *Automatica* **6**, May 1970, pp. 419–430.
5-16. Warren, A. W., "Continuous–Discrete Filtering for Presmoothed Observations," *IEEE Trans* **AC-19** (5), Oct. 1974, pp. 563–567.
5-17. Luenberger, D. G., "Observers for Multivariable Systems," *IEEE Trans* **AC-11** (2), April 1966, pp. 190–197.
5-18. Tse, E., "Observer-Estimators for Discrete-Time Systems," *IEEE Trans* **AC-18** (1), Feb. 1973, pp. 10–16.
5-19. Leondes, C. T., and Novak, L. M., "Reduced-Order Observers for Linear Discrete-Time Systems," *IEEE Trans* **AC-19** (1), Feb. 1974, pp. 42–46.
5-20. Kaminski, P. G., Bryson, A. E., and Schmidt, S. F., "Discrete Square Root Filtering: A Survey of Current Techniques," *IEEE Trans* **AC-16** (16), Dec. 1971, pp. 727–736.
5-21. Bierman, G. J., "Computational Aspects of Discrete Sequential Estimation," JPL Rep. No. 900–661, May 1974.
5-22. Schmidt, S. F., "Application of State-Space Methods to Navigation Problems," *Advan. Control Systems*, **3**. New York: Academic Press, 1966, pp. 293–340.

5-23. Friedland, B., "Treatment of Bias in Recursive Filtering," *IEEE Trans* **AC-14** (4), Aug. 1969, pp. 359–367.
5-24. Danik, B., "Integrated Inertial Velocity-Aided and Position-Aided Aircraft Navigation," presented at 24th ION meeting, Monterey, California, June 21, 1968.
5-25. Albert, A., and Sittler, R. W., "A Method for Computing Least Squares Estimators that Keep Up with the Data," *J. SIAM Control Ser. A* **3** (3), Aug. 1966, pp. 384–417.
5-26. Rust, B., Burrus, W. R., and Schneeberger, C., "A Simple Algorithm for Computing the Generalized Inverse of a Matrix," *Comm. ACM* **9** (5), May, 1966, pp. 381–387.

Chapter 6

Navigation Modes and Applicable Dynamics

6.0

The preceding chapters have established a basis for characterizing the motions encountered in aircraft navigation, conventions for describing these motions, response of inertial instruments to the motion patterns, and a general rationale for processing navigation measurements. All that remains for completion of an integrated nav system concept is definition of specific dynamics for each mode (covered in the present chapter), characterization of measured data for each navaid (Chapter 7), and a full system summary with descriptive examples (Chapter 8). State formulations and dynamic equations are now put into standard form for application of the estimation procedure already described. A thorough understanding of the preceding chapter is assumed in presenting this material. The topical sequence (recall Section 1.1) covers earth nav (e.g., latitude, longitude, altitude above sea level, geographic velocities), point nav (e.g., all aircraft position components defined relative to a designated point, with either a geographic or an alternative azimuth reference), and air-to-air (e.g., all aircraft displacements and derivatives defined relative to a tracked airborne vehicle). Further subdivisions of modes are affected by data window duration and the extent to which the formulation accounts for various error propagation characteristics, as discussed in Section 5.4.2. Also, multiple modes can be conducted simultaneously (e.g., separate vertical channel estimation concurrent with short-term geographic position and velocity determination, during air-to-air tracking).

Formulations for all INS modes here are uniformly applicable to gimballed platform or strapdown mechanizations, with all concepts and notations consistent with the preceding chapters. For illustrative purposes some consideration is given to augmented estimation for both systematic dynamic error sources and fixed navaid bias. Since Chapter 7 treats observations in detail, however, they are generally referred to only symbolically here; present emphasis is on deriving state formulations in accordance with (5-76).

As explained in preceding chapters, the estimation procedure calls for one primary instrumenting source used as the basis for a kinematical model, while all other instruments are associated with navaid observations. The primary sources considered here, in preferential order, are

(1) INS (unless it is of very low quality, in which case it is used only as a short-term data supplement);

(2) a doppler radar (which provides a measure of aircraft velocity; see Chapter 7) plus measured roll, pitch, and heading angles [to enable transformation into geographic coordinates; in the absence of an INS, these angles can be supplied by a heading and attitude reference system (HARS)];

(3) airspeed and angle of attack, discussed in Appendix 4.A, plus the aircraft angles as just described.

When one of the preferred (INS or doppler) instrumenting sources is available it is also used, with the airspeed vector, to compute north and east wind. This provides an initial wind vector for the air data mode, to be described herein, in the event of primary source failure. When both a quality INS and a doppler radar are available, the latter is used as a navaid.

Dynamic models for short-, intermediate-, and long-term earth navigation will now be provided in the first three sections, followed by the long-term formulation as augmented by systematic INS errors (Ref. 6-1). The last two sections then derive modifications for point nav and air-to-air tracking. Only a rudimentary Cartesian form is given for these last two items, intended only to explain the underlying principles without regard to actual system designs. Further applications are then illustrated in Appendixes 6.A and 6.B.

6.1 GEOGRAPHIC POSITION AND VELOCITY ESTIMATION

In this mode a state vector is defined in terms of current estimation uncertainties, in accordance with (5-75), as

$$\mathbf{x}_{4\times 1} \triangleq \begin{bmatrix} x_1 \\ x_2 \\ x_3 \\ x_4 \end{bmatrix} \triangleq \begin{bmatrix} X_1 - \hat{X}_1 \\ X_2 - \hat{X}_2 \\ X_3 - \hat{X}_3 \\ X_4 - \hat{X}_4 \end{bmatrix} \triangleq \begin{bmatrix} \tilde{\lambda} \\ \tilde{\varphi} \\ \tilde{V}_N \\ \tilde{V}_E \end{bmatrix} \triangleq \begin{bmatrix} \lambda - \hat{\lambda} \\ \varphi - \hat{\varphi} \\ V_N - \hat{V}_N \\ V_E - \hat{V}_E \end{bmatrix} \tag{6-1}$$

where the circumflex and the tilde denote estimated or observed values, and the error in these values, respectively. Figure 6-1 shows the uncoupled north channel, in close analogy with the altitude channel already described.

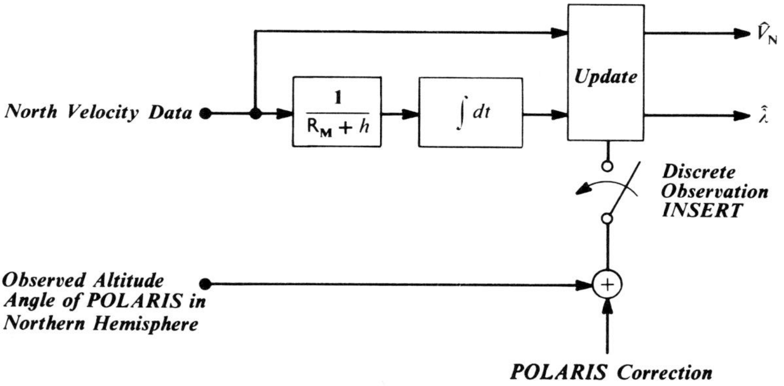

FIG. 6-1. Isolated north channel estimator (see Ex. 6-1).

Each time a scalar measurement is obtained (the mth observation denoted as $\hat{Y}_m$), a residual z is computed,*

$$z_m = \hat{Y}_m - Y_m^{-} \tag{6-2}$$

[where Y_m^{-} is the predicted value of the measurement, based on nav estimates immediately before taking the observation; e.g., see (5-83)] and an updated estimate is obtained by discrete adjustment via (5-85), where

* Although the overall technique is applicable to data *vectors* representing multiple simultaneous scalar observations, this leads to a requirement for matrix inversion operations. The above example is tailored to separate processing of each scalar measurement, subject to the restriction of negligible cross correlation between errors in separate observations. The reader should be able, at this point, to apply the principles of Section 5.4.1 to Fig. 6-1 (Ex. 6-1), and to apply the present discussion to either the north channel or the coupled horizontal channel pair.

the weighting matrix (**W**; in this case a column vector) is computed from recursive estimation theory as

$$\mathbf{W}_m = (\mathscr{H}_m \mathbf{P} \mathscr{H}_m^{\mathrm{T}} + \sigma_m^2)^{-1} \mathbf{P} \mathscr{H}_m^{\mathrm{T}} \tag{6-3}$$

while $\mathscr{H}_m$ (in this case a row matrix) consists of measurement sensitivities; the superscript T denotes transpose; σ_m represents rms random error (sequentially uncorrelated for the immediate example) in the observation at the time t_m when the mth measurement is taken; and **P** is a time-varying uncertainty covariance matrix with the dynamic characteristics as determined below.

The navigation mode corresponding to (6-1) accounts for "white" noise excitation (denoted e_{v1} and e_{v2} for north and east channels, respectively, with autocorrelation times shorter than intervals between observations) for velocity errors, while position errors are kinematically related to velocity errors through (3-23). Thus, for the estimation relation (5-76), it is seen that **A** and **e** take the values (Ex. 6-2)

$$\mathbf{A}_{4\times 4} = \left[\begin{array}{cccc} 0 & 0 & \dfrac{1}{\mathscr{R}} & 0 \\ \left(\dfrac{V_{\mathrm{E}}}{\mathscr{R}}\right) \sec\lambda \tan\lambda & 0 & 0 & \dfrac{\sec\lambda}{\mathscr{R}} \\ \hline \multicolumn{4}{c}{\mathbf{O}_{2\times 4}} \end{array}\right], \qquad \mathbf{e}_{4\times 1} = \begin{bmatrix} e_{v1} \\ e_{v2} \\ 0 \\ 0 \end{bmatrix} \tag{6-4}$$

Expressions are now available to specify the sequential estimation procedure for the mode being considered. The matrix **E** would be assigned only two nonzero elements E_{11} and E_{22}, corresponding to a prescribed error spectral density (denoted η_v; not necessarily white, but defined as an effective average level throughout a noise bandwidth expressed in hertz). An operational specification would dictate a computational cycle as follows: $\mathbf{A}_{4\times 4}$, via (6-4); integration of **P** by (5-49); **W** [using measurement relationships from Chapter 7, for elements of $\mathscr{H}$ in (6-3)]; discrete updating [via (5-85) and (5-50), respectively] of $\hat{\mathbf{X}}$ and **P**, thus preparing for the next integration segment [through (3-20), (3-21), and (5-49), respectively]. The wideband spectral density would correspond to effective data window duration T such that the accumulated contribution of **E** to **P**, from (5-49) integrated for an interval T, produces a statistical residual level (via $\mathscr{H}\mathbf{P}\mathscr{H}^{\mathrm{T}}$) comparable to sensor inaccuracy effects. For example, $(200\ \mathrm{kt}^2/\mathrm{Hz} \times 60\ \mathrm{sec})^{1/2}$ yields approximately 200 ft, comparable to 5-mr azimuth pointing uncertainty at 40 Kft; therefore, the effective averaging time for radar azimuth fixes under these conditions would be one minute (Ex. 6-3).

Customarily, the above concepts are modified to accommodate

systematic error sources encountered in practice. A doppler radar, for example, may have a fixed groundspeed bias b_D; with this as the continuous velocity information source, then an augmented form of the state vector and the dynamic coefficient matrix could take the partitioned form

$$\mathbf{x}_{5\times 1} = \begin{bmatrix} \mathbf{x}_{4\times 1} \\ \hdashline b_D \end{bmatrix}, \qquad \mathcal{A}_{5\times 5} = \left[\begin{array}{c:c} \mathbf{A}_{4\times 4} & \begin{matrix} (1/\mathscr{R})\cos\zeta \\ (\sec\lambda/\mathscr{R})\sin\zeta \\ 0 \\ 0 \end{matrix} \\ \hdashline \multicolumn{2}{c}{\mathbf{O}_{1\times 5}} \end{array}\right] \tag{6-5}$$

where ζ is the ground track angle. In the corresponding augmented version of (5-49), the spectral densities E_{11} and E_{22} would be scaled by geometric factors while including effects of noise in ground track angle as well as in groundspeed. It should be noted that direct frequency calibration would be more natural, and therefore more promising from a performance standpoint, than doppler information expressed in terms of groundspeed and drift angle (see Chapter 7).

With the INS as the continuous velocity information source in position–velocity estimation, the main component of *systematic* error in the dynamics arises from platform tilt. For typical parameters in the mode under immediate attention, however (choice of a 60-sec data window, corresponding to 200 kt^2/Hz in the preceding example, and producing 10 fps rms for a 1-rad/sec noise bandwidth), a few tenths mr tilt would produce an order of magnitude less accumulated error than that caused by the *sequentially uncorrelated* velocity noise. Thus systematic departures could be ignored; the elements in the upper right partition of $\mathcal{A}_{5\times 5}$ in (6-5) would be nulled and in-flight doppler correction could evolve through *navaid* calibration (for fixed observation bias). If at any time in this INS primary mode the doppler radar (now processed as a navaid) became inoperative, the groundspeed bias augmenting state would be removed from the computational algorithm. Obviously there is no in-flight INS calibration in this mode, since even the first-order tilt has been rendered ineffective and azimuth bias is absorbed in the aligning adjustment for the doppler radar and/or the nav update instrument (see the end of Section 7.1.2 for consideration of azimuth uncertainty).

Ground rules at the beginning of this chapter included consideration of a case with neither inertial nor doppler information available. That situation is covered by a natural extension for the present mode; a current wind vector estimate is simply added to the measured airspeed vector (e.g., obtained from measured scalar airspeed and angle of attack at negligible sideslip, with transformation into geographic coordinates through measured roll, pitch, and heading angles). The horizontal components of this essentially continuous estimate for total velocity, substituted into (3-20) and (3-21), provide continuous latitude and longitude estimates. The preceding description of a typical discrete filter adjustment cycle also remains applicable with just a simple change, i.e., the **P**-matrix dynamics are affected by reason of (1) larger values for velocity noise spectral densities, and (2) modified **e** and **A**-matrix properties. Instead of nulling the velocity error dynamic coefficients and forcing function components as in (6-4), the uncertainties (x_3, x_4) are modeled in accordance with the standard band-limited noise relation, (5-61). Thus A_{33} and A_{44} would be set equal to $-1/\tau_{W1}$ and $-1/\tau_{W2}$, indicative of time constants for north and east velocity error components, respectively, while the corresponding spectral densities would be entered as E_{33} and E_{44} in the integration of (5-49).* As pointed out in Ref. 6-2, which uses this excitation noise model, the error states x_3, x_4 include discrepancies in airspeed and aircraft angles as well as wind uncertainty. The implicit assumption of simultaneous but uncoupled vertical channel estimation, providing reasonably accurate knowledge of altitude dynamics [continuously known $\mathscr{R}$, to within required model accuracy, in (6-4)] is acceptable for this air data mode also.

Before the discussion of geographic position and velocity is terminated, the significance of vertical error coupling will be considered. If altitude and vertical velocity uncertainties had been included here, the matrix in (6-4) would have taken the form

$$\mathbf{A} = \left[\begin{array}{c|cc} & V_N/\mathscr{R}^2 & 0 \\ \mathbf{A}_{4\times 4} & (V_E \sec\lambda)/\mathscr{R}^2 & 0 \\ & 0 & 0 \\ & 0 & 0 \\ \hline & 0 & 1 \\ \mathbf{O}_{2\times 4} & 0 & 0 \end{array}\right] \tag{6-6}$$

Comparison of typical contributions of the different state components to total nav error (Ex. 6-5) shows that coupling is loose enough to be ignored in the present mode; uncoupled vertical estimation (Chapter 5) and the spheroidal geographic position-plus-velocity estimation herein can be implemented separately.

6.2 ADDITION OF BASIC INS KINEMATICS

As indicated at the beginning of this chapter, there is an intermediate data window to be considered. During an interval of a few minutes, cumulative effects of INS misorientation can be quite significant, but the

* For a velocity excitation variance of $\sigma_W{}^2$ these spectral densities are easily shown to be $2\sigma_W{}^2/\tau_W$, in acceleration units squared/Hz (Ex. 6-4).

natural kinematics of ψ can be ignored within the data window. The nav update system would then use the same latitude and longitude equations as those already given, supplemented by simple relations for velocity errors,

$$\dot{\tilde{V}}_{\mathrm{N}} = A_{\mathrm{V}} x_6 - A_{\mathrm{E}} x_7 + e_{\mathrm{a}1} \tag{6-7}$$

$$\dot{\tilde{V}}_{\mathrm{E}} = -A_{\mathrm{V}} x_5 + A_{\mathrm{N}} x_7 + e_{\mathrm{a}2} \tag{6-8}$$

where A_{N}, A_{E}, A_{V} represent error-free outputs of north, east, and vertical accelerometers, respectively; x_5, x_6, x_7 represent INS misorientations ψ_{N}, ψ_{E}, ψ_{A}, respectively; and $e_{\mathrm{a}1}$, $e_{\mathrm{a}2}$ represent north and east acceleration errors, respectively. Thus, for the form of (5-76), **A** and **e** for this mode become (Ex. 6-6)

$$\mathbf{A}_{7\times 7} = \left[\begin{array}{c:ccc} & 0 & 0 & 0 \\ & 0 & 0 & 0 \\ \mathbf{A}_{4\times 4} & 0 & A_{\mathrm{V}} & -A_{\mathrm{E}} \\ & -A_{\mathrm{V}} & 0 & A_{\mathrm{N}} \\ \hdashline \multicolumn{4}{c}{\mathbf{O}_{3\times 7}} \end{array}\right], \qquad \mathbf{e}_{7\times 1} = \left[\begin{array}{c} 0 \\ 0 \\ e_{\mathrm{a}1} \\ e_{\mathrm{a}2} \\ \hdashline \mathbf{e}_{\omega} \end{array}\right] \tag{6-9}$$

where $\mathbf{e}_{\omega}$ is the 3×1 vector of random drift that, in combination with the acceleration error level, determines the data window for any given navaid specification. For example a simplified drift model could be postulated, whereby total INS error is represented by effective acceleration noise of $10^{-4}g$ rms, with an autocorrelation time of $\frac{1}{2}$ min. The corresponding spectral density would be

$$\eta_{\mathrm{g}} = (10^{-4})^2(30) \doteq 3(10)^{-7} g^2/\mathrm{Hz} \tag{6-10}$$

and, for an integrated position error of 100 ft per data block,

$$32.2(\tfrac{1}{3}\eta_{\mathrm{g}} T^3)^{1/2} = 100 \text{ ft} \tag{6-11}$$

which would indicate a data window of about one-tenth Schuler period. In this example, then, the value in (6-10) would correspond to the third and fourth diagonal elements of **E** in a 7×7 computation for (5-49); in any case the state and covariance equations pertinent to this mode are now completely defined, and procedures already described remain applicable.

In regard to in-flight calibration it is seen that this mode automatically implies continuous INS information, while no systematic drifts are observable within the data window. Thus all augmenting states [e.g., the ground-speed bias in (6-5)] fall into the category of navaid calibration variables, for which the procedures illustrated elsewhere in this book are applicable.

Just as in the preceding section, vertical coupling could conceivably be introduced here, but is not necessary for this mode of navigation. Vertical channel estimation should be provided separately.

6.3 COMPLETE INS DYNAMICS

For data window intervals exceeding one-tenth Schuler period, observations are averaged over durations sufficiently long for the full effects of kinematic coupling to be taken into account. This section contains the derivation for coupled errors in position, velocity, and angular orientation. Specifically, the outcome is the relation between initial (i.e., at any chosen reference time) and current (i.e., at the time of any navaid measurement) navigation errors, for a system employing a north-slaved Schuler tuned INS. Similar derivations under various conditions have appeared previously in the literature (see, for example, the constant-altitude analysis in Ref. 6-1); the present development includes position, velocity, and INS orientation errors in all three dimensions. Indices 1 to 9 for observed or estimated values (e.g., $\hat{X}_j$) and variations ($x_j = X_j - \hat{X}_j$) here denote latitude, longitude, north and east velocity, INS misorientation about north, east, and downward axes, vertical distance (positive downward), and downward velocity V_V, respectively. Any desired reordering of indices (e.g., for sequences denoting position, velocity, and orientation with respect to $\mathscr{x}$, $\mathscr{y}$, and $\mathscr{z}$ directions, as in the latter portion of Section 6.5) is easily obtained by a simple transformation. Since circumflexed quantities denote *currently* estimated values (i.e., including the benefit of all previous navaid measurements), the condition $\hat{X}_5 = \hat{X}_6 = \hat{X}_7 = 0$ adopted here implies that the gyros are torqued to realign the platform (or, for a strapdown configuration, a computational attitude matrix adjustment is made) immediately after each new observation. This concept merely simplifies covariance matrix extrapolation regardless of mechanization; in practice a modified accumulation of velocity increments can account for estimates $\hat{\boldsymbol{\psi}}$, while trim adjustments are omitted. This item is discussed further after the derivation.

Expressions from Chapters 2 and 3 will now be used to derive the long-term INS error dynamics. The outcome is of course consistent with short-term analysis [i.e., (6-9) is a special case of the result to follow] and, by expressing the complete dynamics in accordance with (5-76), adaptation to the standard overall procedure is naturally achieved.

Error dynamics for the first state variable ($x_1 = \lambda - \hat{\lambda}$) can immediately be obtained by a simple extension of the procedure used in deriving (6-4). By differentiating the latitude expression of (3-23) while allowing for imperfections in $\mathscr{R}$,

$$\dot{x}_1 \doteq (\partial\dot{\lambda}/\partial V_N)x_3 + [\partial\dot{\lambda}/\partial(-\mathscr{R})]x_8 \tag{6-12}$$

so that the elements A_{13} and A_{18} of the canonical dynamics matrix are simply $1/\mathscr{R}$ and $V_N/\mathscr{R}^2$, respectively [note (6-6)]. These are the only nonzero elements in the top row of Table 6-1. An analogous procedure readily

TABLE 6-1 **A**-MATRIX FOR GEOGRAPHIC POSITION, VELOCITY, AND MISORIENTATION

$$
\left[\begin{array}{c:c:c:c:c:c:c:c:c}
0 & 0 & 1/\mathscr{R} & 0 & 0 & 0 & 0 & V_N/\mathscr{R}^2 & 0 \\
\frac{V_E}{\mathscr{R}}\sec\lambda\tan\lambda & 0 & 0 & \sec\lambda/\mathscr{R} & 0 & 0 & 0 & \frac{V_E\sec\lambda}{\mathscr{R}^2} & 0 \\
-V_E\left(\frac{V_E}{\mathscr{R}}\sec^2\lambda + 2\omega_s\cos\lambda\right) & 0 & V_V/\mathscr{R} & 2\varpi_3 & 0 & A_V & -A_E & -\left(\frac{V_E}{\mathscr{R}}\right)^2\tan\lambda + \frac{V_V V_N}{\mathscr{R}^2} & V_N/\mathscr{R} \\
V_N\left(\frac{V_E}{\mathscr{R}}\sec^2\lambda + 2\omega_s\cos\lambda\right) - 2V_V\omega_s\sin\lambda & 0 & -\varpi_3 + \omega_s\sin\lambda & \frac{V_N}{\mathscr{R}}\tan\lambda + \frac{V_V}{\mathscr{R}} & -A_V & 0 & A_N & \frac{V_N V_E}{\mathscr{R}^2}\tan\lambda + \frac{V_V V_E}{\mathscr{R}^2} & \varpi_1 + \omega_s\cos\lambda \\
\omega_s\sin\lambda & 0 & 0 & -1/\mathscr{R} & 0 & \varpi_3 & -\varpi_2 & -V_E/\mathscr{R}^2 & 0 \\
0 & 0 & 1/\mathscr{R} & 0 & -\varpi_3 & 0 & \varpi_1 & V_N/\mathscr{R}^2 & 0 \\
(V_E/\mathscr{R})\sec^2\lambda + \omega_s\cos\lambda & 0 & 0 & \tan\lambda/\mathscr{R} & \varpi_2 & -\varpi_1 & 0 & (V_E/\mathscr{R}^2)\tan\lambda & 0 \\
0 & 0 & 0 & 0 & 0 & 0 & 0 & 0 & 1 \\
2\omega_s V_E\sin\lambda & 0 & -2V_N/\mathscr{R} & -2\varpi_1 & A_E & -A_N & 0 & \frac{-(V_N^2 + V_E^2)}{\mathscr{R}^2} + \frac{2g}{\mathscr{R}} & 0
\end{array}\right]
$$

provides the elements for the second row. The eighth row follows, of course, from identification of vertical velocity as the time derivative of X_8.

Dynamic behavior of the velocity states x_3, x_4, x_9 will next be obtained as follows: For purposes of error analysis the absolute acceleration of (3-26) in geographic coordinates is written in terms of measured outputs of perfectly oriented error-free accelerometers as

$$\frac{d^2\mathbf{R}}{dt^2} = \begin{bmatrix} A_\mathrm{N} \\ A_\mathrm{E} \\ A_\mathrm{V} \end{bmatrix} + \mathbf{G} \tag{6-13}$$

while, with imperfections in both orientation and calibration,

$$\widehat{\frac{d^2\mathbf{R}}{dt^2}} \doteq \mathbf{T}_{\mathrm{P/G}} \begin{bmatrix} A_\mathrm{N} \\ A_\mathrm{E} \\ A_\mathrm{V} \end{bmatrix} - \mathbf{e}_\mathrm{a} + \hat{\mathbf{G}} \tag{6-14}$$

so that (note phasing; e.g., a positive azimuth angle error ψ_A in combination with positive east acceleration produces apparent northern motion)

$$\frac{d^2\mathbf{R}}{dt^2} - \widehat{\frac{d^2\mathbf{R}}{dt^2}} \doteq \begin{bmatrix} 0 & -\psi_\mathrm{A} & \psi_\mathrm{E} \\ \psi_\mathrm{A} & 0 & -\psi_\mathrm{N} \\ -\psi_\mathrm{E} & \psi_\mathrm{N} & 0 \end{bmatrix} \begin{bmatrix} A_\mathrm{N} \\ A_\mathrm{E} \\ A_\mathrm{V} \end{bmatrix} + \mathbf{e}_\mathrm{a} + \mathbf{G} - \hat{\mathbf{G}} \tag{6-15}$$

It is now convenient to combine (6-13) with (3-33),

$$\begin{bmatrix} \dot{V}_\mathrm{N} \\ \dot{V}_\mathrm{E} \\ \dot{V}_\mathrm{V} \end{bmatrix} = \frac{d^2\mathbf{R}}{dt^2} + (\mathbf{g} - \mathbf{G}) + \mathbf{f} \tag{6-16}$$

Aside from altitude divergence, all quantities involving gravity uncertainties can be included in $\mathbf{e}_\mathrm{a}$, and a representation for the velocity error is obtained from the last two equations as

$$\dot{\mathbf{V}} \doteq \left[\frac{\partial \mathbf{f}}{\partial \mathbf{X}}\right] \mathbf{x} + \begin{bmatrix} 0 & A_\mathrm{V} & -A_\mathrm{E} \\ -A_\mathrm{V} & 0 & A_\mathrm{N} \\ A_\mathrm{E} & -A_\mathrm{N} & 0 \end{bmatrix} \boldsymbol{\psi} - 2\frac{g}{\mathscr{R}}\tilde{h} + \mathbf{e}_\mathrm{a} \tag{6-17}$$

This substantiates the third, fourth, and ninth rows of Table 6-1 (Ex. 6-7).

All that remains to complete the derivation of Table 6-1 is an expression for the dynamics of $\boldsymbol{\psi}$. This INS misorientation is driven not only by the rotational drift $\mathbf{e}_\omega$ but also via a nav error feedback $\tilde{\boldsymbol{\varpi}}$,

$$d\boldsymbol{\psi}/dt = \mathbf{e}_\omega + \tilde{\boldsymbol{\varpi}} \tag{6-18}$$

where $\tilde{\boldsymbol{\varpi}}$ is derived from (3-24) thus: The *estimated* nav information $\hat{\mathbf{X}}$ determines the *actual* command vector ($\boldsymbol{\varpi}_{\mathrm{act}}$) for gyro torquing (gimballed platform applications) or for computational geographic frame rate adjustments (strapdown applications), whereas the *desired* command vector $\boldsymbol{\varpi}_{\text{des}}$ is of course determined by the true state $\mathbf{X}$. Thus, if positive $\tilde{\boldsymbol{\varpi}}$ must contribute positively to $d\boldsymbol{\psi}/dt$ as in (6-18), then (Ex. 6-8)

$$\tilde{\boldsymbol{\varpi}} = \boldsymbol{\varpi}_{\text{act}} - \boldsymbol{\varpi}_{\text{des}} \doteq [-\partial\boldsymbol{\varpi}/\partial\mathbf{X}]\mathbf{x} \tag{6-19}$$

In combination with (3-38) and (6-18),

$$\dot{\boldsymbol{\psi}} = [-\partial\boldsymbol{\varpi}/\partial\mathbf{X}]\mathbf{x} - \boldsymbol{\varpi} \times \boldsymbol{\psi} + \mathbf{e}_{\omega} \tag{6-20}$$

The fifth through the seventh rows of Table 6-1 follow readily from this relation and (3-24). Thus, the complete result in the form of (5-76) has been achieved with $\mathbf{e}$ defined as the 9×1 vector having components $(0, 0, e_{a1}, e_{a2}, e_{\omega 1}, e_{\omega 2}, e_{\omega 3}, 0, e_{a3})$. The remainder of this section provides further insight into this mode.

PHASING

Certain algebraic signs in the preceding derivation will now be examined:

(1) Since a platform is driven to counter measured angular rates, a positive component of $\mathbf{e}_{\omega}$ (e.g., a low gyro reading in the presence of a positive rate) produces a negative rotation from actual to desired orientation (i.e., a positive contribution to $\dot{\boldsymbol{\psi}}$).

(2) A low estimate of north velocity $(+x_3)$ causes a tilt angle corresponding to a positive rotation about the east axis $(+x_6)$; thus the (6, 3) element of Table 6-1 is positive.

(3) In the northern hemisphere, a low estimate of east velocity $(+x_4)$ implies that the axis of the rotation from desired (north) to actual platform heading will be about the downward vertical $(+x_7)$; the sign of A_{74} thus provides appropriate navigation error coupling, as a function of hemispheric position.

RELATION TO PREVIOUS DEVELOPMENTS

Various special cases of the dynamics just derived are now recalled. First, for shorter intervals it is permissible to uncouple the vertical and ignore variations in $\boldsymbol{\psi}$ while setting $V_{\text{N}} = V_{\text{E}} = V_{\text{V}} = \omega_{\text{s}} = 0$, in which case the dynamics matrix of (6-9) is obtained. If in addition the latitude were set to zero while conditions of (3-41) were imposed, but the torquing error of (3-87) were reinserted without adding drift, results would conform to (1-44) plus similar relations for east tilt and north velocity error. As one generalization of this case, nonzero values can be reintroduced for the latitude and the sidereal rate. If the upper 7×7 partition of Table 6-1 were then diagonalized (see the opening discussion of Appendix II.3), the eigenvalue equation would produce the roots derived on pp. 144–146 of Ref. 6-3:

$$\Lambda(\Lambda^2 + \omega_s^2)[\Lambda^2 + (W + \omega_s \sin \lambda)^2][\Lambda^2 + (W - \omega_s \sin \lambda)^2] = 0 \quad (6\text{-}21)$$

VERTICAL COUPLING

Unlike the case for preceding sections, horizontal and vertical channel interaction is significant in the long term. The reason is traceable to the (3, 9) and (4, 9) elements of Table 6-1. It is easily shown that the error terms associated with these elements, while appreciably smaller than the accompanying error terms associated with the basic platform coupling (Ex. 6-9),

$$\frac{V}{\mathscr{R}} \tilde{V}_V \div (A_V \psi) \doteq \frac{1}{30} \text{(typically)} \quad (6\text{-}22)$$

are nevertheless comparable in magnitude to the error terms introduced in the long-term data mixer dynamics; e.g., from position error coupling,

$$\frac{V}{\mathscr{R}} \tilde{V}_V \div \left(\frac{V^2}{\mathscr{R}} \tilde{\lambda}\right) = \frac{\tilde{V}_V/V}{\tilde{\lambda}} \quad (6\text{-}23)$$

and from velocity error coupling (with V_H denoting horizontal velocity),

$$\frac{V}{\mathscr{R}} \tilde{V}_V \div \left(\frac{V}{\mathscr{R}} \tilde{V}_H\right) = \frac{\tilde{V}_V}{\tilde{V}_H} \quad (6\text{-}24)$$

Interestingly, a similar investigation regarding relative significance of *altitude* error coupling, in comparison with errors associated with the other elements of $\mathbf{A}_{9 \times 9}$, leads to smaller error contribution ratios, which could be neglected (Ex. 6-10).

MECHANIZATION

It is recalled that discrete estimation makes two different uses of the state dynamics, i.e., extrapolation of the state itself and extrapolation of the estimation error covariance matrix between measurements. The former often uses a more detailed dynamic model than the latter, e.g., compare (3-20) and (3-21) with (3-23). Another consideration is the assumption, stated before this analysis, of platform trim by gyro torquing (or attitude computer adjustment in strapdown systems) after each navaid measurement. This omits a small dynamic coupling (involving ψ_N, ψ_E, and ψ_A) from Table 6-1. While this simplification is permissible for covariance extrapolation in this mode, it does not imply that actual systems must be mechanized accordingly. In practice the torquing commands may be kept independent of data mixer refinements [tantamount to accepting the presence of known orientation errors internal to the INS, while corrections are applied *after* (3-32)], or replaced by small-angle transformations of accelerometer data (analogous to substitution of computation for refinement of the torquing commands). This latter procedure is possible only when accelerometer information is made accessible to the Kalman estimator data interface.

In closing this section, attention is drawn to basic differences between the present formulation for error dynamics and the analysis of Section 3.4.2. Certain restrictive conditions in Section 3.4.2 have been generalized here, most notably the full geographic reference frame rotation [completion of $\tilde{\boldsymbol{\varpi}}$, instead of the partial omission implied by (3-87)] and fuller kinematical coupling. On the other hand, the present formulation is limited to situations in which systematic INS error accumulation is much smaller than the effects of random excitation $\mathbf{e}$ prescribed for a data window. The next section is addressed to a general case including systematic INS errors.

6.4 IN-FLIGHT INS CALIBRATION

For data blocks of extended duration, or for moderate data windows with significant INS bias, systematic errors can be included as augmenting state components. This necessitates a decision as to whether a separate identification of individual degradations (Chapter 4) should be attempted. For many applications, such efforts could be thwarted by similarities between observed effects of superimposed error sources (Ex. 6-11). In principle, however, total biases could be decomposed with sufficient variations in flight conditions. A subdivision is also theoretically possible for bias components with specified autocorrelation properties (e.g., initial state augmentation to separate abnormal drifts during gyro warmup), but time variations and/or uncertainties in spectral characteristics, as well as the "augmentation principles" of Section 5.4.2, must be kept in mind. No single rule can cover all cases; simple or elaborate in-flight calibration provisions are prescribed for each system individually.

Since expressions for separate errors have already been presented, and since multiple calibration procedures are extensions of a single operation, only a total bias determination (e.g., $\langle \mathbf{N}_a \rangle$, $\langle \mathbf{N}_\omega \rangle$) need be demonstrated here. An example sometimes cited in the literature uses a 12-dimensional augmented state, the first seven identified with x_1–x_7 as previously defined and the other five identified as total systematic INS errors with specified time constant τ, for the three gyros and the two level accelerometers.

This state vector of course will conform to (5-76) for a dynamics matrix (note: $\mathscr{A}$ is the upper left 7 × 7 partition of $\mathbf{A}_{9\times 9}$):

$$\boldsymbol{\mathfrak{a}}_{12\times 12} = \left[\begin{array}{c|c} \mathscr{A} & \begin{array}{c} \mathbf{O}_{2\times 5} \\ \hline \mathbf{I}_{5\times 5} \end{array} \\ \hline \mathbf{O}_{5\times 7} & -\mathbf{T}_{5\times 5}^{-1} \end{array}\right] \tag{6-25}$$

and a forcing vector $\boldsymbol{\varepsilon}_{12\times 1}$ having the same first seven components as those in (6-9), while the last five elements drive the biases, each at its own specified time constant. Addition of vertical channel and/or navaid bias calibration, as well as the aforementioned inertial instrument bias decompositions, follows from material already presented (Ex. 6-12). With typical values of tilt (ψ_N, ψ_E) and model imperfections, the accelerometer biases are often considered unobservable, and consequently omitted from the augmented state.

6.5 MODIFICATIONS FOR POINT NAV MODE

In the vicinity of a designated "target" (e.g., reference position on a runway), aircraft nav operations are defined with respect to this point, irrespective of its latitude λ_0, longitude φ_0, or altitude h_0, as previously explained. The relative distances, and aircraft excursions involved, subtend angles of at most a few milliradians at earth center. The subtended angles, as well as differential latitudes $(\lambda - \lambda_0)$ and longitudes $(\varphi - \varphi_0)$, easily conform to small-angle approximations, which lead to simplified dynamics. It follows that a point nav mode can be initiated when close to destination so that short or intermediate data block conditions apply. This section specifies the modifications required in position–velocity estimation (Section 6.1) and intermediate-duration INS update (Section 6.2). Each of these subdivides into applications independent of earth-related orientation (e.g., targeting) and those retaining a specific azimuth reference (e.g., landing). Present discussion begins with estimation of position and velocity alone with arbitrary azimuth reference, and all other cases follow as straightforward extensions.

From the time t_0 of point nav mode commencement, the computer is directed to begin a dynamic covariance computation with initialization at $\mathbf{P}_0$, and to begin calculating continuously current values for Cartesian components of position ($\hat{S}_1$, $\hat{S}_2$, $\hat{S}_3$) and velocity ($\hat{S}_4$, $\hat{S}_5$, $\hat{S}_6$), where the $\hat{S}_j$ denote estimated values of position or velocity components expressed in target coordinates (defined with the origin at the designated point). With no special azimuth reference the target coordinate frame is defined with the

$+z$ axis downward along the line from (λ_0, φ_0), and the x axis defined such that the x-z plane contains the aircraft velocity (positive forward) at reference time t_0. Although the small rotation of the local vertical is naturally taken into account in INS implementation, it is otherwise ignored for point navigation. Thus, the orthogonal transformation $[\hat{\zeta}_0]_z$ from geographic to target coordinates is equal to the transformation matrix for initial indicated ground track angle:

$$[\hat{\zeta}_0]_z = [\text{Arctan}\ \hat{V}_{E(t_0)}/\hat{V}_{N(t_0)}]_z \tag{6-26}$$

With the origin translated from the earth center to the target, the initial position and velocity vectors of the aircraft, expressed in target coordinates, are

$$\begin{bmatrix} S_1 \\ S_2 \\ S_3 \end{bmatrix} = [\hat{\zeta}_0]_z(-\mathbf{T}_0 + \mathbf{R}), \qquad \begin{bmatrix} S_4 \\ S_5 \\ S_6 \end{bmatrix} = [\hat{\zeta}_0]_z \begin{bmatrix} V_N \\ V_E \\ V_V \end{bmatrix} \tag{6-27}$$

where the target position vector $\mathbf{T}_0$, expressed in geographic coordinates, is written as an earth vector transformed into geographic coordinates:

$$\mathbf{T}_0 = (R_E + h_0)\left[-\frac{\pi}{2} - \lambda\right]_y [\varphi]_z \begin{bmatrix} \cos\lambda_0 \cos\varphi_0 \\ \cos\lambda_0 \sin\varphi_0 \\ \sin\lambda_0 \end{bmatrix} \tag{6-28}$$

and the same procedure applied to current aircraft position (Ex. 6-13) simply produces (1-25). It is noted that this procedure effectively eliminates initial ground track angle uncertainty [i.e., substitution of estimated components into the velocity transformation of (6-27) produces initial local x, y, z elements equal to estimated horizontal velocity, zero, and vertical velocity, respectively, regardless of imperfections in $\hat{\zeta}_0$]; only the small time variations in azimuth orientation error are effective.

When the last three equations are combined with the previously discussed small-angle approximations,

$$\begin{bmatrix} S_1 \\ S_2 \\ S_3 \end{bmatrix} = [\hat{\zeta}_0]_z \begin{bmatrix} R_E(\lambda - \lambda_0) \\ (R_E \cos\lambda_0)(\varphi - \varphi_0) \\ -(h - h_0) \end{bmatrix} \tag{6-29}$$

in agreement with standard definitions. Computational procedures for position and velocity in the point nav mode can now be defined as follows: With the incremental velocity since time t_0 and its time integral denoted

$$\mathbf{q} \triangleq [\hat{\zeta}]_z\{\hat{\mathbf{V}} - \hat{\mathbf{V}}_{(t_0)}\}, \qquad \mathbf{Q} \triangleq \int_{t_0}^{t} \mathbf{q}\, dt \tag{6-30}$$

the instantaneous velocity in target coordinates as determined by the nav system is written

$$\begin{bmatrix} \hat{S}_4 \\ \hat{S}_5 \\ \hat{S}_6 \end{bmatrix} = \begin{bmatrix} \hat{S}_{4(t_0)} \\ \hat{S}_{5(t_0)} \\ \hat{S}_{6(t_0)} \end{bmatrix} + \mathbf{q} \tag{6-31}$$

and, by integration, the instantaneous aircraft position in target coordinates is

$$\begin{bmatrix} \hat{S}_1 \\ \hat{S}_2 \\ \hat{S}_3 \end{bmatrix} = \begin{bmatrix} \hat{S}_{1(t_0)} \\ \hat{S}_{2(t_0)} \\ \hat{S}_{3(t_0)} \end{bmatrix} + (t - t_0)\begin{bmatrix} \hat{S}_{4(t_0)} \\ \hat{S}_{5(t_0)} \\ \hat{S}_{6(t_0)} \end{bmatrix} + \mathbf{Q} \tag{6-32}$$

For the mode under present consideration these last two equations, of course, constitute the transformation of the state* as mechanized by (5-82). Two important features of the present formulation are now recalled, i.e., (1) insensitivity to initial ground track angle uncertainty, and (2) cancellation of INS velocity errors from (6-30) for short-term applications, so that $\mathbf{q}$ and $\mathbf{Q}$ can be considered *known* in the evaluation of errors in $\hat{\mathbf{S}}$: From (6-31) and (6-32) it is readily deduced that

$$\mathbf{s}_{6\times 1} \triangleq \begin{bmatrix} S_1 - \hat{S}_1 \\ S_2 - \hat{S}_2 \\ S_3 - \hat{S}_3 \\ S_4 - \hat{S}_4 \\ S_5 - \hat{S}_5 \\ S_6 - \hat{S}_6 \end{bmatrix} = \mathbf{s}_{(t_0)} + (t - t_0)\begin{bmatrix} s_{4(t_0)} \\ s_{5(t_0)} \\ s_{6(t_0)} \\ 0 \\ 0 \\ 0 \end{bmatrix} \tag{6-33}$$

Obviously this fits the canonical form for nav error dynamics with the transition matrix of (5-90) and, if velocity and acceleration noise sources with constant spectral densities are included, (5-92) is applicable (Ex. 6-14).

This procedure for estimation of velocity and position with respect to a designated point is immediately extended to applications requiring a north azimuth reference. The matrix $[\hat{\zeta}_0]_\times$ is simply replaced by the 3×3 identity, and all relations presented in this section thus far remain applicable. Also, the air data mode of Section 6.1 is easily adapted to point navigation by straightforward modification of the above formulation, with or without the north reference.

* The output of this dynamic transformation in this mode has another usage, i.e., automatic cursor drive. When the position reference is the center of a designated cell on a display grid, a cursor can be positioned for implementing (5-85) via discrete INSERT operations. With this automatic drive provision, any wandering of the cursor from the reference cell location is attributable to imperfections in the state estimate; therefore, only a small corrective adjustment is required prior to INSERT.

As the data window approaches one-tenth the Schuler period, the previously described cancellation of velocity errors from (6-30) may not be complete. Still, the horizontal velocity variations may be modeled as linear ramps proportional to platform tilts, which remain essentially fixed within the data block duration. This can be viewed as either (1) an extension of the previous example to account for acceleration errors introduced through $\boldsymbol{\psi}$, or (2) a simplification of the dynamics implied by Table 6-1, with all slow variations ignored. Either viewpoint leads to the same result; the present nine-state exposition will begin with the case of a north azimuth reference, introducing an error state vector $\mathbf{x}_G$, related to $\mathbf{x}$ of Section 6.3 with the following redefinitions and rearrangements of components, in order: north distance (instead of latitude); east distance (instead of longitude); downward displacement; north, east, and downward velocity; and the elements of $\boldsymbol{\psi}$, respectively. If for the short term all slow dynamic effects are ignored it follows from Table 6-1 that $\mathbf{x}_G$ conforms to (5-76) with the dynamics matrix,

$$\mathbf{A}_G \triangleq \left[\begin{array}{c|c|c} & \mathbf{I}_{3\times 3} & \mathbf{O}_{3\times 3} \\ \cline{2-3} \mathbf{O}_{9\times 3} & & \mathbf{G} \\ \cline{3-3} & \mathbf{O}_{6\times 3} & \mathbf{O}_{3\times 3} \end{array}\right] \tag{6-34}$$

in which

$$\mathbf{G} \triangleq \begin{bmatrix} 0 & A_V & -A_E \\ -A_V & 0 & A_N \\ A_E & -A_N & 0 \end{bmatrix} \tag{6-35}$$

and the components of the error forcing function (denoted here as $\mathbf{e}_G$) are correspondingly redefined also (e.g., rearrange the components of $\mathbf{e}$ listed in Section 6.3). This formulation generalizes the type of analysis applicable to (3-53), but with instrument biases replaced by zero mean random errors.

All that remains to complete the nine-state point nav formulation is to apply the above development to the case of arbitrary azimuth reference. To this end, the matrix

$$\mathscr{L} \triangleq \left[\begin{array}{c|c|c} [\hat{\zeta}_0]_z & \mathbf{O}_{3\times 3} & \\ \cline{1-1} & [\hat{\zeta}_0]_z & \mathbf{O}_{6\times 3} \\ \cline{3-3} \mathbf{O}_{6\times 3} & \mathbf{O}_{3\times 3} & \mathbf{I}_{3\times 3} \end{array}\right] \tag{6-36}$$

is introduced, to perform the transformation of $\mathbf{x}_G$ into initial ground track coordinates,

$$\mathbf{s}_G = \mathscr{L}\mathbf{x}_G \tag{6-37}$$

and it is noted also that a similar transformation applies to $\mathbf{e}_G$. Furthermore, by combination of (6-37) and (5-76) noting that $\mathscr{L}$ is constant and orthogonal, the dynamics for $\mathbf{s}_G$ can be written

$$\dot{\mathbf{s}}_G = \mathscr{Z}\mathbf{A}_G \mathscr{Z}^T\mathbf{s}_G + \mathscr{Z}\mathbf{e}_G \tag{6-38}$$

The dynamics matrix for $\mathbf{s}_G$ is similar to $\mathbf{A}_G$ in (6-34) except that $\mathbf{G}$ is premultiplied by $[\zeta_0]_z$. Before acceptance of this definition, however, recall that initial azimuth error has been nulled and the relatively small variation in this misorientation angle is ineffective for moderate durations. Thus the state to be estimated is an 8×1 vector $\mathbf{s}$ that consists of the first eight components of $\mathbf{s}_G$. From the preceding analysis the appropriate dynamics can be expressed in terms of an 8×8 dynamics matrix that, for periods of constant acceleration, would give a transformation matrix of the form

$$\mathbf{\Phi}_s = \begin{bmatrix} \mathbf{I}_{3\times 3} & (t-t_0)\mathbf{I}_{3\times 3} & \frac{1}{2}(t-t_0)^2\mathbf{F} \\ & \mathbf{I}_{3\times 3} & (t-t_0)\mathbf{F} \\ \mathbf{O}_{5\times 3} & \mathbf{O}_{2\times 3} & \mathbf{I}_{2\times 2} \end{bmatrix} \tag{6-39}$$

where, with χ defined as the ratio

$$\chi \triangleq A_V/([\hat{V}_{N(t_0)}]^2 + [\hat{V}_{E(t_0)}]^2)^{1/2} \tag{6-40}$$

the 3×2 matrix $\mathbf{F}$ is written

$$\mathbf{F} = \begin{bmatrix} -\chi\hat{V}_{E(t_0)} & \chi\hat{V}_{N(t_0)} \\ -\chi\hat{V}_{N(t_0)} & -\chi\hat{V}_{E(t_0)} \\ A_E & -A_N \end{bmatrix} \tag{6-41}$$

Although these relations are expressed in terms of nav mode switching time t_0, the transition properties of course have the same form for recursion between navaid observations.

Obviously the *instantaneous* ground track convention of Fig. 2-2 could have been used in the above formulation instead of a reference based on an initial value. For some applications the resulting time-variations in $\mathscr{Z}$ would constitute a significant disadvantage. A re-expression in terms of instantaneous ground track coordinates is easily obtained at any time, by transforming the vectors in (6-31) and (6-32) through a z axis rotation of

$$\Delta\zeta = \arctan(\hat{V}_E/\hat{V}_N) - \hat{\zeta}_0 \tag{6-42}$$

6.6 MODIFICATIONS FOR AIR-TO-AIR TRACKING

At initiation (again denoted by t_0) of the air-to-air mode, position and velocity vector estimates (with Cartesian components again denoted $\hat{S}_1$–$\hat{S}_6$) between tracking aircraft and tracked object may be obtained [e.g., by straightforward computation from range, LOS, range rate, and LOS rate at the time of mode switching, using the initial ground track azimuth reference of (6-26)], while tracking aircraft velocity and position thereafter are governed by (6-31) and (6-32), respectively. To use a fixed acceleration Z_T example for the tracked object (with initial velocity V_{T0}), its position time history following mode initiation evolves simply as $\{V_{T0}(t - t_0) + Z_T(t - t_0)^2/2\}$. With V_T now absorbed into the relative velocity states, the displacement vector from the tracking aircraft to the object can be written

$$\begin{bmatrix} S_1 \\ S_2 \\ S_3 \end{bmatrix} = \begin{bmatrix} S_{1(t_0)} \\ S_{2(t_0)} \\ S_{3(t_0)} \end{bmatrix} + \begin{bmatrix} S_{4(t_0)} \\ S_{5(t_0)} \\ S_{6(t_0)} \end{bmatrix}(t - t_0) - \mathbf{Q} + \boldsymbol{Z}_T \frac{(t - t_0)^2}{2} \tag{6-43}$$

With high short-term accuracy for $\mathbf{Q}$ throughout a data window ($\tilde{\mathbf{Q}} = \mathbf{0}$), the uncertainties in (1–3) relative position vector, (4–6) relative velocity vector, and (7–9) constant acceleration vector of the tracked object in this example are expressible as (Ex. 6-15)

$$\mathbf{s} = \begin{bmatrix} \mathbf{I} & (t - t_0)\mathbf{I} & \frac{1}{2}(t - t_0)^2\mathbf{I} \\ \mathbf{O} & \mathbf{I} & (t - t_0)\mathbf{I} \\ \mathbf{O} & \mathbf{O} & \mathbf{I} \end{bmatrix} \mathbf{s}_{(t_0)} \tag{6-44}$$

As a nonhomogeneous extension, each channel of the above system can be assumed to experience its own separate disturbance [e.g., note that the x axis state variables in (6-44), uncoupled from the states associated with the other two axes, can be expressed by (5-76) with the random vector $\mathbf{e}$ containing velocity and acceleration excitation with fixed spectral densities, the latter being caused by variations in the tracked object's acceleration vector]. By integrating (5-49) analytically, the contribution of noise excitation to uncertainty covariances during a data window could then be obtained (Ex. 6-16) by an extension of (5-92). Effects of INS errors ($\tilde{\mathbf{Q}} \neq \mathbf{0}$) can be considered as included in the aforementioned spectral densities. Note, therefore, that the lenient requirements for knowledge of own-ship motion enable usage of doppler or air data (plus HARS) in this mode, when INS information is unavailable.

A more common practice in air-to-air tracking is to base the dynamics of $\boldsymbol{Z}_T$ on (5-61). Description of that scheme now follows from previous developments. Just as (6-44) is a three-dimensional re-expression of (5-64), the matrices in (5-62) and (5-63) can be applied to all three reference frame directions (Ex. 6-17). Standard calculus readily provides a closed-form solution to (5-49) under those conditions but, for applications with high radar data rates, approximate dynamics (Ex. 6-16a) may suffice for covariance extrapolation.

The relations presented here apply in general to an inertially oriented or a slowly rotating reference frame, regardless of radar mechanization approaches (e.g., gyro stabilization of a rotating antenna dish, or computational stabilization of an electronically steerable array). To cite an example, let (2-67) characterize a reflective antenna beam, computationally slaved to a geographic reference via frequent adjustment of $\mathscr{A}_D$ and $\mathscr{E}_D$ based on instantaneous gimbal angle readings (Φ, Θ, and Ψ, in addition to the antenna deflections). The indicated sightline from an error-free radar would then correspond to relative position states, as resolved along beam-referenced axes:

$$(S_1{}^2 + S_2{}^2 + S_3{}^2)^{-1/2} \mathscr{T}^{\mathrm{T}} \mathbf{T}_{A/G} \begin{bmatrix} S_1 \\ S_2 \\ S_3 \end{bmatrix} \doteq \begin{bmatrix} 1 \\ \epsilon_A \\ -\epsilon_E \end{bmatrix} \tag{6-45}$$

where ϵ_A, ϵ_E conform to Fig. 6-2.

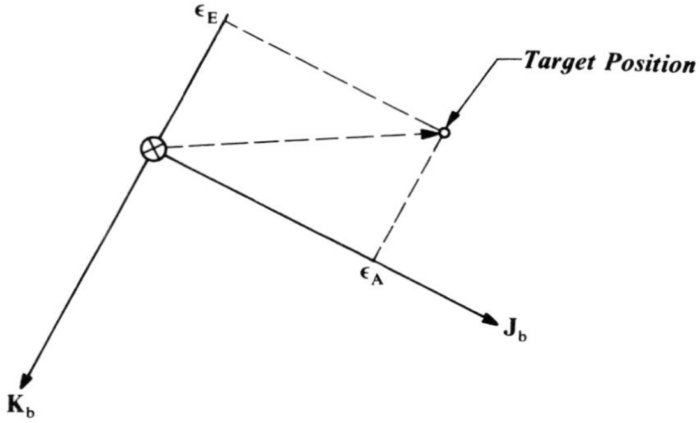

FIG. 6-2. LOS error convention (beam axis $\mathbf{I}_b$ is the inward normal to the plane of the figure).

APPENDIX 6.A: REMARKS CONCERNING NONGEOGRAPHIC REFERENCE MODES

Although much of the development in this chapter can easily be modified for wander azimuth or for nonrotating reference frames, certain portions (particularly Section 6.3) are more firmly based on the geographic reference. In the wander azimuth case, direct modification based on Section 3.6 is complicated by the need for a position *state* formulation in terms of three variables [rather than four, using (3-71), or nine, as in (3-63)]. While a four- or nine-parameter formulation is quite acceptable for navigation between updates, it cannot be used in the update equation itself (5-85) nor for covariance extrapolation (5-49). For those operations, velocity and reference orientation error equations would have to be re-expressed as explicit functions of the three variables chosen to define position.

One possible set of position states, immune to singularity, would be the Cartesian components of $\mathbf{R}_I$. The method of computing these components and the corresponding velocities, for both the gimballed platform and the strapdown implementation, was described for a nonrotating reference in Section 3.1. Unlike the geographic and wander azimuth systems, however, the nonrotating reference approach has not yet been examined in this book for error growth characteristics (e.g., presence of Schuler oscillations). The following development summarizes these characteristics.

ERROR PROPAGATION IN SPACE STABLE SYSTEMS

The well-known equivalence of natural dynamics for errors in locally and inertially stabilized systems will now be demonstrated. Consider an attempt to compute $\ddot{\mathbf{R}}_I$ from (3-2), and from (3-29) re-expressed as

$$\mathbf{G} = \mathbf{g} + \boldsymbol{\gamma}, \qquad \boldsymbol{\gamma} \triangleq \boldsymbol{\omega}_s \times (\boldsymbol{\omega}_s \times \mathbf{R}) \tag{6-46}$$

on the basis of an apparent specific force $\hat{\mathbf{A}}_I$, in the presence of unknown deviation between apparent $(\hat{\mathbf{I}}_I, \hat{\mathbf{J}}_I, \hat{\mathbf{K}}_I)$ and true inertial reference axes:

$$\hat{\ddot{\mathbf{R}}}_I = \mathbf{T}_{\hat{I}/I}\hat{\mathbf{A}}_I + \hat{\mathbf{T}}_{I/G}\hat{\mathbf{g}} + \hat{\boldsymbol{\gamma}}_I, \qquad \hat{\boldsymbol{\gamma}}_I = \boldsymbol{\omega}_E \times (\boldsymbol{\omega}_E \times \hat{\mathbf{R}}_I) \tag{6-47}$$

From combination of (3-2) with these expressions (Ex. 6-18) and

$$\hat{\mathbf{g}} = \mathbf{g} - \tilde{\mathbf{g}}, \qquad \hat{\mathbf{A}}_I = \mathbf{A}_I - \boldsymbol{n}_a \tag{6-48}$$

navigation error here conforms to the differential equation

$$\tilde{\ddot{\mathbf{R}}}_I \doteq (\mathbf{T}_{I/G} - \hat{\mathbf{T}}_{I/G})\mathbf{g} + \mathbf{T}_{I/G}\tilde{\mathbf{g}} + \tilde{\boldsymbol{\gamma}}_I + (\mathbf{I} - \mathbf{T}_{\hat{I}/I})\mathbf{A}_I + \boldsymbol{n}_a \tag{6-49}$$

The first two terms on the right of this equation contain the Schuler and the altitude divergence phenomena, respectively, since (Ex. 6-19)

$$g\left(\begin{bmatrix}\mathbf{I}_I\cdot\mathbf{K}_G\\ \mathbf{J}_I\cdot\mathbf{K}_G\\ \mathbf{K}_I\cdot\mathbf{K}_G\end{bmatrix} - \begin{bmatrix}\widehat{\mathbf{I}_I\cdot\mathbf{K}_G}\\ \widehat{\mathbf{J}_I\cdot\mathbf{K}_G}\\ \widehat{\mathbf{K}_I\cdot\mathbf{K}_G}\end{bmatrix}\right) = g\left(\frac{-\mathbf{R}_I}{|\mathbf{R}_I|} - \frac{(-\hat{\mathbf{R}}_I)}{|\hat{\mathbf{R}}_I|}\right)$$

$$\doteq \frac{-g}{|\mathbf{R}_I|}\{\tilde{\mathbf{R}}_I - (\tilde{\mathbf{R}}_I\cdot \boldsymbol{K})\boldsymbol{K}\}, \qquad \boldsymbol{K} \triangleq \frac{-\mathbf{R}_I}{|\mathbf{R}_I|} \tag{6-50}$$

and, with vertical deflections and gravity anomalies included in $\boldsymbol{n}_a$ instead of $\tilde{\mathbf{g}}$,

$$g\begin{bmatrix}\mathbf{I}_I\cdot\mathbf{K}_G\\ \mathbf{J}_I\cdot\mathbf{K}_G\\ \mathbf{K}_I\cdot\mathbf{K}_G\end{bmatrix}\left(\frac{2\tilde{\mathbf{R}}_I\cdot \boldsymbol{K}}{|\mathbf{R}_I|}\right) = \frac{2g}{|\mathbf{R}_I|}(\tilde{\mathbf{R}}_I\cdot \boldsymbol{K})\boldsymbol{K} \tag{6-51}$$

In substituting these relations into (6-49), the vertical and horizontal position uncertainties are denoted

$$\tilde{\mathbf{R}}_{\mathrm{IV}} = (\tilde{\mathbf{R}}_{\mathrm{I}} \cdot \boldsymbol{K})\boldsymbol{K}, \qquad \tilde{\mathbf{R}}_{\mathrm{IH}} = \tilde{\mathbf{R}}_{\mathrm{I}} - \tilde{\mathbf{R}}_{\mathrm{IV}} \tag{6-52}$$

respectively, while directional uncertainty is represented by

$$\mathbf{T}_{\hat{\mathrm{I}}/\mathrm{I}} = \mathbf{I} - (\boldsymbol{\upsilon} \times \quad) \tag{6-53}$$

and the nav error differential equation becomes

$$\ddot{\tilde{\mathbf{R}}}_{\mathrm{I}} - \tilde{\gamma}_{\mathrm{I}} \doteq \frac{-g}{|\mathbf{R}_{\mathrm{I}}|}\tilde{\mathbf{R}}_{\mathrm{IH}} + \frac{2g}{|\mathbf{R}_{\mathrm{I}}|}\tilde{\mathbf{R}}_{\mathrm{IV}} + \boldsymbol{\upsilon} \times \mathbf{A}_{\mathrm{I}} + \boldsymbol{n}_{\mathrm{a}} \tag{6-54}$$

Unlike preceding analyses, involving rotating reference frames, the Schuler phenomenon appears wholly within the translational equation here, and is absent from the rotational relation (Ex. 6-20)

$$d\boldsymbol{\upsilon}/dt = \dot{\boldsymbol{\upsilon}} = \boldsymbol{n}_{\omega} \tag{6-55}$$

Regardless of mathematical viewpoint, the first two terms on the right of (6-54) demonstrate the benefit to horizontal navigation, and the degradation of the vertical channel, associated with gravity.

At nonpolar latitudes, the left of (6-54) can represent imperfect knowledge of the quantity (Ex. 6-21)

$$\ddot{\mathbf{R}}_{\mathrm{I}} - \gamma_{\mathrm{I}} = \mathbf{T}_{\mathrm{I/G}}(\mathbf{A} + \mathbf{G} - \gamma) = \mathbf{T}_{\mathrm{I/G}}(\dot{\mathbf{V}} - \mathbf{f}) \tag{6-56}$$

so that, by transformation into the geographic frame,

$$\dot{\tilde{\mathbf{V}}} - \tilde{\mathbf{f}} + \frac{g}{|\mathbf{R}|}\begin{bmatrix} \tilde{x}_{\mathrm{N}} \\ \tilde{x}_{\mathrm{E}} \\ -2\tilde{x}_{\mathrm{V}} \end{bmatrix} = \mathbf{T}_{\mathrm{G/I}}(\boldsymbol{\upsilon} \times \mathbf{A}_{\mathrm{I}} + \boldsymbol{n}_{\mathrm{a}}) \tag{6-57}$$

with $\tilde{x}_{\mathrm{N}}, \tilde{x}_{\mathrm{E}}, \tilde{x}_{\mathrm{V}}$ denoting position uncertainty resolved along $(\mathbf{I}_{\mathrm{G}}, \mathbf{J}_{\mathrm{G}}, \mathbf{K}_{\mathrm{G}})$, respectively. This relation emphasizes the similarity between space stable and locally slaved (Section 6.3) system error propagation, extending even to variations in **f**, for the natural (zero forcing function) dynamics. Note, however, two implications involving the forcing function in (6-57):

(1) In general, all instruments on an inertially stabilized platform experience an appreciable component of the lift force, while rotation-sensitive errors can be ignored.

(2) Constant components of inertial instrument errors, defined here in the stabilized coordinate frame, are transformed into time-varying effects via $\mathbf{T}_{\mathrm{G/I}}$ in (6-57). The long-term result is detrimental since rates that drive this transformation [recall (2-29) and (3-24)] are closely connected with the natural dynamics on the left of (6-57) (recall the description of natural

frequencies in Section 6.3). While correlation between forcing function and natural frequencies produces resonance, large amplitudes of long-period oscillations cannot develop within the time allotted to any mode here (Section 1.5). For further investigation of resonance, the reader is referred to pp. 159–163 of Ref. 6-3.

As with all modes, then, navigation with an inertial reference frame is a straightforward modification of other approaches already described. The same comment applies to a strapdown mechanization with the nonrotating reference, developed in Ref. 6-5* or easily adapted from methods presented earlier in this book.

APPENDIX 6.B: SUBOPTIMAL DAMPING

Consider an INS augmented with a separate source of indicated velocity $\hat{\mathbf{V}}_{\text{ref}}$, in which the velocity estimate to be integrated is adjusted by weighing its departure from this reference source; e.g., let (3-36) be replaced by

$$\dot{\hat{\mathbf{V}}} = (\mathbf{I} - \boldsymbol{\psi} \times \;)\hat{\mathbf{A}} + \hat{\mathbf{g}} + \hat{\mathbf{f}} + \mathbf{B}(\hat{\mathbf{V}}_{\text{ref}} - \hat{\mathbf{V}}) \tag{6-58}$$

where $\mathbf{B}$ denotes the suboptimal weighting matrix. By subtraction from (3-33), while using (3-34) and noting that

$$\hat{\mathbf{V}}_{\text{ref}} - \hat{\mathbf{V}} \equiv \hat{\mathbf{V}}_{\text{ref}} - \mathbf{V} + \mathbf{V} - \hat{\mathbf{V}} = \tilde{\mathbf{V}} - \tilde{\mathbf{V}}_{\text{ref}} \tag{6-59}$$

the error equation becomes

$$\dot{\tilde{\mathbf{V}}} + \mathbf{B}\tilde{\mathbf{V}} = \boldsymbol{\psi} \times \mathbf{A} + \mathbf{n}_a + \tilde{\mathbf{g}} + \tilde{\mathbf{f}} + \mathbf{B}\tilde{\mathbf{V}}_{\text{ref}} \tag{6-60}$$

which permits immediate recognition of dynamic modifications in any nav mode. With a diagonal matrix of fixed damping coefficients, for example, constants could be subtracted from A_{33} and A_{44} of Table 6-1, while another term would be added to north and east velocity error forcing functions. Specific examples can illustrate the effectiveness and the limitations of damping, or generalized conditions (e.g., cross-axis coupling and/or transfer functions in $\mathbf{B}$) lead to a more involved analysis (e.g., pp. 188–193 of Ref. 6-3 illustrate damping of the azimuth error and of the 84-min oscillations). For present purposes, however, it suffices to note that (1) damping can be included in any nav mode, by straightforward adjustments of the dynamics matrix and the forcing function, and (2) the resulting

* While Ref. 6-5 provided extensive coverage of INS errors, gravity effects were included in the horizontal channel only (i.e., the Schuler phenomenon was present, but not altitude divergence); altitude was assumed known independently. Also, since gradual rotation of the local reference was ignored, results were only approximate even for durations much shorter than the specified maximum (about 2 hr).

change in dynamic behavior is exemplified by adapting (6-60) to (4-1), under conditions of a constant diagonal damping coefficient matrix and

$$\tilde{\mathbf{f}} = \tilde{\mathbf{g}} = \mathbf{0}, \qquad \boldsymbol{\varpi} = \mathbf{0} \tag{6-61}$$

so that

$$\begin{bmatrix} \dot{\tilde{V}}_{\mathrm{N}} \\ \dot{\tilde{V}}_{\mathrm{E}} \\ \dot{\psi}_{\mathrm{N}} \\ \dot{\psi}_{\mathrm{E}} \end{bmatrix} = \begin{bmatrix} -B_{\mathrm{N}} & 0 & 0 & -g \\ 0 & -B_{\mathrm{E}} & g & 0 \\ 0 & -1/\mathscr{R} & 0 & 0 \\ 1/\mathscr{R} & 0 & 0 & 0 \end{bmatrix} \begin{bmatrix} \tilde{V}_{\mathrm{N}} \\ \tilde{V}_{\mathrm{E}} \\ \psi_{\mathrm{N}} \\ \psi_{\mathrm{E}} \end{bmatrix} + \begin{bmatrix} n_{\mathrm{a}1} + B_{\mathrm{N}} \tilde{V}_{\mathrm{R}1} \\ n_{\mathrm{a}2} + B_{\mathrm{E}} \tilde{V}_{\mathrm{R}2} \\ n_{\omega 1} \\ n_{\omega 2} \end{bmatrix} \tag{6-62}$$

which readily produces equations in classical form, e.g., if the forcing functions are differentiable (Ex. 6-22),

$$\ddot{\tilde{V}}_{\mathrm{N}} + B_{\mathrm{N}} \dot{\tilde{V}}_{\mathrm{N}} + \frac{g}{\mathscr{R}} \tilde{V}_{\mathrm{N}} = g\dot{n}_{\omega 2} + \dot{n}_{\mathrm{a}1} + B_{\mathrm{N}} \dot{\tilde{V}}_{\mathrm{R}1} \tag{6-63}$$

The aforementioned reference (Ref. 6-3) noted that the accuracy of $\hat{\mathbf{V}}_{\mathrm{ref}}$ must rival that of the INS for effective damping. This can be illustrated by considering large degradations for the last term in (6-63); that term would then be dominant (so that $\tilde{V}_{\mathrm{N}}$ would approach $\tilde{V}_{\mathrm{R}1}$) unless the coefficient B_{N} were small enough to weaken the damping term on the *left* of the equation.

EXERCISES

6-1 Draw an analogy between Figs. 6-1 and 1-14. Relate to (3-20) and Ex. 1-6d. Explain why variations in observation accuracy and scheduling are easily accommodated. Show that uncoupling of the east channel is less natural. Would the correction in Fig. 6-1 be needed if the sightline to Polaris were exactly parallel to $\mathbf{K}_{\mathrm{E}}$?

6-2 (a) Use (3-23), (5-76), and (5-77) to derive (6-4). Recall Ex. 5-33a.

(b) Draw an extended version of Fig. 6-1 for coupled north and east channels, using Fig. 5-4 as a guide. Let (3-20) and (3-21) be the model for (5-82).

6-3 (a) Show that the white noise spectral density $(200)(6080)^2/(3600)^2$ $(\mathrm{fps})^2/\mathrm{Hz}$ produces about 200 ft rms position uncertainty after 1 min. Compare versus position error contributed from 0.3-mr platform tilt.

(b) Apply the methods of Section 5.3.1 to generalize (6-4) for a doppler radar with groundspeed bias. If $\zeta = 90°$, why does the bias appear in the east channel?

6-4 Write a scalar version of (5-48) with a transition "matrix" of the form $\exp\{-(t-\tau)/\tau_W\}$, obtained from (5-61) and (II-55). Use a lower limit of $-\infty$ in the integral term, to verify the last footnote of Section 6.1.

6-5 (a) Show how the positive downward convention has affected the derivation of (6-6).

(b) Compute the effects of a 500-ft altitude uncertainty at 45° latitude in (6-6).

6-6 (a) Discuss the derivation of (6-9). Compare versus the type of analysis represented by (3-53), in regard to coordinates used, errors considered, and dynamic conditions assumed.

(b) Show that (6-11) implies a 500-sec data window.

6-7 (a) Relate altitude divergence error in (6-17) to Ex. 5-31e. Show that the uncertainty in centripetal acceleration from sidereal rotation has not been neglected.

(b) Demonstrate the influence of a positive downward convention on the derivation of Table 6-1.

(c) Verify rows 1–4 and 8–9 of Table 6-1 by differentiation.

(d) Compare (6-17) versus (4-78), in regard to types of errors considered.

6-8 (a) Interpret (6-19) according to the pertinent discussion given in Section 1.4.

(b) Compare (6-20) versus (4-73), and explain differences by conditions assumed for each.

(c) Extend Ex. 6-2b to include INS misorientation states. Which discussion in Section 5.4.2 is related to usage of $\mathscr{R}$ throughout Table 6-1?

6-9 Verify (6-22)–(6-24).

6-10 Multiply the eighth column of Table 6-1 by a 400-ft altitude uncertainty. Show that, at moderate latitudes, the effect is overshadowed by a 1-min latitude uncertainty, 1-fps horizontal velocity uncertainty at 1000 fps, and 0.2-mr misorientation,

$$\tilde{h}/\mathscr{R} \ll \tilde{\lambda}, \qquad \tilde{h}/\mathscr{R} \ll \tilde{V}_H/V_H, \qquad \tilde{h}/\mathscr{R} \ll \psi$$

6-11 (a) Consider the combined effect of gyro scale factor, misalignment, and mass unbalance drifts, in cruise segments separated by sharp turns. Recall Ex. 4-19c,d and discuss prospects of identifying each drift source separately.

(b) If six augmenting states could in principle represent gyro misalignment, how many more augmenting states could theoretically be introduced for accelerometer misalignment? Use (4-49) to derive an answer.

(c) Recall Ex. 4-17b. Could in-flight calibration detect biases remaining from imperfect laboratory procedures? Even after both of these measures are taken, could some residual mass unbalance, misalignment, scale factor error, etc., remain uncompensated? How does this residual effect interact with maneuvers designed to identify separate drift sources?

(d) On the basis of Ref. 6-4, p. 18, sequentially correlated acceleration errors of order $10^{-4}g$ can arise from mechanization of (3-32), since vertical deflections can be 20 $\widehat{\text{sec}}$. How would this affect in-flight calibration of level accelerometers? Discuss also the implications of $3(10^{-5})g$ gravity anomalies in the vertical channel.

6-12 (a) Extend Ex. 6-8c to include in-flight INS calibration states as well as vertical channel coupling. Compare conditions applicable to (6-25) versus (4-4).

(b) Recall the discussion regarding the fitting of data to a model, prior to Section 5.4.1. Does a finite time constant assumption offer greater flexibility than a fixed bias model?

(c) Consider the closing statement of Section 6.4. Use (3-48) to support that statement, under cruise conditions.

(d) Recall the discussion of units at the end of Section 5.4.2. Should unit conversions be considered for augmenting states in (6-25)?

6-13 (a) Apply (6-28) to **R**. Compare versus (1-25).

(b) Substitute estimated velocity components into (6-27) and assess the effect of uncertainty in the initial ground track angle.

(c) Apply the principles of Ex. 5-35 to the point nav mode with 5-min data window.

6-14 (a) If errors in **q** and **Q** are ignored in deriving (6-33), are all INS imperfections assumed to be negligible?

(b) Apply (5-90), (5-91), and (5-92) to (6-33).

6-15 (a) Rewrite (6-43) with circumflexes to denote estimated quantities. Perform a subtraction to obtain (6-44).

(b) Identify a difference between (6-43) and (1-10); explain by choice of origins.

6-16 (a) Extend (5-91) to obtain a closed-form solution for (5-49) with the transition matrix in (6-44).

(b) What are the prospects for in-flight INS calibration in the air-to-air track mode? Explain by applying (5-61) to components of $\boldsymbol{Z}_{\mathrm{T}}$.

(c) Reinterpret the tracking operation with air data used for ownship motion. Are the velocity states then associated with relative airspeed?

Do the acceleration states then represent target maneuvers with respect to a moving air mass? Is this a critical item?

6-17 (a) Extend (6-44) with a three-dimensional generalization of (5-63). With (5-91) extended to include random maneuvers governed by (5-61), use Ex. 6-4 to derive a mean squared acceleration of the form $(\eta_3 \tau_3/2000)g^2$.

(b) Are the dynamics in Section 6.6 linear? Would linearization be required for re-expression in spherical coordinates referenced to the relative position vector? Could the transformed equations involve expressions in a form similar to (2-37) and (4-58)?

(c) Let $(\mathbf{I}_A, \mathbf{J}_A, \mathbf{K}_A)$ deviate from radar gimbal mechanical axes (RGMA), and redefine $\mathscr{T}$ in (6-45) as the transformation between RGMA and radar beam coordinates. Should $\mathbf{T}_{A/G}$ then be replaced by the transformation between INS and RGMA reference axes?

(d) Show how the analysis in Section 6.6 easily reduces to tracking from a ship, or from an air traffic control (ATC) observation point. Discuss also a possible extension to track-while-scan (TWS) operation, wherein a time-shared radar beam illuminates multiple objects in repeated succession.

6-18 Show that $\mathbf{K}_G$ is always well defined, even when $\mathbf{I}_G$ and $\mathbf{J}_G$ are not. Recall the direction of gravity. Establish the generality of (6-47) and (6-49) accordingly. Verify (6-49).

6-19 (a) Relate (6-50) to the first term on the right of (6-49), using the positive downward convention and the bottom row of the matrix in (2-3).

(b) Express the approximation

$$\frac{\hat{\mathbf{R}}_I}{|\hat{\mathbf{R}}_I|} \doteq \frac{\mathbf{R}_I - \tilde{\mathbf{R}}_I}{|\mathbf{R}_I|\{1 - 2\mathbf{R}_I \cdot \tilde{\mathbf{R}}_I/|\mathbf{R}_I|^2\}^{1/2}}$$

as a Taylor series to obtain the right of (6-50). What vector in that equation, resolved along inertial axes, points along the geodetic vertical?

(c) Use Ex. 3-21a to justify (6-51).

6-20 Draw an analogy between (6-53) and (2-15). Compare (6-55) versus (6-20).

6-21 Use (3-33) and (3-79) to obtain (6-56).

6-22 (a) Discuss differences between (6-17) and (6-60).

(b) Could the left of (6-63) be described in terms of a damping ratio and a natural frequency, as in (4-55)? Would this procedure facilitate the analysis of fixed errors in reference velocity?

(c) Can (II-55) be solved in closed form, for the dynamics matrix in (6-62)? With fixed spectral densities for $\mathbf{n}_a$, $\mathbf{n}_\omega$, and $\tilde{\mathbf{V}}_{\text{ref}}$, could (5-48) be used to analyze random error effects?

(d) Would a reference velocity source, far more accurate than the INS, seem practical and economical?

(e) Does the system of Fig. 1-14 need velocity damping?

REFERENCES

6-1. Richman, J., and Friedland, B., "Design of Optimum Mixer-Filter for Aircraft Navigation Systems," presented at 19th Ann. NAECON Conf., Dayton, Ohio, May 15–17, 1967.

6-2. Bryson, A. E., and Bobick, J. C., "Improved Navigation by Combining VOR/DME Information and Air Data," *AIAA J. Aircraft* **9** (6), June 1972, pp. 420–426.

6-3. Broxmeyer, C., *Inertial Navigation Systems*. New York: McGraw-Hill, 1964.

6-4. Kayton, M., and Fried, W. R. (eds.), *Avionics Navigation Systems*. New York: Wiley, 1969.

6-5. Farrell, J. L., "Analytic Platforms in Cruising Aircraft," *AIAA J. Aircraft* **4** (1), Jan.–Feb. 1967, pp. 52–58.

Chapter 7

Navigation Measurements

7.0

Modern aided inertial systems combine short-term INS accuracy with measurements obtained from navaids, such as airborne receivers or optical sensing devices. Mathematical representation and typical accuracies will now be given for navaids commonly used in the discrete-update estimation scheme. Although some existing systems serve as examples, restriction to specific mechanizations is generally avoided. It follows that bias error sources are not fully analyzed here. Also deemphasized or omitted in this chapter are

(1) inertial measurements, except where they appear as observables (e.g., from antenna-mounted gyros);
(2) air data (discussed briefly in preceding chapters and in Appendix III);
(3) observations requiring contour matching (e.g., magnetic or terrain profiles), for an excursion of some length over the earth;
(4) errors in *processing* navaid data (which, in a well-designed system, should not be a dominant source of degradation); and
(5) delays due to sensor lag or data processing (which can limit control functions *using* nav information, but are not regarded as errors here).

In general, an explicit characterization of observations in terms of the state is stressed. This is appropriate, not only for evaluating partial derivatives (Ex. 7-1), but also for properly assessing the information content of each measurement. A few counterexamples, with "pseudomeasurements" formed from quantities actually observed, are shown to be at a disadvantage. Radar observations and landing aids are described here in terms of local state variables (denoted by **S**, **s**). All other communication signals, as well as celestial, magnetic, and external initialization data, are associated with earth nav mode variables (denoted by **X**, **x**). A few illustrative partial derivatives for estimation are also presented.

7.1 LOCALLY REFERENCED OBSERVATIONS

Radar measurements are applicable to landmarks of known latitude and longitude, as well as designated points or tracked objects of unknown location relative to earth. Chosen for illustration in this section are velocities resolved along axes that are independent of north direction, and position with respect to a fixed or moving point, irrespective of latitude and longitude. ILS observations are similarly characterized, with a runway centerline used as the reference azimuth direction. Local state variables for point nav consist of position coordinates for an airborne IMU at the location in Fig. 7-1

$$\boldsymbol{R}_S \triangleq \begin{bmatrix} S_1 \\ S_2 \\ S_3 \end{bmatrix} \tag{7-1}$$

and velocity states defined by

$$\dot{\boldsymbol{R}}_S = \begin{bmatrix} S_4 \\ S_5 \\ S_6 \end{bmatrix} = \begin{bmatrix} \dot{S}_1 \\ \dot{S}_2 \\ \dot{S}_3 \end{bmatrix} \tag{7-2}$$

A succinct description of each information source is provided in this text. Applicable radar theory is summarized in Appendices 7.A–7.C, and more fully presented in Ref. 7-1.

7.1.1 Doppler Radar

Of all sensors considered in this chapter, only the doppler radar can provide *either* continuous velocity data *or* updates in an estimation scheme. Mathematical definition of a doppler measurement is expressed here in terms

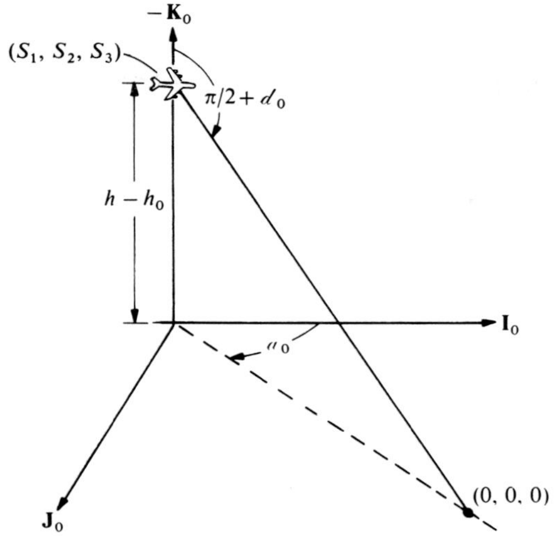

FIG. 7-1. Aircraft position with respect to reference origin.

of *closing range rate** along an antenna beam axis. For a gimballed antenna, with azimuth and depression angles denoted $\mathscr{A}$, $\mathscr{D}$, respectively, a unit vector along the beam for the jth doppler measurement is

$$[\mathscr{A}_j]_x^T[\mathscr{D}_j]_y \begin{bmatrix} 1 \\ 0 \\ 0 \end{bmatrix} = \begin{bmatrix} \cos \mathscr{D}_j \cos \mathscr{A}_j \\ \cos \mathscr{D}_j \sin \mathscr{A}_j \\ \sin \mathscr{D}_j \end{bmatrix} \tag{7-3}$$

as expressed in aircraft coordinates. The same vector, transformed into the $(\mathbf{I}_0, \mathbf{J}_0, \mathbf{K}_0)$ frame used for Fig. 7-1, is denoted

$$\boldsymbol{I}_j \triangleq \begin{bmatrix} I_{j1} \\ I_{j2} \\ I_{j3} \end{bmatrix} \triangleq \mathbf{T}_{0/A}[\mathscr{A}_j]_x^T[\mathscr{D}_j]_y \mathbf{1}_1, \qquad \mathbf{1}_1 \triangleq \begin{bmatrix} 1 \\ 0 \\ 0 \end{bmatrix} \tag{7-4}$$

and the jth closing range rate measurement Y_j is

$$Y_j = \dot{\boldsymbol{R}}_S{}^T \boldsymbol{I}_j = I_{j1} S_4 + I_{j2} S_5 + I_{j3} S_6 \tag{7-5}$$

A velocity vector can therefore be determined by linear estimation from three noncoplanar beams. Whereas avoidance of the horizon and of the area directly beneath the aircraft precludes orthogonality of the three beams, directional spread should be as great as the constraints allow (Ex. 7-2).

Doppler navigation systems have customarily been mechanized using

* Range is the distance between an antenna and the surface it illuminates. Closing range rate is the rate at which this range is decreasing, i.e., the component of relative velocity along the beam.

continuous-wave (CW) transmission, through three or four simultaneously generated beams, with outputs expressed in terms of groundspeed V_H and drift angle:

$$V_H = (S_4{}^2 + S_5{}^2)^{1/2} \tag{7-6}$$

$$\text{drift angle} = \arctan\left(\frac{S_4 \sin \hat{\zeta}_0 + S_5 \cos \hat{\zeta}_0}{S_4 \cos \hat{\zeta}_0 - S_5 \sin \hat{\zeta}_0}\right) - \Psi \tag{7-7}$$

Although estimation can be implemented in this form, the added transformations introduce unnecessary complexity and potential degradations, while nonlinearities in (7-6) and (7-7) complicate the error statistics. Also, when used to correct INS errors, doppler observations need not be continuous. An occasional three-direction pointing *sequence*, obtained by scanning a time-shared radar beam, is adequate for update purposes (Refs. 7-2 and 7-3).

MBC energy dependence on backscatter reflectivity and distance suggests that doppler data can be lost under conditions of specular reflection, high altitude, or shallow antenna depression. The information is commonly lost when flying high over calm seas. Also, at shallow depression angles, unsymmetrical energy density within the illuminated area can introduce appreciable frequency offset. Additional possible error sources include quantization of angle data used in (7-4), mechanical misalignments (e.g., traceable to gimbal pickoff mounting), electrical misalignments (e.g., radome refraction), and oscillator bias in a frequency track loop. Despite degradations, horizontal velocity accuracies to within a few fps are typical.

7.1.2 Conventional Radar Fixes

Radar systems transmit electromagnetic energy through an antenna and monitor the characteristics of reflections received. One possible observation is the range rate just described. An antenna beam can be oriented along chosen *directions* according to (7-4), or directed at a designated *point* such as the origin in Fig. 7-1. The following discussion pertains to a beam orientation that is nominally collinear with $\boldsymbol{R}_S$.

RANGE AND RANGE RATE OBSERVATIONS

With a radar beam pointed at the origin in Fig. 7-1, a range Y_R can be measured,

$$Y_R = |\boldsymbol{R}_S| = (S_1{}^2 + S_2{}^2 + S_3{}^2)^{1/2} \tag{7-8}$$

and the rate at which this range is decreasing is denoted here by the observable closing range rate (Ex. 7-3),

$$Y_d = -\dot{\boldsymbol{R}}_S{}^T \boldsymbol{R}_S / |\boldsymbol{R}_S| = -(S_1 S_4 + S_2 S_5 + S_3 S_6)/(S_1{}^2 + S_2{}^2 + S_3{}^2)^{1/2} \tag{7-9}$$

The negative sign in (7-9) is consistent with accepted convention for closing doppler, and with the projection of relative velocity along a line from aircraft to origin (Ex. 7-4). In contrast with point nav, air-to-air tracking conventions locate the position reference at the tracking aircraft. Velocity states, however, are then defined in terms of the tracked object relative to the tracking vehicle. Thus, closing doppler as well as range observations, as expressed above, are applicable to both Sections 6.5 and 6.6.

LINE OF SIGHT (LOS) OBSERVATIONS

Just as replacement of $\boldsymbol{I}_j$ by $(-\boldsymbol{R}_S/|\boldsymbol{R}_S|)$ in (7-5) produces (7-9), the antenna gimbal angles $\mathscr{A}$ and $\mathscr{D}$ become position observables Y_A, Y_D when the beam is pointed at the origin in Fig. 7-1. With $\boldsymbol{I}_j$ thus replaced in (7-4), the result can be expressed in aircraft coordinates as

$$\begin{bmatrix} \cos Y_D \cos Y_A \\ \cos Y_D \sin Y_A \\ \sin Y_D \end{bmatrix} = \frac{-1}{(S_1{}^2 + S_2{}^2 + S_3{}^2)^{1/2}} \mathbf{T}_{A/0} \begin{bmatrix} S_1 \\ S_2 \\ S_3 \end{bmatrix} \tag{7-10}$$

For air-to-air tracking, with the *antenna* at the origin, the nominal beam direction is along $+\boldsymbol{R}_S/|\boldsymbol{R}_S|$; also, the LOS observables are then indicated *departures* of the tracked object off the beam, already illustrated in Fig. 6-2. It is then convenient to define a beam-referenced coordinate frame $(\mathbf{I}_b, \mathbf{J}_b, \mathbf{K}_b)$ such that $\mathbf{I}_b$ coincides with the antenna axis and $\mathbf{J}_b$ points along the elevation gimbal axis (i.e., opposite the axis for the depression angle). The transformation between $(\mathbf{I}_b, \mathbf{J}_b, \mathbf{K}_b)$ and $(\mathbf{I}_0, \mathbf{J}_0, \mathbf{K}_0)$ is then expressed as

$$\mathbf{T}_{b/0} = [\mathscr{D}]_y^T [\mathscr{A}]_z \mathbf{T}_{A/0} \tag{7-11}$$

and the almost linear LOS observation/state relationship can be obtained from (Ex. 7-5)

$$\begin{bmatrix} \epsilon_A \\ \epsilon_E \end{bmatrix} \doteq \begin{bmatrix} 0 & 1 & 0 \\ 0 & 0 & -1 \end{bmatrix} \mathbf{T}_{b/0} \boldsymbol{R}_S/|\boldsymbol{R}_S| \tag{7-12}$$

Note that, in applying these measurements to (5-84) and (5-85), the predicted observation $\mathbf{Y}_m^-$ should be determined by substituting a priori position state estimates into (7-12).

Gyros mounted with sensitive axes along $\mathbf{J}_b$ and $\mathbf{K}_b$ can measure absolute angular rates (ω_{be}, ω_{ba} respectively) of the range vector $\boldsymbol{R}_S$, when the beam is pointed toward the origin in Fig. 7-1. As expressed in the $(\mathbf{I}_0, \mathbf{J}_0, \mathbf{K}_0)$ frame, the unit vector $\boldsymbol{I}$ along $\mathbf{I}_b$ is then equal to $-\boldsymbol{R}_S/|\boldsymbol{R}_S|$, which, upon differentiation, yields

$$\mathbf{T}_{0/b} \begin{bmatrix} 0 \\ -\omega_{ba} \\ \omega_{be} \end{bmatrix} = \frac{\dot{\boldsymbol{R}}_S}{|\boldsymbol{R}_S|} - \frac{\dot{\boldsymbol{R}}_S \cdot \boldsymbol{R}_S}{|\boldsymbol{R}_S|^3} \boldsymbol{R}_S \tag{7-13}$$

or

$$\begin{bmatrix} \omega_{\text{be}} \\ \omega_{\text{ba}} \end{bmatrix} = \begin{bmatrix} 0 & 0 & 1 \\ 0 & -1 & 0 \end{bmatrix} \mathbf{T}_{\text{b/0}} \left(\frac{\dot{\boldsymbol{R}}_{\text{S}}}{|\boldsymbol{R}_{\text{S}}|} - \frac{\dot{\boldsymbol{R}}_{\text{S}} \cdot \boldsymbol{R}_{\text{S}}}{|\boldsymbol{R}_{\text{S}}|^3} \boldsymbol{R}_{\text{S}} \right) \tag{7-14}$$

With the beam nominally parallel to $\boldsymbol{R}_{\text{S}}$, the term proportional to $\boldsymbol{R}_{\text{S}}$ can be ignored (Ex. 7-6), and the result is written in terms of observables $Y_{\omega\text{e}}$, $Y_{\omega\text{a}}$ as

$$\begin{bmatrix} Y_{\omega\text{e}} \\ Y_{\omega\text{a}} \end{bmatrix} \doteq \frac{1}{(S_1{}^2 + S_2{}^2 + S_3{}^2)^{1/2}} \begin{bmatrix} 0 & 0 & 1 \\ 0 & -1 & 0 \end{bmatrix} \mathbf{T}_{\text{b/0}} \begin{bmatrix} S_4 \\ S_5 \\ S_6 \end{bmatrix} \tag{7-15}$$

TIME-SHARED RADAR MECHANIZING CONSIDERATIONS

A multimode radar (MMR) can provide position in spherical coordinates from (7-8) and (7-10) and/or velocity information from (7-9) and (7-15). Cartesian velocity components can also be determined, on the basis of (7-5). The full set of measurements need not be simultaneous; in fact, economy dictates an observational sequence with one radar beam, subjected to computer control of its parameters and beam steering commands. Furthermore, a tracking operation using (7-8), (7-9), and (7-12) can be interleaved with the point nav mode for an electronically steerable array.

With wavelengths chosen for specific applications, radar information is less sensitive to conditions (e.g., adverse weather) that could deactivate optical sensors. Rain can still be a potential degradation, as well as clutter from terrain or sea below. Electronic countermeasures (ECM) or other sources of false target information can render the radar temporarily inoperative.

Wideband pulsed radar can provide range accuracies to within a hundred feet or less, while narrowband FM ranging may be in error by several hundred feet. Range rates are commonly accurate to within a few fps. These accuracies are obtainable only when the ambiguities described in Appendices 7.A and 7.B are resolved.

LOS rate observations can be of varying quality, depending on gyro drift characteristics (Chapter 4). A similar statement applies to sightline information, with errors of less than 1 mr to more than 1°. For fine measurements of angular rate (e.g., obtained from high quality gyros) or angles (e.g., exploiting steep gradients or nulls in antenna directivity patterns), effectiveness can be limited by resolution of gimbal pickoffs (Ex. 7-7).

Plurality of attitude and sensor gimbal angles that influence radar observations will hamper complete in-flight calibration. Mounting misalignments sometimes combine additively, such as $\tilde{\Psi}$ and $\tilde{\mathscr{A}}$ in level flight, but conditions for combining bias errors will not be consistent in general.

7.1.3 Imaging Systems

When MMR operation includes provisions for synthetic aperture radar (SAR) mapping, its characteristics in that mode permit classification as an imaging device. Other instruments in this class include television (TV) and infrared (IR) devices, collectively referred to here as electro-optical (E/O) sensors. Imaging systems effectively combine signal intensity time histories with a "scan function," which converts the time reference into position on a surface. Because different wavelengths affect both the type of information provided and the extent of nav system interaction, E/O and SAR imaging will now be discussed separately.

SAR

A plot of relative signal intensity versus position on a grid can be displayed as a map image. Radar imaging (Ref. 7-4), with grid cells smaller than the area illuminated by the beam, uses precise characterization of rf phase relationships. Typical parameter values are a few centimeters and a few seconds for rf wavelength and SAR frame time, respectively. It follows that, in marked contrast to most INS applications, SAR mapping depends critically on *short-term* nav system accuracy to perform its basic function. For years there was also strict dependence on flight path and beam stabilization, but digital implementation (Ref. 7-5) removes these constraints and permits mapping with variable azimuth orientation during maneuvers (see Appendix 7.C for a basic theoretical explanation of SAR imaging).

For an operational definition of updating with SAR, attention is again drawn to Fig. 7-1. Consider the origin located at a consistently identifiable object, at which the radar display is also centered. The nav system, while providing essentially continuous position and velocity data, can simultaneously drive a cursor toward the display center. At the same time, as rapid changes in aircraft/map area geometry occur during flight, airborne processing algorithms repeatedly compute range and range rate corresponding to each image cell. On any SAR frame, the cursor may be repositioned at the origin on the display. The amount of horizontal and vertical adjustment required for repositioning (Ex. 7-8) constitutes a 2×1 residual data vector based on (7-8) and (7-9), with accuracy determined by the distance across an image cell. Whereas range increments are fixed by design (e.g., ratio of map size to the number of cells), the doppler spread depends upon conditions such as speed and azimuth beam orientation.

E/O SENSOR DATA

Pointing of an optical beam at the origin in Fig. 7-1, by manual or automatic means, can be associated with the following types of E/O observations:

(1) laser ranging, in conformance with (7-8),
(2) angular rates, in conformance with (7-15),
(3) LOS, in conformance with (7-10),
(4) LOS with a reversed gimbal order.

When the gimbal order in (7-3) is reversed, an observable/state relation can be expressed in terms of azimuth $\mathcal{A}$ and elevation $\mathcal{E}$ angles as (Ex. 7-9)

$$[\mathcal{E}]_y^{\mathrm{T}}[\mathcal{A}]_x^{\mathrm{T}}\,\mathbf{1}_1 = \begin{bmatrix} \cos Y_a \cos Y_e \\ \sin Y_a \\ -\cos Y_a \sin Y_e \end{bmatrix} = \frac{-1}{(S_1{}^2 + S_2{}^2 + S_3{}^2)^{1/2}}\,\mathbf{T}_{A/O} \begin{bmatrix} S_1 \\ S_2 \\ S_3 \end{bmatrix} \tag{7-16}$$

Sensor configurations can also employ reflective devices (Section 2.4).

COMPARISON OF OPTICAL AND RADAR IMAGERY

The night visibility characteristic of IR devices is not provided by TV sensors, except those of the "low intensity level" variety. SAR can function at night and in adverse weather.

Wavelengths of E/O devices ensure an intrinsic fine resolution. SAR can subdivide an area illuminated by a radar beam, only through precise knowledge of rf phase relationships.

Intrinsic high resolution of E/O devices generally produces LOS accuracies limited by pickoff resolution, but SAR fix accuracy depends directly on image cell size. This resolution is controlled by few parameters (e.g., the number of cells, map size, aircraft/map area geometry, speed), but recognition is subject to a host of other influences (e.g., static and transient INS misorientations, velocity quantization, attitude pickoff quantization,* directivity variations from antenna pointing transients; wandering of a phase center within an antenna, shock mount deformation or other vibratory relative motion between IMU and antenna). If degradations from these latter influences are acceptably low, and if the (in general, nonorthogonal) scan coordinates do not excessively distort the image, objects can be recognized within a swath of adequate size.

* Pickoff quantization effects grow more serious as the IMU/antenna displacement is increased. This degradation is greater for gimballed platforms than for strapdown systems. The opposite is true, however, for gyro misalignment and scale factor errors (Section 4.3.1), which affect the crucial short-term INS performance.

7.1.4 Instrument and Microwave Landing Systems (ILS, MLS)

During runway approach, a nav system estimates aircraft position with respect to a reference point such as Ⓞ in Fig. 7-2. The aircraft can then be guided to a prescribed descent path in a vertical plane containing the runway centerline. Typically the desired path ends in a straight segment with slight inclination, before a gradual leveling "flareout" arc prepares the aircraft for touchdown between Ⓞ and the *localizer* antenna line.

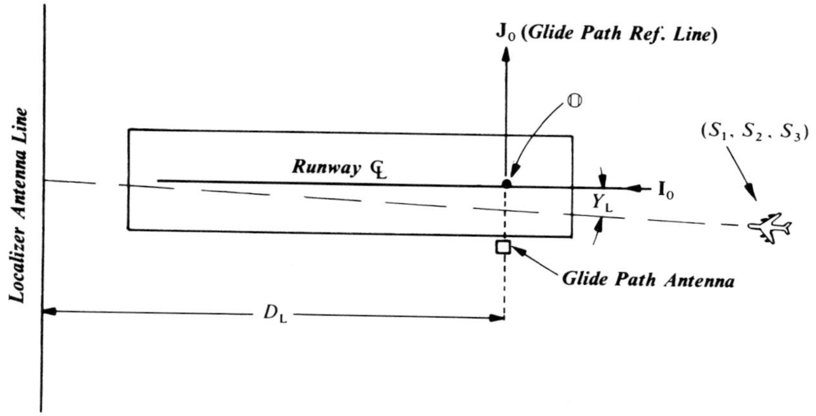

FIG. 7-2. Plan view of aircraft before landing.

Azimuth and elevation measurements obtained from landing aids are defined in terms of a local $(\mathbf{I}_0, \mathbf{J}_0, \mathbf{K}_0)$ coordinate frame. The azimuth reference $\mathbf{I}_0$ is now chosen coincident with the runway centerline. The origin Ⓞ is the intersection of $\mathbf{I}_0$ with a lateral line through the *glide path* antenna position. With the localizer antenna displacement beyond Ⓞ denoted by D_L, the azimuth measurement Y_L is expressible as

$$Y_L = \text{Arctan}\{S_2/(D_L - S_1)\}, \qquad |Y_L| \le Y_{L(\max)}, \qquad S_1 < 0 \qquad (7\text{-}17)$$

whereas, beyond a "proportional coverage limit" $Y_{L(\max)}$, only the algebraic sign of S_2 can be observed. The deviation Y_G from a desired descent path inclination is similarly restricted to "proportional" coverage between limits $\pm Y_{G(\max)}$; e.g., for a 2.5° nominal glide slope (Ex. 7-10),

$$Y_G = \text{Arctan}(S_3/S_1) - 2.5\pi/180, \qquad |Y_G| \le Y_{G(\max)} \qquad (7\text{-}18)$$

Aside from mechanization, principal differences between ILS and MLS are (Ref. 7-6)

(1) *Coverage.* $Y_{G(\max)}$ and $Y_{L(\max)}$ are about one or two degrees for

ILS, whereas MLS provides coverage over a sector that is an order of magnitude broader.

(2) *Accuracy.* ILS localizer and glide path measurements are accurate to within a fraction of a degree. MLS, however, must enable landing under conditions of extremely poor visibility. Typical objectives for positions near touchdown are lateral and vertical accuracies of order 10 and 5 ft, respectively.

(3) *Additional observations.* MLS information includes distance measurements (Ex. 7-11) in a form similar to (7-8). ILS installations sometimes include a *compass locator* antenna at a known position (D_{C1}, D_{C2}, 0), for measurements of angle Y_C between $\mathbf{I}_A$ and the sightline

$$Y_C = \text{Arccos}\left\{[(S_1 - D_{C1})^2 + (S_2 - D_{C2})^2 + S_3{}^2]^{-1/2}\mathbf{1}_1{}^{\mathrm{T}}\mathbf{T}_{A/0}\begin{bmatrix} D_{C1} - S_1 \\ D_{C2} - S_2 \\ -S_3 \end{bmatrix}\right\} \tag{7-19}$$

while ILS *marker beacons* provide direct observations of horizontal position. At the observed flyover time for a specified marker (Ex. 7-12),

$$Y_B = S_1 \tag{7-20}$$

Compass locator and marker beacon fixes, typically measured to within 2°, are less accurate than localizer and glide path observations. As another limitation of marker beacons, only two or three updates are generally obtainable, and each update requires passage through a wedge emanating upward from a fixed position. Beacon coverage broader than the 2° resolution is achieved through gradients in tone volume, which can be perceived during flight through the wedge. Nevertheless, compass locator and marker beacon fixes cannot yield the same performance as MLS distance measurements, continuously available to within hundreds of feet.

For localizer and glide path data accuracy, MLS superiority over ILS can be explained in terms of rf signal considerations. This refers not to modulating waveforms, but to the *ultra high frequency* (UHF) carrier; ILS uses the *very high frequency* (VHF) band, with greater degradation from ground reflection. Also, a scanned beam or an array antenna allows MLS coverage over a wide angular sector. ILS is restricted to a single glide slope for descent, observed with an antenna pattern null* formed by differencing two overlapping beams.

* A null between two positions of high directivity facilitates pointing an antenna in a desired orientation. Compass locator measurements are obtained using *automatic direction finding* (ADF) equipment, which exploits the null in a cardioid antenna pattern.

While pilots have reasonably versatile training for different landing conditions and instruments, there need not be separate procedures for every possible combination of navaids and accuracies. Data and their accuracy levels can be displayed to a pilot in standardized forms, enabling him to make prompt and reliable decisions. The accuracy, much more than the content, of the displayed information will depend upon the instrumenting configuration. This standardization is fully consistent with the integrated nav concept, which is as readily adaptable to a complete MLS airport installation as to intermittent fixes from a partial ILS without a localizer.

7.1.5 Altimeter

Height above local terrain or sea can be determined, typically to within a few percent, by radar altimeters. While above-surface height is extremely important at low altitudes, most operations are concerned ultimately with another reference datum (such as point Ⓞ of Fig. 7-2). As an example involving three different datum levels, consider a landing strip near the edge of a jungle. A radar altimeter indicates a safe height above the trees, while altitude with respect to mean sea level is furnished by barometric data. Known elevation at the landing strip is then used to adjust the barometric information, to determine aircraft altitude with respect to the desired touchdown point. Random error of order 50-ft rms in barometric observations can be appreciably reduced by the scheme of Fig. 1-14, but there could be a bias of 1 ft per 0.001 in.-Hg uncertainty in current local barometric pressure.

7.2 EARTH-BASED OBSERVATIONS

This section expresses quantities obtained from radio, celestial, magnetic, and externally measured sources, in terms of instantaneous aircraft latitude X_1 and longitude X_2. Where permissible, measurements are expressed with the influence of aircraft altitude and/or earth curvature ignored. Communication aids are described in terms of position with respect to earth coordinate locations such as

$$\mathbf{P}_k = (R_\mathrm{E} + h_k)\begin{bmatrix} \cos\lambda_k \cos\varphi_k \\ \cos\lambda_k \sin\varphi_k \\ \sin\lambda_k \end{bmatrix} \tag{7-21}$$

for the kth station at latitude λ_k, longitude φ_k, and altitude h_k; R_E denotes earth radius.

7.2.1 VOR/DME/TACAN

VHF "omnidirectional range" or "omnirange" (VOR) information, previously discussed in Section 1.3.1, expresses bearing from a station at $\mathbf{P}_k$ as measured off a magnetic azimuth reference. In Fig. 7-3a, $\mathbf{I}_{\mathrm{M}}$ points along the horizontal component of the local geomagnetic field and η_k represents local magnetic variation, i.e., eastward angular deflection from $\mathbf{I}_{\mathrm{G}}$ to $\mathbf{I}_{\mathrm{M}}$. The VOR bearing is then essentially

$$Y_{\mathrm{V}} \doteq \arctan\{(X_2 - \varphi_k) \cos \lambda_k/(X_1 - \lambda_k)\} - \eta_k \tag{7-22}$$

if earth curvature may be ignored and equal magnetic variation throughout the reception range of a station may be assumed. Degradations introduced by these simplifications, and uncertainties in η_k, are generally smaller than typical ratings for VOR data as received (e.g., 2°).

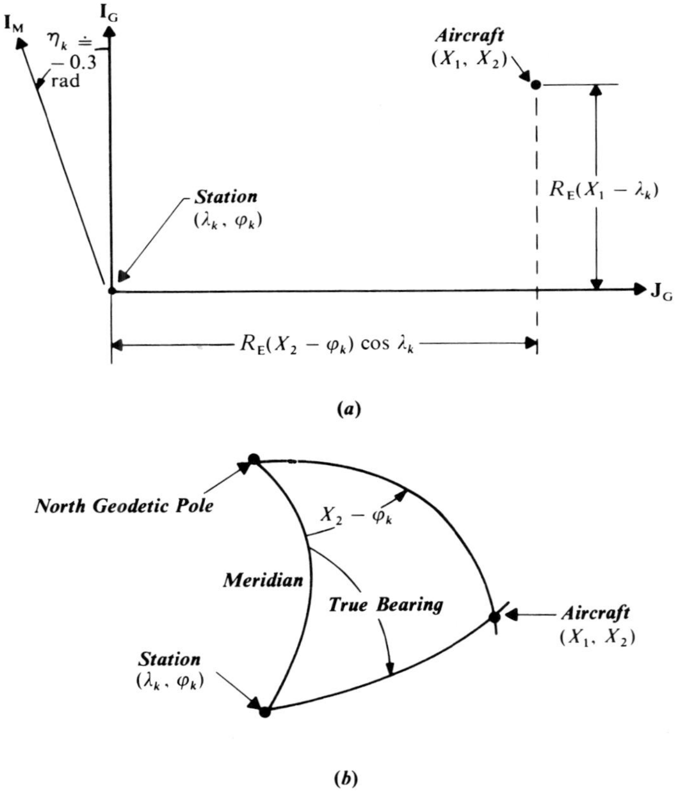

FIG. 7-3. Aircraft/station geometry; (a) planar representation, (b) spherical representation.

Many VOR stations are furnished with distance measuring equipment (DME). A microwave *transponder*, in response to a signal emanating from the aircraft, transmits rf information that is converted into distance Y_P from the station at $\mathbf{P}_k$. With aircraft altitude denoted by h (Ex. 7-13),*

$$Y_P = |\mathbf{P}_k - \mathbf{T}_{E/G}\,\mathbf{R}| \doteq [R_E{}^2(X_1 - \lambda_k)^2 + R_E{}^2(X_2 - \varphi_k)^2 \cos^2 \lambda_k + (h - h_k)^2]^{1/2} \tag{7-23}$$

where the flat earth approximation is again used. DME observations are typically accurate to within 1000 ft.

Some of the simplified expressions used here could be generalized for longer range or more precise measurements. Directional data processing, for example, could account for magnetic variation at $(\hat{X}_1, \hat{X}_2)$ and the prediction Y_V^- could be based on Fig. 7-3b. From the spherical law of sines, aircraft bearing relative to the station meridian is

$$\text{brg} = \text{Arcsin}\left\{\frac{\cos X_1 \sin(X_2 - \varphi_k)}{\sin \xi}\right\} \tag{7-24}$$

where ξ, representing the angle subtended by aircraft and station, is obtained from the spherical law of cosines,

$$\cos \xi = \sin X_1 \sin \lambda_k + \cos(X_2 - \varphi_k) \cos X_1 \cos \lambda_k \tag{7-25}$$

Even in many cases calling for the above precise prediction, partial derivatives can still be computed from (7-22); e.g., (Ex. 7-14),

$$\frac{\partial Y_V}{\partial X_2} \doteq \frac{\cos^2(Y_V + \eta_k) \cos \lambda_k}{X_1 - \lambda_k} \doteq \frac{(X_1 - \lambda_k) \cos \lambda_k}{(X_1 - \lambda_k)^2 + (X_2 - \varphi_k)^2 \cos^2 \lambda_k} \tag{7-26}$$

Aside from uncertainties in station location, VOR measurement accuracies are limited by rf signal characteristics. One limiting feature is precision in the transmitter, which modulates a VHF carrier and an adjacent subcarrier with the same tone. The signals are radiated in all directions, but with a direction-dependent phase difference between the modulating tones. The phase angle observed in an airborne receiver is then identified with Y_V as defined in this section.

The TACtical Air Navigation (TACAN) system provides the same type of observations just described, with increased precision. In addition to the previously defined DME measurements, and UHF communications for furnishing Y_V, TACAN supplies a "vernier" tone with several degrees phase

* At distances exceeding 25 nm, R_E must be represented to within a few tenths of one percent in (7-23). From (3-19) it follows that aircraft and station locations should be taken into account when applying (7-23) to (5-83). The representation need not be exact, but approximation errors exceeding measurement uncertainties are unacceptable.

difference per degree of bearing. The level of precision theoretically obtainable is limited by control of modulating signals and, though to a lesser extent with UHF, refraction.

Geographic coverage for VOR, DME, and TACAN is discussed in Ref. 7-7. Often in the absence of stations on a flight segment, nav information is re-expressed in terms of equivalent "RHO–THETA" observations from a "phantom station" at a nominal on-course position (Ex. 7-15). Such transformations of integrated nav system outputs are readily accommodated for display to flight crew personnel in any desired form.

7.2.2 Hyperbolic Navigation

Differences in distance from two or more specified geographic locations can be obtained from mobile station pairs, or from stationary systems such as LOng RAnge Navigation (LORAN) or OMEGA (Ref. 7-8). Consider a *master* station that controls rf signal transmission, and a *slave* station at another known location with its signal transmission keyed to the master. *Ground waves* characterize the preferred mode of propagation, i.e., the signal paths can follow the earth's curvature far beyond the horizon. With known propagation speed c, differences in time of signal reception from master and slave stations can be re-expressed as path length difference measurements,

$$\begin{aligned} Y_{\mathrm{H}} &= |\mathbf{P}_k - \mathbf{T}_{\mathrm{E/G}}\mathbf{R}| - |\mathbf{P}_j - \mathbf{T}_{\mathrm{E/G}}\mathbf{R}| \\ &\doteq \mathscr{K}_k \operatorname{Arccos}\{\sin\lambda_k \sin X_1 + \cos\lambda_k \cos X_1 \cos(X_2 - \varphi_k)\} \\ &\quad - \mathscr{K}_j \operatorname{Arccos}\{\sin\lambda_j \sin X_1 + \cos\lambda_j \cos X_1 \cos(X_2 - \varphi_j)\} \end{aligned} \tag{7-27}$$

in which altitude differences are ignored and (7-25) is used for two stations located at (λ_k, φ_k) and (λ_j, φ_j); $\mathscr{K}_j$, $\mathscr{K}_k$ can be set equal to a mean local geoid radius for the jth and kth arcs, respectively, plus half the estimated altitude. In any case, irrespective of airborne computations, much insight is gained from the following simplified analysis: Let Fig. 7-4 represent a locally level plane with the origin chosen midway between two stations. With the *base line* segment of length $2\mathscr{L}$ defining x axis direction, the time difference for two synchronized signal epochs arriving at the aircraft position designated (x, y) would be

$$t_k - t_j = \frac{1}{c}\left([(x+\mathscr{L})^2 + y^2]^{1/2} - [(x-\mathscr{L})^2 + y^2]^{1/2}\right) \tag{7-28}$$

which can be re-expressed as

$$(\mathscr{L}x + \mathscr{P}^2)^2 = \mathscr{P}^2[(x+\mathscr{L})^2 + y^2], \qquad \mathscr{P} \triangleq \frac{c}{2}(t_k - t_j) \tag{7-29}$$

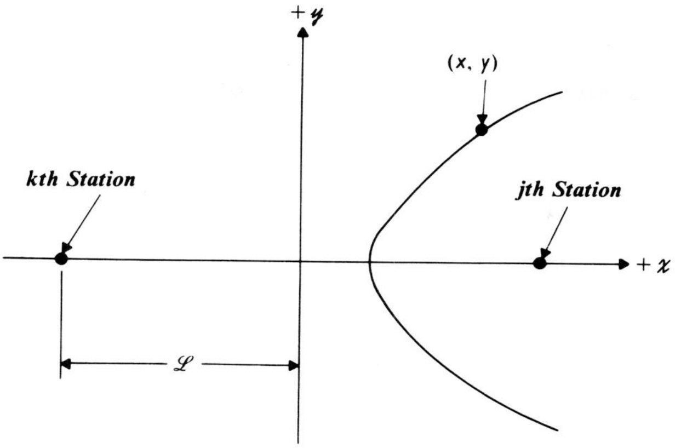

FIG. 7-4. Planar representation of hyperbolic navigation.

or

$$\frac{x^2}{\mathscr{P}^2} - \frac{y^2}{\mathscr{L}^2 - \mathscr{P}^2} = 1 \tag{7-30}$$

Although this equation defines a complete hyperbola, only the cusp conforming to the observed reception sequence is significant. Potential ambiguities exist for aircraft positions near the y axis, where the time difference is on the order of timing uncertainties. At long range, such measurements are *ill-conditioned*; even with the ambiguity resolved, time differences with that geometry are relatively insensitive to position changes along the x direction. Similarly, at long ranges near the base line extension, the measurements are insensitive to position changes along y.

Other possible sources of ambiguity, unlike the above geometric ill-conditioning, are not apparent from the foregoing mathematical description. Even with perfect timing and ideal rf propagation, the identification of synchronized stations with like signals could be transposed. Also, common usage of phase differences to represent time differences creates whole families of hyperbolic cusps; each *lane* corresponds to a 2π-rad phase angle separation. Furthermore, *sky waves* are useful outside ground wave range, only if they are properly identified and if compensated as necessary. Typical accuracies considered here (e.g., a few microseconds error in the absence of propagation corrections, with a possible order of magnitude improvement applicable to ground wave corrections) presuppose successful rf signal acquisition and resolution of all ambiguities. The implications upon the nav system may include independent approximate knowledge of velocity (so that station tuning can take doppler effects into account) and of position (for cycle or lane identification). Also, wherever asynchronous transmission is used, known "triggering" delays between master and slave stations could be compensated in either the receiver or the nav data processor.

Direct range LORAN measurements could be characterized by one term on the right of (7-27) or, for short distances, by (7-23).

Comparison of different hyperbolic systems will show significant contrast in coverage, electromagnetic propagation, power, rf carrier, bandwidth, signal coding, and parameters. In general, however, low-frequency communication is favored for all-weather operation and ground wave characteristics. With wide geographic coverage, stations can be tied to the Universal Time reference (Section 1.2.1).

The desirability of processing actual observations, without transformation in the sensor, comes to attention once again. With reception from multiple hyperbolic nav stations, conventional systems have often produced outputs of latitude and longitude instead of range differences. These "measured" coordinates have been obtained, computationally or graphically, as intersections of quasi-hyperbolic arcs corrected for earth curvature. For full data utilization, however, each "partial fix" observation should be processed on the basis of (7-27) (Ex. 7-16).

7.2.3 Satellite Navaids

The range difference measurements just described can be applied in three dimensions to signals emanating from multiple high-altitude satellites. In practice, satellite position vectors are obtained from earth-based tracking, expressed in the $(\mathbf{I}_\text{I}, \mathbf{J}_\text{I}, \mathbf{K}_\text{I})$ frame. Following transformation into earth coordinates, $\mathbf{P}_j$ and $\mathbf{P}_k$ in this section represent the jth and kth satellite locations in the $(\mathbf{I}_\text{E}, \mathbf{J}_\text{E}, \mathbf{K}_\text{E})$ frame, irrespective of geodetic conventions. Tracking is carried out in a *geocentric* reference, and aircraft position $\boldsymbol{P}_\text{E}$ in earth coordinates must also be expressed here with respect to earth center. In terms of the ellipsoid and its radius of curvature R_P described in Chapter 3, Eq. (10) of Ref. 7-9 gives polar and equatorial distance projections corresponding to geodetic latitude and altitude. The resulting aircraft position is (Ex. 7-17)

$$\boldsymbol{P}_\text{E} = \begin{bmatrix} (\mathsf{R}_\text{P} + h)\cos X_1 \cos X_2 \\ (\mathsf{R}_\text{P} + h)\cos X_1 \sin X_2 \\ \{\mathsf{R}_\text{P}(1 - e_\text{E}^2) + h\}\sin X_1 \end{bmatrix} \tag{7-31}$$

which is substituted into satellite distance differences of the form

$$Y_\text{S} = |\mathbf{P}_k - \boldsymbol{P}_\text{E}| - |\mathbf{P}_j - \boldsymbol{P}_\text{E}| \tag{7-32}$$

Since each range difference corresponds to a three-dimensional surface, multiple observations are needed for a full position fix. Each scalar observation per (7-32) should nevertheless be processed separately.

A more accurate range difference measurement, in the presence of significant refraction and ephemeris errors, would be available from a receiver *pair*. An airborne receiver, and another receiver at known location in Fig. 7-2 (denoted here by T_E), would detect any given satellite signal asynchronously. Range differences would then be of the form

$$|\mathbf{P}_j - T_E| - |\mathbf{P}_j - T_E - (P_E - T_E)|$$

Taylor expansion readily shows reduction of satellite position uncertainty effects, in this case by a factor of order

$$(\text{distance to airport runway}) \div (\text{distance to satellite})$$

Refraction effects are similarly reduced and, with coverage from three non-coplanar satellite sightlines, the aircraft-to-runway position vector estimate is then as accurate as the timing mechanization will allow.

Transponders could in principle be used to obtain aircraft-to-satellite range observations, but that would require transmission of signals from the aircraft. Range difference signals can be controlled by orbiting transmitters, so that only passive airborne reception is required.

High-altitude satellites are visible over a wide field of view, but measurements from lower altitudes are more sensitive to aircraft position. A compromise is therefore in order, and it is easily seen that a similar conclusion holds for other (e.g., doppler or doppler difference) satellite measurements.

7.2.4 Star Trackers

Stellar observations afford an opportunity to correct orientation or position errors, as will now be demonstrated. According to the conventions in Section 1.2.1, a star with known declination (DEC) and sidereal hour angle (SHA) has the unit sightline vector $\mathbf{c}_E$ in earth coordinates

$$\mathbf{c}_E = [\Xi]_z \begin{bmatrix} \cos(\text{DEC})\cos(\text{SHA}) \\ -\cos(\text{DEC})\sin(\text{SHA}) \\ \sin(\text{DEC}) \end{bmatrix} \tag{7-33}$$

and, from (2-14) and (2-15), the sightline in the $(\mathbf{I}_P, \mathbf{J}_P, \mathbf{K}_P)$ frame is

$$\mathbf{c}_P = [\mathbf{I} - (\boldsymbol{\psi} \times \quad)]\left[-\frac{\pi}{2} - X_1\right]_y [X_2]_z \mathbf{c}_E \tag{7-34}$$

For a star tracker mounted along platform axes, the azimuth and elevation angles are

$$Y_{\mathrm{AZ}} = \arctan(c_{\mathrm{P}2}/c_{\mathrm{P}1}) \tag{7-35}$$

$$Y_{\mathrm{EL}} = \mathrm{Arcsin}(-c_{\mathrm{P}3}) \tag{7-36}$$

Instead of expanding (7-34) to express the last two observations in terms of scalars, all computations can be programmed in this abbreviated form; partial derivatives are also expressible in vector and matrix form, e.g.,

$$\frac{\partial \mathbf{c}_{\mathrm{P}}}{\partial X_2} \doteq \left[-\frac{\pi}{2} - X_1\right]_y \begin{bmatrix} -\sin X_2 & \cos X_2 & 0 \\ -\cos X_2 & -\sin X_2 & 0 \\ 0 & 0 & 0 \end{bmatrix} \mathbf{c}_{\mathrm{E}} \tag{7-37}$$

and

$$\frac{\partial \mathbf{c}_{\mathrm{P}}}{\partial \psi_{\mathrm{N}}} \doteq \begin{bmatrix} 0 & 0 & 0 \\ 0 & 0 & 1 \\ 0 & -1 & 0 \end{bmatrix} \left[-\frac{\pi}{2} - X_1\right]_y [X_2]_z \mathbf{c}_{\mathrm{E}} \tag{7-38}$$

which can be substituted into expressions such as

$$(\sec^2 Y_{\mathrm{AZ}}) \frac{\partial Y_{\mathrm{AZ}}}{\partial c_{\mathrm{P}2}} = \frac{1}{c_{\mathrm{P}1}} \tag{7-39}$$

While preceding expressions determine required airborne computations for systems employing star trackers, some insight regarding performance can be gained from restrictive approximations. When (7-34) is written

$$\mathbf{c}_{\mathrm{P}} = [\mathbf{I} - (\boldsymbol{\psi} \times \quad)][-\tilde{x}_1]_y \left[-\frac{\pi}{2} - \hat{X}_1\right]_y [\hat{X}_2 + \Xi + \mathrm{SHA}]_z [\tilde{x}_2]_z [\mathrm{DEC}]_y \mathbf{1}_1 \tag{7-40}$$

the information obtained with certain geometries becomes obvious. A pole star, for example, would have a sightline vector

$$\mathbf{c}_{\mathrm{P}} \doteq \begin{bmatrix} 1 & \psi_{\mathrm{A}} & \tilde{x}_1 - \psi_{\mathrm{E}} \\ -\psi_{\mathrm{A}} & 1 & \psi_{\mathrm{N}} \\ \psi_{\mathrm{E}} - \tilde{x}_1 & -\psi_{\mathrm{N}} & 1 \end{bmatrix} \begin{bmatrix} \cos \hat{X}_1 \\ 0 \\ -\sin \hat{X}_1 \end{bmatrix}, \qquad \mathrm{DEC} = \frac{\pi}{2} \tag{7-41}$$

so that, on the equator, $(Y_{\mathrm{AZ}}, Y_{\mathrm{EL}})$ would essentially reduce to $(-\psi_{\mathrm{A}}, \tilde{x}_1 - \psi_{\mathrm{E}})$, respectively. For an overhead star, however, these gimbal angles would be uninformative:

$$\mathbf{c}_{\mathrm{P}} \doteq \begin{bmatrix} \psi_{\mathrm{E}} - \tilde{x}_1 \\ -\psi_{\mathrm{N}} - \tilde{x}_2 \cos(\mathrm{DEC}) \\ -1 \end{bmatrix}, \qquad \mathrm{DEC} = \hat{X}_1, \quad \hat{X}_2 + \Xi + \mathrm{SHA} = 0 \tag{7-42}$$

The perturbed elevation measurement, to first order, is still a right angle, and Y_{AZ} would be ill-defined for (7-42). A star with a near-horizontal sight-line, in view of (7-25), would conform to the relation [recall (1-36)]

$$\sin \hat{X}_1 \sin(\text{DEC}) + \cos(\hat{X}_2 + \Xi + \text{SHA}) \cos \hat{X}_1 \cos(\text{DEC}) \doteq 0 \quad (7\text{-}43)$$

There are several ways for this to happen, e.g.,

$$\mathbf{c}_P \doteq \begin{bmatrix} 1 & \psi_A & \tilde{x}_1 - \psi_E \\ -\psi_A & 1 & \psi_N \\ \psi_E - \tilde{x}_1 & -\psi_N & 1 \end{bmatrix} \begin{bmatrix} 0 & 0 & 1 \\ 1 & 0 & 0 \\ 0 & 1 & 0 \end{bmatrix} \begin{bmatrix} \cos(\text{DEC}) \\ -\tilde{x}_2 \cos(\text{DEC}) \\ \sin(\text{DEC}) \end{bmatrix}$$

$$\hat{X}_1 = 0, \qquad \hat{X}_2 + \Xi + \text{SHA} = -\pi/2 \quad (7\text{-}44)$$

so that the measurements would become

$$Y_{AZ} \doteq \arctan \frac{\cos(\text{DEC}) - \psi_A \sin(\text{DEC})}{\sin(\text{DEC}) + \psi_A \cos(\text{DEC})} \quad (7\text{-}45)$$

$$Y_{EL} \doteq (\psi_N + \tilde{x}_2)\cos(\text{DEC}) + (\tilde{x}_1 - \psi_E)\sin(\text{DEC})$$

These examples illustrate the capabilities and the limitations of stellar observations. Whereas ψ_A is often obtainable from a single measurement, other observations contain combined effects of tilt and position errors. A 30-$\widehat{\text{sec}}$ star tracker accuracy could fix a position coordinate to within $\frac{1}{2}$ nm if attitude were known; alternatively it could correct orientation to within ~0.15 mr if position were known. With uncertainties present in both attitude and position, a stellar observation at any time provides only a partial correction of each.

Modifications for other star tracker mounting conventions, augmenting states for misalignment, etc., are straightforward. Bias correction prospects will depend on accuracy (which may be on the order of a few arc seconds) and data volume. Stellar observations can be obtained during day or night, if unobstructed by clouds.

7.2.5 Initialization with Magnetic Heading and Externally Derived Data

The magnetic compass is a highly significant instrument for flight training and operations. With a compass needle pointing along $\mathbf{I}_M$ (recall Fig. 7-3a),* and magnetic variation η accurately known at any given position, both heading Ψ and azimuth offset ψ_A could in principle be determined.

* Guided by magnetic force, a level needle seeks an orientation along the horizontal component of the local geomagnetic field. Although referenced to a known pole location at approximately (78.9°N, 70.1°W), field direction is a highly irregular function. Effective usage of compass readings is enabled by route charts based on a magnetic reference, and by extensive tabulation of η.

The determination of nav quantities, once again, calls for a different perspective than in-flight operating considerations. By integrated system standards, magnetic heading information is quite crude (Ex. 7-18) for all purposes except backup and initialization. A "magnetic data drive phase," with ψ_A determined as described above, can be inserted between the leveling and gyrocompassing procedures discussed near the end of Section 3.2. Initialization is thus expedited, since the slow earth rate detection is begun with nearly correct INS orientation.

Since magnetic modeling and instrumenting errors may be a degree or more, other means of providing an azimuth reference are sometimes used. One possible backup is a directional gyro, which also requires initial orientation as well as occasional resets to counteract drift. Other alternatives involve azimuth orientation from external sources, introduced after initialization of position (e.g., latitude and longitude of takeoff point, which may be known to within seconds of arc) and geographic velocity (which may be nulled before takeoff from land). After leveling as described in Section 3.2, azimuth fix information could be physically transported to the aircraft, or introduced through rf transmission or an optical aligning procedure. To be useful, external azimuth data should be more accurate than magnetic heading; however a 1-mr fix would seem a limiting goal in any case.

APPENDIX 7.A: FUNDAMENTAL CHARACTERISTICS OF AIRBORNE RADAR

This appendix discusses radar theory, at a level sufficient for understanding position measurements and error sources. A far more comprehensive development appears in Ref. 7-1. Emphasis is placed here upon airborne (rather than ground-based) radar, since this chapter in general deals with information measured on board an aircraft (Ex. 7-19).

Radar systems transmit electromagnetic energy through an antenna, and monitor the properties (e.g., frequency, time, modulation) of the reflected "echo." To introduce the basic principles, consider a radio frequency (rf) transmitter, radiating energy uniformly in all directions. With rf wavelength and the speed of light denoted as λ_{rf} and c, respectively, the *carrier frequency* is

$$f = c/\lambda_{rf} \tag{7-46}$$

so that a *continuous-wave* (CW) transmitter with power level P would have an output waveform proportional to*

$$F_a(t) = (2P)^{1/2} \cos 2\pi ft \tag{7-47}$$

Average power per unit area received in any direction at a distance R from the source would then be $P/(4\pi R^2)$. Of greater interest here is the case of a *directive* antenna, which concentrates most of the transmitted energy into a *main beam*, represented by a cone with an apex angle of a few degrees. The antenna is then characterized by a *one-way power gain* denoted G, so that the power per unit area received along the direction $\mathbf{I}_b$ of the beam axis is $PG/(4\pi R^2)$. With P_I denoting the total power incident upon a target object along $\mathbf{I}_b$, with projected cross-sectional area A_T,

$$P_I = PGA_T(4\pi R^2) \tag{7-48}$$

In reflecting a fraction f_T of this incident energy, the illuminated object acts as a source, while the antenna is a receiving device with effective capture area $G\lambda_{rf}^2/(4\pi)$. The power returned from the object is then

$$P_R = \{f_T P_I/(4\pi R^2)\}G\lambda_{rf}^2/(4\pi) \tag{7-49}$$

Customarily the product $f_T A_T$ is expressed as an effective radar cross section σ_T, while the last two equations are combined and corrected by a factor Δ for various losses,

$$P_R = \frac{PG^2\lambda_{rf}^2 \sigma_T \Delta}{(4\pi)^3 R^4} \tag{7-50}$$

The inverse fourth power relationship, shown in this radar range equation, explains the dramatic reduction in radar signal detectability at long distances. In addition to receiver noise, interference from the following sources may obscure the radar signal:

(1) Ground or sea *clutter*, i.e., radar energy reflected from the earth below. When the antenna beam is pointed below the horizon, the region illuminated by the main beam produces *main beam clutter* (MBC). Also, since the antenna cannot concentrate all energy into the main beam, there is *side lobe clutter* reflected from the vast surface area outside the main beam.†

* Initially this development assumes ideal conditions; degradations are introduced later. Also, in the analysis to follow, only *relative* energy levels will be of interest. Thus the impedance factor applicable to (7-47) can be ignored, and $F_a(t)$ can be interpreted as the "normalized" instantaneous transmitter output waveform.

† Even a narrow beam antenna collects some energy from each direction. Unfortunately, small objects in the main beam do not always produce more energy than large objects lying outside the main beam.

(2) Radar energy reflected from rain and other extraneous objects in the atmosphere (Ex. 7-20).

(3) Other communication signals at or near frequency f. *Electronic countermeasure* (ECM) signals, actively generated to help a target object escape detection, can produce particularly heavy suppression of a radar echo (Ex. 7-21).

Influence of signal characteristics will now be described for radar position measurements.

7.A.1 Radar Range

A conceptually simple distance measurement is obtained by observing elapsed time t from radar signal transmission to echo reception. Suppose that the waveform represented by (7-47) were switched on intermittently, with a duration τ_G given to each pulse. Then, since t represents the *two-way* transit time of the rf signal.

$$R = c\mathrm{t}/2 \tag{7-51}$$

and range measurements of this type are normally accurate to within a nominal resolution of $c\tau_G/2$. CHIRP radars, while using pulses of long duration, achieve fine range resolution through a pulse compression technique (Ref. 7-10).

Ability to perform the range measurements just described is influenced by the pulse repetition frequency (PRF) and by the detection range, i.e., the distance at which nominal-sized reflectors are marginally detectable. With the *interpulse period* denoted T,

$$\mathrm{PRF} = 1/T \tag{7-52}$$

the maximum *unambiguous* range is $cT/2$. The detection range of a low-PRF radar is less than $cT/2$, so that t in (7-51) can never exceed T. Thus each echo received in a low-PRF radar can be reliably attributed to the most recent transmitted pulse. For convenient mechanization, T is chosen as an integral multiple of τ_G, and the radar receiver can divide each interpulse period into *range gates* of width τ_G. Detections are then made on the basis of rf energy in each gate, with the following implications:

(1) Desired radar echoes are obscured by only those interfering signals lying within the same range gate.

(2) The echo from a target at an arbitrary range will not, in general, be confined to one gate. A measure of relative energy within adjacent gates can be used to determine a range "centroid," and thus enhance accuracy of the observation (Ex. 7-22).

The foregoing remarks are not restricted to low PRF radars, but added procedures are necessary when the detection range exceeds $cT/2$. Range ambiguities are customarily resolved through independent knowledge of approximate distance and/or PRF switching.

As an alternate means of range determination, frequency modulation (FM) can be applied. The waveform represented in (7-47) can be replaced by a function proportional to

$$F_T(t) \triangleq \cos(2\pi ft + \beta_f \sin \omega_f t) \tag{7-53}$$

which, after the transit time, would produce a received echo proportional to

$$F_R(t + 2R/c) \triangleq \cos(2\pi ft + \beta_f \sin \omega_f t) \tag{7-54}$$

At the time of echo reception, the transmitter function is

$$F_T(t + 2R/c) = \cos\{2\pi f(t + 2R/c) + \beta_f \sin \omega_f(t + 2R/c)\} \tag{7-55}$$

At most ranges of interest, with a modulating frequency of a few hundred rad/sec,

$$\cos(2\omega_f R/c) \doteq 1, \qquad \sin(2\omega_f R/c) \doteq 2\omega_f R/c \tag{7-56}$$

and (7-55) reduces to

$$F_T(t + 2R/c) \doteq \cos\{2\pi f(t + 2R/c) + \beta_f \sin \omega_f t + (2\beta_f \omega_f R/c) \cos \omega_f t\} \tag{7-57}$$

By multiplying the instantaneous transmitted and received waveforms, a separable difference frequency (Ex. 7-23) is obtainable. From (7-54) and (7-57) the difference function is

$$\cos\{4\pi fR/c + (2\beta_f \omega_f R/c) \cos \omega_f t\}$$

and thus the observed frequency deviation is proportional to range.

FM ranging is applicable to pulsed as well as CW transmission and can employ modulation characteristics differing from those just considered. Accuracies on the order of several hundred feet are typical.

7.A.2 Radar Line of Sight (LOS)

Antenna axis orientation, at the time midway between transmission and reception of a pulse, indicates the approximate direction corresponding to any echo received from that pulse. Multiple detections from near-coincident beam positions can also be centroided to provide LOS resolution finer than the antenna beamwidth (Ex. 7-24).

As an illustration of refined angular measurements, consider a beam intentionally pointed slightly to the left, and slightly to the right, of the estimated sightline to an object. If either beam position should produce a stronger echo than the other, estimated relative position should be revised accordingly (Ex. 7-25). This technique can be applied to each of two perpendicular directions, in the plane normal to the beam axis, by various methods such as:

(1) *Four-quadrant sequencing.* The beam is allowed to dwell temporarily in each of four fixed positions, placed symmetrically off the estimated sightline direction.

(2) *Conical lobing.* The beam continuously circumscribes the estimated sightline direction, which is re-evaluated from amplitude modulation on the received echo.

(3) *Null seeking.* Two separate overlapping beams are simultaneously formed. The ratio of their energy difference to their sum is used to generate an antenna pointing adjustment signal. Because it does not require repeated transmission, this *monopulse* technique is less sensitive to *amplitude scintillation*, i.e., fluctuation in the radar cross section.

LOS observations based on energy comparison techniques can be accurate to within a few mr, under favorable conditions. This potential accuracy can be inhibited, however, when the radar echo is competing with interference signals (noise, clutter, ECM) of comparable strength. Particularly misleading, from the standpoint of energy assessment, are "discretes" (i.e., highly concentrated objects with large radar cross section) lying in an off-sightline direction. Furthermore, the desired radar echo itself represents an object of finite size; within 0.5-nm range, a 30-ft object subtends an angle of nearly 10 mr. Finally, transmitted and received signals can be refracted by the atmosphere and by a protective surface called a *radome.* Refraction can produce LOS errors of several mr, if left uncompensated.

APPENDIX 7.B: DOPPLER RADAR

When an airborne radar antenna is pointed toward the forward right quadrant of the earth surface (Fig. 7-5), a family of loci can be drawn for various directivity levels. These elliptical loci are ground plane intersections for a family of coaxial cones with various apex angles, all centered on the beam axis. For narrow beam antennas, directivity decreases rapidly with increasing apex angles; thus the smallest ellipses have the heaviest concentration of incident energy. Also, by reason of the inverse square law for incident power received at a distance, energy density within an elliptical ring is greatest for the points closest to the aircraft.

In reflecting a portion of the incident energy, the surface area acts as a source, while the directivity pattern again applies in the role of a *receiving*

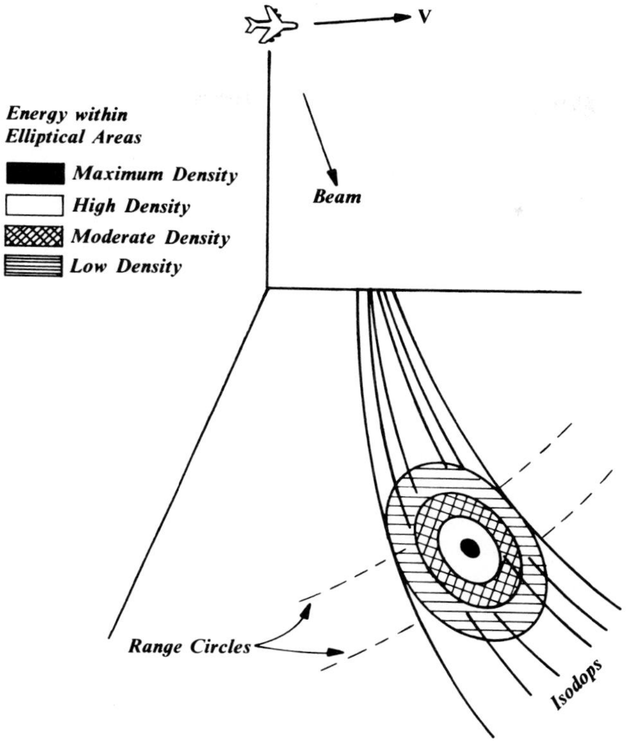

FIG. 7-5. MBC geometry.

antenna. The result is a main beam clutter (MBC) accumulation from several, unequally illuminated, reflectors. Further MBC characterization calls for the analysis of relative velocity effects, which follows.

EFFECTS OF RELATIVE MOTION

In Fig. 7-5 let an arbitrary infinitesimal reflector be chosen from within any ellipse and let $\boldsymbol{i}$ denote its instantaneous unit sightline vector from the aircraft. The closing range rate between aircraft and reflector is nominally $v_i = \mathbf{V}_A \cdot \boldsymbol{i}$, where $\mathbf{V}_A$ is the aircraft velocity (Ex. 7-26). The radiation received by the reflector has an energy level e_i inversely proportional to the square of its distance R_i from the aircraft, and a frequency determined as follows: The ith reflector response $F_i(t)$ to a transmitted waveform (7-47) will be delayed by the *one-way* transit time, so that

$$F_i(t + R_i/c) = \mathsf{e}_i^{1/2} \cos(2\pi f t) \tag{7-58}$$

where f and c denote the carrier frequency and propagation speed for the transmitted radar signal, respectively. By shifting the received signal time reference while expressing R_i as $R_0 - v_i t$,

$$F_i(t) = e_i^{1/2} \cos\left\{2\pi\left(f + \frac{v_i}{\lambda_{rf}}\right)t - \frac{2\pi R_0}{\lambda_{rf}}\right\} \tag{7-59}$$

where $\lambda_{rf} = c/f$ denotes rf wavelength in units used for $\mathbf{V}_A$. Obviously $R_0 \gg \lambda_{rf}$, so that the last term above represents an arbitrary phase angle.

The "backscatter" reflection from this ith point back to the antenna produces an energy level e_i' proportional to e_i/R_i^2 and a frequency that, by reapplying the procedure above, is again shifted by v_i/λ_{rf} Hz. Thus e_i' varies as R_i^{-4} and the total *doppler shift* f_i can be equated to $2v_i/\lambda_{rf}$. More commonly, with $\mathbf{V}_A$ in fps and wavelength λ_{cm} in centimeters,

$$f_i \doteq 61 v_i/\lambda_{cm} \tag{7-60}$$

Obviously, not all ground reflectors will produce the same doppler shift. From the definition of v_i, the locus of points corresponding to a given doppler frequency is obtained from a cone having an axis along $\mathbf{V}_A$. Ground plane intersections of cones with various apex angles conform to the hyperbolic loci (called "isodops") in Fig. 7-5. With this geometry the MBC spectrum approximates the form of Fig. 7-6. Spectral parameters are given

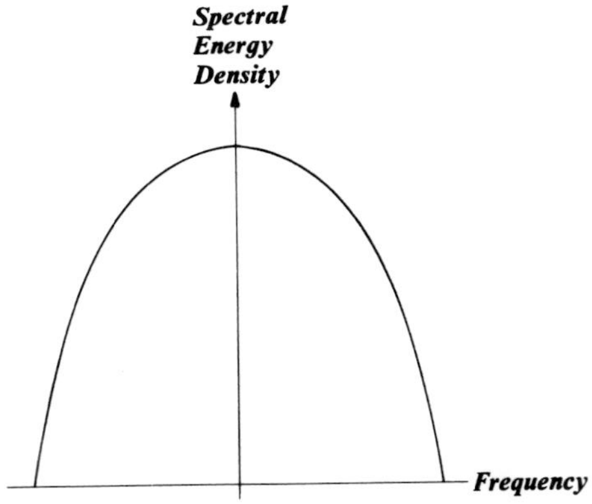

FIG. 7-6. Hypothetical MBC spectrum.

in Ref. 7-11 as simple functions of aircraft speed, antenna orientation, etc. Typical conditions can produce several hundred hertz doppler spread across the antenna main beam,* and the aforementioned asymmetry within elliptical rings causes spectral skew, depicted in Fig. 7-7.

* Total frequency spread also includes effects of finite dwell time for the antenna beam over the illuminated area. Ramifications of this finite duration, plus further extensions of the analysis in Ref. 7-11, are treated in Ref. 7-12. Additional extensions could include superimposed "discretes" (Appendix 7.A.2) and reflector motion described in Ref. 7-13.

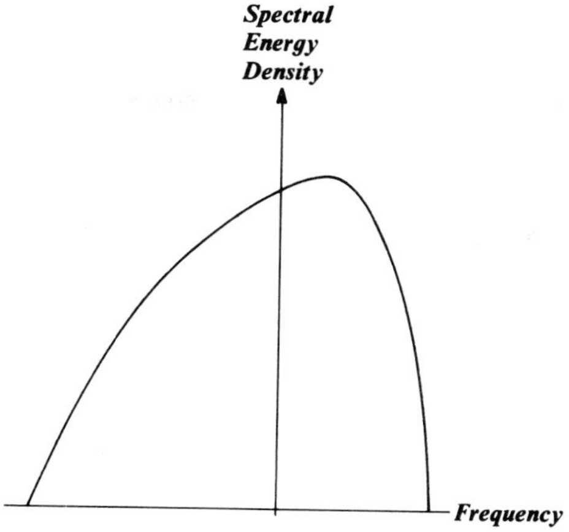

FIG. 7-7. Skewed MBC spectrum.

PULSED TRANSMISSION

Figure 7-8 exemplifies a *pulse doppler* radar transmitter output. The waveform $F_c(t)$ is *coherent*, i.e., expressible as a product $F_a(t) \cdot F_b(t)$ so that rf phase relationships are preserved across silent periods. The spectrum of $F_c(t)$ is repetitive in the frequency domain, with a spacing of $1/T$ Hz between adjacent clusters. Theoretically then, spectral bands of width $2f_{max}$ are nonoverlapping if $1/T > 2f_{max}$. For a high-PRF pulse doppler radar, the PRF is greater than twice the maximum doppler shift, by an amount exceeding spectral spread from signal modulation and finite time effects. Whereas range measurements are unambiguous for a low-PRF radar, doppler observations are unambiguous for a high-PRF radar. A medium-PRF radar produces some ambiguity in both range and doppler measurements (Ex. 7-27).

Distribution of MBC energy among range gates corresponds to distance along the beam in Fig. 7-5. Radar echoes from the area between the two concentric arcs, shown dotted, could produce the spectrum of Fig. 7-7. The component of aircraft velocity along the beam would then be surmised from (7-60), with f_i replaced by a "center frequency" in Fig. 7-7. A spectral average, accurate to within a few knots, can be obtained by applying a frequency track servo loop to range-gated MBC.

MULTIBEAM VERSUS SCANNED SINGLE BEAM DOPPLER

Conventional doppler navigation systems employ three or four antenna beams with separate continuous frequency tracking for each. Simultaneous tracking of multiple beams is appropriate for a primary source of continuous velocity data, as in Section 6.0, item 2. However, when doppler information

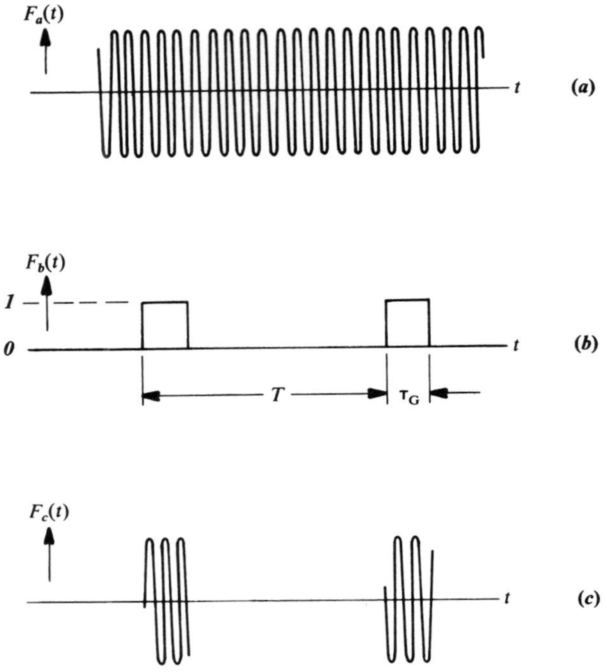

FIG. 7-8. Gated rf waveform; (a) rf sinusoid, (b) unit gate function, (c) gated sinusoid.

is used only to correct INS errors, only intermittent fixes are needed. Every few minutes, a time-shared radar beam can be successively pointed along three noncoplanar directions. Dwell time in each direction is influenced by the frequency track loop pull-in response. Doppler fixes then consume only a few percent of the time-shared radar duty cycle.

LIMITATIONS OF DOPPLER NAVIGATION

Reflection of electromagnetic rays from smooth surfaces is highly *specular*, i.e., characterized by equal angles of incidence and reflection as in Fig. 2-6. Doppler navigation, however, relies on prominence of MBC, which requires *backscattering* from a more diffuse reflecting phenomenon. At shallow beam depression angles over wet terrain or calm seas, MBC energy is not always readily distinguishable from side lobe clutter or other rf interference. The use of shallow depression angles is further discouraged by the fourth power law and by horizon avoidance (Ex. 7-28). At the other extreme, however, large antenna depression angles would also be impractical.

Even when mechanical constraints do not restrict downward orientation, reflection from the "zero-doppler" surface area below the aircraft could tax the receiver dynamic range. Specular effects, which cause weak echoes at shallow depression angles, produce very high reflectivity at normal incidence.

Among the sources of bias in doppler navigation are misalignments (including refraction) and offsets incurred through frequency tracking (spectral skew, servo lag, oscillator drift). Many of these bias effects can be counteracted. Radome refraction can be measured for subsequent computational correction. Spectral asymmetry and servo following errors are also partly predictable. Oscillator frequency can be recalibrated through comparison with other data sources (Ref. 7-2).

Some of the unbiased random doppler navigation errors are also reduced (e.g., averaging of the instantaneous frequency in the tracker), while others are maintained at acceptable levels by design specifications (e.g., resolution of pickoffs affecting apparent beam orientation). The doppler information is corrupted by another false relative velocity component, whenever ocean currents are encountered. Despite all degradations, doppler nav data can be accurate to within a few knots, *provided* the aforementioned ambiguities are resolved.

DOPPLER MEASUREMENTS FROM DISCRETE OBJECTS

Range rate observations indicate components of velocity along an antenna axis, relative to moving objects for tracking, or to objects that are fixed on the earth for navigation. By time-sharing a beam for sequential measurements, doppler information can be supplied from the same *multimode radar* (MMR) used to provide the aforementioned range and LOS observations. Typically the MMR will employ coherent pulsed transmission, at least during doppler measurements. In the pulse doppler mode, total rf energy in each range gate can be subdivided into spectral *cells*. The echo from a given object is then obscured by only those interfering signals at or near the same range gate *and* doppler cell. Assessment of energy from multiple adjacent doppler cells enhances the previously described centroiding operation, for both range and LOS refinements.

Indicated range rate discrepancies will be in direct proportion to uncertainties associated with observed frequency. Catastrophic errors include frequency shifts caused by moving parts within a tracked object (e.g., aircraft engines), loss of echo (e.g., obscuration by MBC during air-to-air tracking; Ex. 7-29), and unresolved PRF ambiguities. In the absence of such degradations, range rate accuracy is limited by doppler cell resolution or by finite-duration spectral spread (nominally the reciprocal of the beam dwell time), whichever is greater.

APPENDIX 7.C: SYNTHETIC APERTURE RADAR (SAR)

One of the many functions commonly performed by a radar is terrain mapping, particularly when visibility is hampered by darkness or adverse weather. From a display of relative energy levels received at various sub-horizon beam positions, a trained operator can recognize coastlines and other prominent features. Azimuth resolution of the displayed image, corresponding to the *real* aperture of an antenna, is normally comparable to an outer ellipse of Fig. 7-5. To resolve contrasting features *within* the area illuminated by a beam, known relations between doppler frequency and angular position can be exploited. Following is a basic explanation of how such a *synthetic* aperture is obtained. The aforementioned range increment $c\tau_G/2$ is used for illustration, but the technique remains applicable to CHIRP systems with equivalent resolution.

Consider an aircraft on the flight path in Fig. 7-9, with a long series of range gates always timed so that the origin ⓞ is centered. After a long succession of transmitted pulses, imagine each separate range gate time sequence applied to a bank of contiguous narrowband filters, or to a discrete

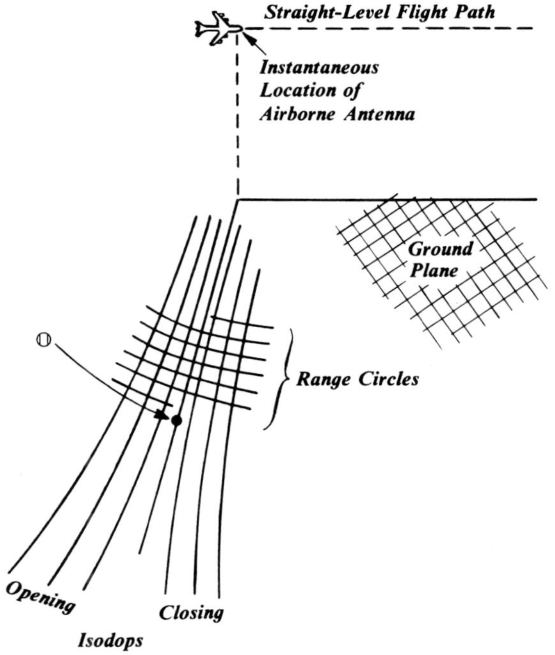

FIG. 7-9. Side-looking radar geometry.

Fourier transform (DFT). Based on (7-60), an association could be made between each doppler frequency and a position angle,

$$\vartheta_i = \text{Arccos}\{\lambda_{\text{cm}} f_i/(61|\mathbf{V}_A|)\} \tag{7-61}$$

so that each spectral amplitude denotes intensity for a particular range–LOS *combination*. The full set of intensity levels, properly processed, defines a two-dimensional image.

SAR imaging is a far-reaching topic (Ref. 7-4), but its influence on navigation requirements can be exhibited here. To maintain the range gate center at any point and perform the subsequent phase adjustments successfully requires precise knowledge of *dynamic variations in antenna location.** Without provisions for monitoring irregular variations, imaging could be performed only on simple flight paths (e.g., straight and level) in the absence of severe gusts. General motion compensation, applicable to variable azimuth "squint" as well as side-look geometries, is achieved in the following scheme devised by J. H. Mims (Refs. 7-5 and 7-14).

THEORY OF SAR MOTION COMPENSATION

Determination of the spectrum for each separate range gate, to form the image just described, allows a simple interpretation in terms of rf phasing. The instantaneous phase shift corresponding to a reflector at range R_i, for one-way transmission as in (7-59), is $-2\pi R_i/\lambda_{\text{rf}}$. To describe the radar information received, the transmission/reception roles of reflector and antenna are again reversed, to provide a total (two-way) phase shift function of the form

$$\exp(-\imath 4\pi R_i/\lambda_{\text{rf}}) = \exp(-\imath 4\pi|\boldsymbol{R}_R - \boldsymbol{P}_i|/\lambda_{\text{rf}}) \tag{7-62}$$

where $\boldsymbol{R}_R$ and $\boldsymbol{P}_i$ represent position vectors of the radar antenna and the ith reflector, respectively. The effect of antenna translation can therefore be taken into account through multiplying the received waveforms by functions of the form $\exp(+\imath 4\pi R_i/\lambda_{\text{rf}})$.

For the general situation depicted in Fig. 7-10, the range to an arbitrary reflector cannot be expressed as any simple function of time. Nevertheless,

* For those with no prior introduction to SAR theory, it might seem that no practical system could be sensitive to displacements far smaller than 1 cm. Typical uncertainties in location between an airborne antenna and a mapped area will far exceed the rf wavelength, so that phase adjustment signals will contain several whole cycles of error. The image, however, is comprised of relative intensities for relative grid positions, established by rf phase variations. Static errors in doppler as well as position are associated with translational shift, rather than degradation, of an image; this is shown in the ensuing analysis, which is taken from Refs. 7-5 and 7-14. Unfortunately a host of printing errors in Ref. 7-14 necessitated a subsequent letter of correction (Ref. 7-15). The right-hand column on p. 965 of Ref. 7-15 also contains false information, which should be disregarded.

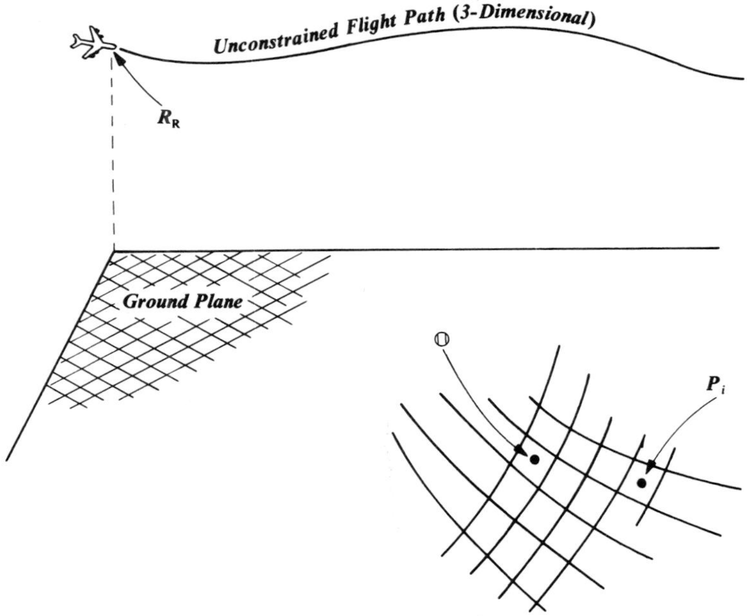

FIG. 7-10. SAR geometry with azimuth squint and maneuvers.

translational motion of the antenna could be taken into account by functions expressible as (Ex. 7-30)

$$\exp(+\iota 4\pi R_i/\lambda_{rf}) \equiv \exp(\iota 4\pi |\boldsymbol{R}_R - \boldsymbol{R}_{Ref}|/\lambda_{rf}) \cdot \exp(\iota 4\pi d_i/\lambda_{rf}) \quad (7\text{-}63)$$

in which

$$d_i \triangleq |\boldsymbol{R}_R - \boldsymbol{P}_i| - |\boldsymbol{R}_R - \boldsymbol{R}_{Ref}| \quad (7\text{-}64)$$

where $\boldsymbol{R}_{Ref}$ is a chosen reference location such as nominal image center $\mathbb{O}$. The product expression in (7-63) corresponds to two stages of motion compensation. First, the distance $|\boldsymbol{R}_R - \boldsymbol{R}_{Ref}|$ governs placement of all range gates as well as an instantaneous phase shift of $4\pi|\boldsymbol{R}_R - \boldsymbol{R}_{Ref}|/\lambda_{rf}$ rad. The second stage performs multiplication with the phasing function

$$\exp(\iota 4\pi d_i/\lambda_{rf}) = \exp(\iota 2\pi f_i t) \quad (7\text{-}65)$$

where f_i represents an *incremental* frequency shift associated with a particular range-doppler cell (Ex. 7-31):

$$f_i = 2\dot{d}_i/\lambda_{rf} \quad (7\text{-}66)$$

Association of the second phase compensation stage with a Fourier transform is evident from (7-65); when the output of the first stage for a given range gate has a spectral representation,

$$F_g(t) = \sum_j \mathscr{C}_j \exp(-\imath 2\pi f_j t) \tag{7-67}$$

then a discrete Fourier transform (DFT) would extract, for the ith image cell, a time average

$$\text{av}\{F_g(t) \exp(\imath 2\pi f_i t)\} = \mathscr{C}_i \tag{7-68}$$

which yields the proper energy level $|\mathscr{C}_i|^2$ for that cell intensity on a SAR map display.

Although the necessary computations vary rapidly with time and differ for different image cells, modern data processing techniques and algorithms afford practical computing requirements. References 7-5 and 7-14 discuss overlapping fast Fourier transform (FFT) blocks with interpolation, and Taylor expansions about nominal aircraft as well as image ($\boldsymbol{R}_{\text{Ref}}$) positions. Also, large *swath* size (i.e., mapped area) and fine resolution can be achieved through compensation of the *range curvature* phenomenon. With large swaths and/or narrow image cells, the locus of ground reflectors lying within any range gate varies (e.g., consider rotation of a circular arc, pivoting about point $\mathbb{O}$ in Fig. 7-10 as the aircraft advances along its path). This phenomenon is not inherently detrimental to the image, as long as interpolations are performed to redistribute energy among adjacent range gates (Ref. 7-5).

By comparison with earlier practice, the procedures just described significantly enhance SAR capability. High-resolution mapping in the past generally implied side-look radar (SLR); even a nominal fixed azimuth *squint* was achieved only with an appreciable increase in complexity. Furthermore, a tight autopilot was needed to ensure a restrictive flight path (usually straight and level), while common implementing provisions (lenses and film) hampered operations; there was also a blind "settling time" delay after a turn. All of these difficulties are eliminated through computationally path-adaptive motion compensation. At the same time, radar parameters can readily be programmed in flight, provided that basic compatibilities between PRF, beamwidth, data processing provisions, etc., are satisfied.

The foregoing description applies to a known instantaneous location $\boldsymbol{R}_R$ of the radar antenna. Although uncertainty in location can be tolerated for range gate placement and for the *second* stage of rf phase adjustment, performance degradations must now be identified for inexact phase shifts of the form

$$\exp(\imath 4\pi |\hat{\boldsymbol{R}}_R - \boldsymbol{R}_{\text{Ref}}|/\lambda_{\text{rf}}) \equiv \exp(\imath \tilde{\psi}) \cdot \exp(\imath 4\pi |\boldsymbol{R}_R - \boldsymbol{R}_{\text{Ref}}|/\lambda_{\text{rf}}) \tag{7-69}$$

in which $\tilde{\psi}$ denotes phase error,

$$\tilde{\psi} \triangleq 4\pi(|\hat{\boldsymbol{R}}_{\mathrm{R}} - \boldsymbol{R}_{\mathrm{Ref}}| - |\boldsymbol{R}_{\mathrm{R}} - \boldsymbol{R}_{\mathrm{Ref}}|)/\lambda_{\mathrm{rf}} \tag{7-70}$$

EFFECTS OF IMPERFECT PHASE ADJUSTMENT

Consider a range gate subjected to one stage of motion compensation, with imperfections as just described. Instead of $F_{\mathrm{g}}(t)$, the output of this first stage will then have the form $F_{\mathrm{g}}(t)\exp(\imath\tilde{\psi})$. A *correlating function* $E_i(t)$ can then be defined,

$$E_i(t) \triangleq \exp\{\imath(2\pi f_i t + \tilde{\psi})\} \tag{7-71}$$

so that, instead of (7-68), the time average extracted for the ith image cell is

$$F_i(t) = \mathrm{av}\{F_{\mathrm{g}}(t)E_i(t)\} \tag{7-72}$$

The effect can be demonstrated for any spectral component $\tilde{V}_{\mathrm{b}}$ of range rate error with amplitude $\tilde{V}_0$ and frequency f_0,

$$\tilde{V}_{\mathrm{b}} = \tilde{V}_0 \cos 2\pi f_0 t \tag{7-73}$$

which produces an rf phase error of

$$(4\pi/\lambda_{\mathrm{rf}}) \int_0^t \tilde{V}_{\mathrm{b}}\, d\ell = (2\tilde{V}_0/\lambda_{\mathrm{rf}}) \frac{\sin 2\pi f_0 t}{f_0} \tag{7-74}$$

Substitution into (7-71), under the condition

$$2\tilde{V}_0/(f_0 \lambda_{\mathrm{rf}}) \ll 1 \tag{7-75}$$

produces the approximate phase-modulated correlating function

$$E_i(t) \doteq \left(1 + \imath \frac{2\tilde{V}_0}{f_0 \lambda_{\mathrm{rf}}} \sin 2\pi f_0 t\right) \exp(\imath 2\pi f_i t) \tag{7-76}$$

Superimposed on the desired spectral amplitude from (7-72), then, will be an attenuated response to image elements with $\pm f_0$ Hz frequency difference. A corner reflector, at the position shown in Fig. 7-11, can significantly hamper recognition of an image object between the designated isodops.

For lower frequency components of motion compensation error in (7-73), inequality (7-75) is inapplicable. The phenomenon just described is then restricted to nearby image cells, causing defocus or blur. Image degradation is then quantified by a two-term expansion of the sine in (7-74), and subsequent formulation of a quadratic phase shift. With $f_0 = 0$ in (7-74), there is *no* image degration; $\tilde{V}_0$ would then produce a fixed doppler shift, detectable in nav update (Ex. 7-32).

The influence of frequency upon motion compensation error has led to widespread usage of acceleration formulations. By generalizing (7-73) to

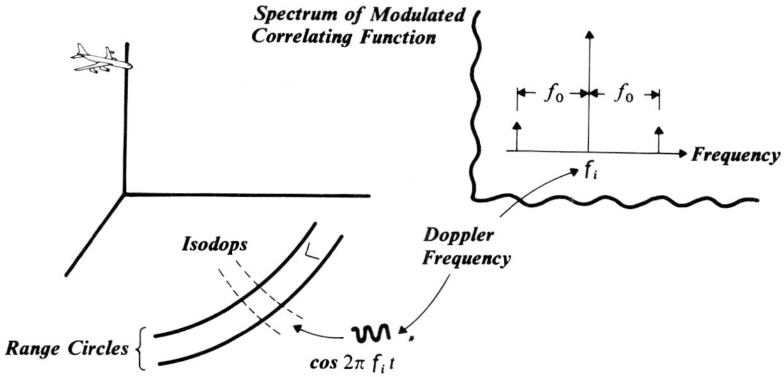

FIG. 7-11. SAR image degradation with corner reflector in remote doppler region.

include multiple spectral components, velocity error along the beam has a derivative of the form

$$\tilde{Z}_b = \sum_k \tilde{Z}_k \sin 2\pi f_k t \tag{7-77}$$

and the relative energy level for each remote sideband (Ex. 7-33) is

$$20 \log_{10}\{\tfrac{1}{2}(2\tilde{V}_k)/(f_k \lambda_{rf})\} = 20 \log_{10}\{\tilde{Z}_k/(2\pi f_k^2 \lambda_{rf})\} \text{ dB} \tag{7-78}$$

NAV SYSTEM IMPLICATIONS

The relation between motion compensation and navigation requirements can be approached by expressing instantaneous radar antenna position as

$$\boldsymbol{R}_R = \boldsymbol{R}_S + \mathbf{T}_{0/A}\,\mathbf{d}_{N/R} \tag{7-79}$$

where $\mathbf{d}_{N/R}$ represents the displacement between IMU and antenna phase center, expressed in aircraft coordinates (recall Ex. 2-1). To control range gate centering and the relatively critical first stage of rf phase adjustment, both $\boldsymbol{R}_S$ and $\mathbf{T}_{0/A}$ are needed at each transmitted pulse. This often leads to special provisions for up-to-date information; the ratio A_{max}/v discussed in Section 3.3 may be comparable to the PRF (typically several hundred hertz), which generally exceeds the standard nav computer iteration rate. An accumulation, reset at each iteration, can provide current knowledge of velocity; addition of integrated velocity to transformed displacement $\mathbf{T}_{0/A}\,\mathbf{d}_{N/R}$ calls for current attitude also.* Effects of lags in this information

* Some systems have used additional antenna-mounted accelerometers, processed through high-pass filters and combined with low-pass INS data. That scheme can experience degradations from filter transients, however, in a general (maneuvering) situation.

as well as other important degradation sources, are identified by considering small variations in the above equation, as follows:

$$\tilde{R}_R \doteq \tilde{R}_S + \tilde{\mathbf{T}}_{0/A}\,\mathbf{d}_{N/R} + \mathbf{T}_{0/A}\,\tilde{\mathbf{d}}_{N/R} \tag{7-80}$$

The first term on the right contains integrated effects of INS misorientation, plus all gravity model uncertainties and accelerometer errors described in Section 4.3.2, as well as velocity quantization (recall Ex. 4-35). The second term includes quantization of attitude data and kinematical interactions connected with rotational uncertainties. The last term includes dynamic (vibratory) as well as fixed uncertainties in antenna phase center location. Static uncertainties in $\mathbf{d}_{N/R}$ are shown to be comparatively unimportant in Ref. 7-14, where other effects just discussed are also investigated.

EXERCISES

7-1 (a) Let the azimuth and depression angles a_0 and d_0 in Fig. 7-1 be denoted by state variables X_2 and X_3, respectively, while X_1 is defined as the distance $(S_1{}^2 + S_2{}^2 + S_3{}^2)^{1/2}$. Show that, for an altitude measurement Y_h,

$$\partial Y_h/\partial X_1 = \sin X_3$$

(b) Consider an alternate state definition, in which X_1 and X_2 retain the meanings just given, while X_3 now represents altitude. Show that, with this convention,

$$\partial Y_h/\partial X_1 = 0$$

despite the fact that the measurement and the first state variable are unchanged. Give an explanation, using the definition of a partial derivative.

(c) For the extended Kalman filter of Section 5.4.1, how should measurements be defined?

(d) In a forward/right/down coordinate convention, what would be the algebraic signs of the position states shown in Fig. 7-1?

7-2 (a) For straight and level flight with the beam pointed at Ⓞ in Fig. 7-1, shown that a_0 and d_0 could replace $\mathscr{A}$ and $\mathscr{D}$ in (7-4). Show how (7-4) could be modified for an RGMA displacement, as in Ex. 6-17c.

(b) Show that (5-27) could be applied to three range rate measurements defined by (7-5), only if the three beam directions are noncoplanar. What would happen if the third beam were only 1° out of the plane of the other two?

(c) Consider usage of doppler observations in the estimation scheme of Chapter 5. Would linearization error be introduced from (7-5)? From

(7-6) and (7-7)? Make a case for acceptability of *linear* pseudomeasurements (e.g., range rate is directly proportional to observed doppler frequency).

(d) Establish agreement between (7-6) and (7-7) and (1-38)–(1-40).

7-3 (a) Draw an analogy between (7-5) and (7-9). Identify the proper algebraic sign in (7-9) for the case of an aircraft flying past (rather than toward) the origin.

(b) Discuss application of (7-1), (7-2), and (6-43) to an object being tracked by airborne radar.

(c) Show that differentiation of (7-8) produces (7-9) without the negative sign.

7-4 (a) With shallow antenna depression in near-level flight, would doppler measurements be a highly informative source of vertical velocity data?

(b) If simultaneous measurement of three velocity vector components were required, could one radar beam be time shared?

7-5 (a) Use small angle transformation methods of Chapter 2 to derive (7-12). Show that ϵ_A and ϵ_E would vanish for a beam pointed directly at a tracked object.

(b) With coincidence between $(\mathbf{I}_0, \mathbf{J}_0, \mathbf{K}_0)$ and the geographic frame, discuss a relation between (6-45) and (7-12).

(c) Write an explicit "observable versus state" relation for (7-12), using Y and S notation. Show that, if $|\boldsymbol{R}_S|$ can be assumed independent of small cross-range variations, the applicable **H**-matrix is independent of the state. Justify the simplification, using the Pythagorean theorem and material from Section 5.4.

(d) Show that the statement following (7-12) enables accounting for known departures (e.g., due to antenna servo lag) of $\mathbf{I}_b$ from the direction of $\hat{\boldsymbol{R}}_S$.

7-6 (a) Express $\boldsymbol{I}$ as a constant vector premultiplied by a matrix. Use the method of (2-28) and product differentiation to derive (7-13).

(b) For a unit beam axis vector $\boldsymbol{I}$ pointed at the origin, re-express the right of (7-13) as

$$\{\dot{\boldsymbol{R}}_S - (\dot{\boldsymbol{R}}_S{}^T\boldsymbol{I})\boldsymbol{I}\}/|\boldsymbol{R}_S| \equiv \boldsymbol{I} \times (\dot{\boldsymbol{R}}_S \times \boldsymbol{I})/|\boldsymbol{R}_S|$$

Also re-express the left of (7-13) as $\mathbf{T}_{0/b}(\mathbf{1}_1 \times \boldsymbol{\omega}_b)$. Realizing that $\mathbf{T}_{0/b}\,\mathbf{1}_1 = \boldsymbol{I}$, use Ex. 4-4b to show that the LOS rate in the $(\mathbf{I}_0,\ \mathbf{J}_0,\ \mathbf{K}_0)$ frame is $\boldsymbol{R}_S \times \dot{\boldsymbol{R}}_S/|\boldsymbol{R}_S|^2$. Would the addition or subtraction of a rotation component about $\boldsymbol{I}$ affect the rates experienced along $\mathbf{J}_b$ and $\mathbf{K}_b$?

(c) With $\boldsymbol{I}$, $\dot{\boldsymbol{R}}_S$, and $\mathbf{K}_0$ coplanar, show that the azimuth LOS rate vanishes.

(d) With $\boldsymbol{I}$ normal to $\dot{\boldsymbol{R}}_S$, express the LOS rates resolved along $\mathbf{J}_0$ and $\mathbf{K}_0$ in terms of the beam depression angle.

(e) Consider a beam that deviates from the origin by a few mr. Use the results of part (b) to determine the pointing error effect. Also discuss the slow rotation of the $(\mathbf{I}_0, \mathbf{J}_0, \mathbf{K}_0)$ frame.

7-7 (a) Consider a 1-nm range vector $\boldsymbol{R}_S$ normal to a 1000-fps velocity vector $\dot{\boldsymbol{R}}_S$. Show that, on the basis of Ex. 7-6e, the high-quality gyros considered in Section 3.4.2 are not required for LOS rate measurement.

(b) In analogy with Ex. 7-6e, describe the influence of a 1-mr pickoff resolution for inertial platform gimbal angles and for $\mathscr{A}$ and $\mathscr{D}$. Would efforts to refine radar gimbals be very fruitful, without refinement of platform pickoffs as well?

7-8 (a) Find the discussion regarding SAR cursor laying operation in Section 6.5, and identify the cursor drive equations.

(b) Time differentiation of the range rate expression would provide a basis for SAR focus adjustments. If the second derivative of range were available for observation could large initial range rate uncertainties be reduced more quickly? Does there appear to be a critical need for focus observations? Find a related discussion in Section 5.3.1.

(c) Pattern recognition methods could enable fast automatic cursor adjustments. If adopted, would more than one independent focus observation be available for each frame? Use the discussion in Section 5.4.4 to answer this question.

7-9 Sightline depression is independent of the azimuth angle in (7-3). How does this differ from (7-16), and from the "rolled beam frame" of (2-67)?

7-10 (a) Show that lateral displacements can be ignored for glide path observations when $Y_{L(\max)}$ is a small angle. If inserted into (7-18), should S_2 be offset by half the runway width?

(b) Under what conditions can the inverse tangent relationship be simplified in (7-17) and (7-18)? Could deviations from proportionality, and uncertainties in antenna locations, be included in total measurement error?

(c) From the discussion immediately following (7-17), explain why MLS enables a more natural landing approach than ILS.

7-11 (a) Formulate a measurement of distance from $\mathbb{O}$ in Fig. 7-2. Use a modified form of (7-8), to include a lateral displacement of a DME antenna, off point $\mathbb{O}$.

(b) Consider a substantial displacement between DME and glide path antenna. How would this displacement affect the determination of altitude above the runway, in a direct triangulation scheme? In an integrated estimation scheme? How should antenna position uncertainties be regarded?

7-12 (a) Instead of a wedge-shaped beam emanating upward from a marker beacon position, consider a cone of elliptical cross section. Let $(D_{B1}, D_{B2}, 0)$ denote known coordinates of the beacon and, by generalization of (7-20), describe implementation of (5-84) with a data vector residual having the form

$$\begin{bmatrix} D_{B1} - \hat{S}_1^- \\ D_{B2} - \hat{S}_2^- \end{bmatrix}$$

with (6-32) used for extrapolation between observations. With reduced accuracy of the lateral position indication, discuss possible emergency landing with no localizer information available.

(b) Show that distance from the runway can be determined with reasonable accuracy by low altitude passage over a marker beacon, but there are undesirable path constraints and limitations on available data.

7-13 (a) Uncoupling of vertical from horizontal navigation is assumed for (7-23). What modification would be appropriate for DME measurements in Sections 6.3 and 6.4?

(b) Compare (7-21) versus (1-24). What concept do (6-29) and (7-22) have in common?

7-14 (a) Find a discussion near the end of Section 5.4.2 that justifies (7-26).

(b) Show how (7-24) and (7-25) can produce (7-22).

(c) Use (7-22) and trigonometric identities to obtain (7-26).

7-15 Consider the R-nav scheme mentioned near the beginning of Section 1.3, with no VOR stations on a particular course leg. Suppose that a pilot nevertheless wants up-to-date values of DME distance ("RHO") and VOR bearing ("THETA") that *would* have been obtained from a specified on-course location. Explain that integrated nav estimation can proceed unchanged, while (7-22) and (7-23) can be applied to derive display outputs. Relate to the discussion at the end of Section 7.1.4.

7-16 Consider a second base line inclined 5° off the x axis in Fig. 7-4. Would a hyperbola, centered about the inclined base line, provide a well-defined intersection (x, y) in the presence of timing errors? Should (7-30) be used to convert time differences to pseudomeasurements of position coordinates, to be substituted into (5-81), (5-83), and (5-84)? How can timing information from a master and only one slave station be used?

7-17 (a) Recall from Appendix 3.A that **R** does not pass through the center of the ellipsoid. Use (3-19) to show that substitution of (1-24), instead of (7-31), into (7-32) could produce errors of several miles in (5-83). Would approximation of (5-81) be permissible?

(b) Discuss the contrast between part (a) and Ex. 3-23, which involves maximum errors introduced into *correction* terms through sphericity assumptions.

(c) From (7-31) show that $|\boldsymbol{P}_E| \doteq \boldsymbol{R}_P(1 - e_E{}^2 \sin^2 X_1) + h$. Use (3-80) in the ellipse equation, and (3-22), to verify the result at zero altitude. Identify the approximation used to obtain this result. Use first-order Taylor expansions for the difference of two square roots, and demonstrate a maximum error of order 100 ft, at 45° latitude.

(d) Consider Cartesian components of $\mathbf{T}_{I/E}\boldsymbol{P}_E$, used as position states in the space stable formulation of Appendix 6.A. Would (7-31) be used in the dynamic equations?

(e) Describe range information obtainable from a precise airborne clock, if exact satellite transmitting schedules were known.

7-18 (a) Compute the effect of 1° azimuth error on the effective drift in (3-46). Could magnetic heading be used for an austere nav system, considered in item (2) or (3) of Section 6.0?

(b) If Ψ and η were known, could magnetic heading be displayed to the pilot?

7-19 Recall Ex. 6-17d and discuss possible usage of observations from ground communications.

7-20 If highly reflective material were ejected from an object being tracked by radar, what might be expected to happen?

7-21 In contrast to the *passive* ECM of Ex. 7-20, suppose that signals were generated from a remote source outside the main beam, but at the same range as the tracked object. Let P_S denote the remote source power and G_S represent the radar antenna directivity along the sightline to the remote source. Employ the principles used to derive (7-48)–(7-50) to show dominance of remote source energy if

$$(4\pi/\Delta)(R^2/\sigma_T) > (P/P_S)(G^2/G_S)$$

Note the prospects for masking reflections from long range, by a low-power source.

7-22 Suppose that two adjacent range gates have received energy levels far exceeding all other gates. Should range be estimated at a position midway between the two strongest gates, when their relative energy levels are not balanced?

7-23 (a) Multiply (7-54) by (7-57) and express the product in terms of sum and difference frequencies. Note that the sum frequency term can be removed by a low-pass filter.

(b) Differentiate the phase argument on the right of (7-54). The result is an *instantaneous frequency*. Show that the maximum instantaneous frequency in hertz is $f + \beta_f \omega_f/(2\pi)$. The quantities β_f and $\beta_f \omega_f/(2\pi)$ are, respectively, the *modulating index* and the *frequency deviation* of the FM waveform in (7-54). What is the frequency deviation of the difference function following (7-57)?

(c) All other things being equal, a larger frequency deviation implies both a greater bandwidth for the transmitted signal and a better measure of range via FM observations. Recall the text accompanying (7-51); does finer range resolution again correspond to greater bandwidth?

(d) *Without* the simplification (7-56), obtain the difference function corresponding to a product of (7-54) with (7-55). Obtain the effective amplitude of the function

$$\beta_f\{\sin \omega_f t[\cos(2\omega_f R/c) - 1] + \cos \omega_f t \sin(2\omega_f R/c)\}$$

and use trigonometric identities to derive an observed index,

$$2\beta_f \sin(\omega_f R/c)$$

When does this approximately equal the modulating index for the expression following (7-57)?

7-24 Consider the TWS scheme of Ex. 6-17d. With no refinements, angular resolution would be on the order of the antenna beamwidth. Suppose, however, that strong but *unequal* energy levels are received from two adjacent scan directions, nominally a beamwidth apart. Is there any reason to believe that a detectable object may be located slightly *off* midscan position?

7-25 (a) Apply the reasoning of Ex. 7-24 to a single target track mode.

(b) Consider an amplitude scintillation much slower than the rate of conical lobing. Discuss the effect on LOS observations.

(c) Amplitude scintillation is accompanied by *glint*, i.e., variation in the effective center of reflection on or near an object. Discuss the effect of a 20-ft shift on range and LOS measurements.

7-26 (a) If $\dot{\boldsymbol{R}}_S = \mathbf{T}_{0/A} \mathbf{V}_A$, show that v_i equals a scalar product of $\dot{\boldsymbol{R}}_S$ with a transformed unit vector $\mathbf{T}_{0/A}\boldsymbol{i}$. More precisely, with displacements $\mathbf{d}_{C/N}$ and $\mathbf{d}_{N/R}$ from aircraft mass center to IMU, and IMU to radar antenna, respectively, use (4-102) to express range rate between antenna and reflector as

$$\boldsymbol{i} \cdot \{\mathbf{V}_A + \boldsymbol{\omega}_A \times (\mathbf{d}_{C/N} + \mathbf{d}_{N/R})\}$$

(b) From (7-51) deduce the one-way transit time; combine with (7-46) and a time-shifted form of (7-58) to obtain (7-59).

(c) Describe the radar information obtained by superposition of several individual reflector echoes. Relate to (4-104).

(d) Describe the radar echo from a reflector with a doppler shift, in the presence of transmitter FM as in (7-53).

7-27 (a) Sketch a number of similar spectral clusters, separated by $(1/T)$ Hz. Show that adjacent clusters with bandwidth $(2f_{\max})$ would overlap if $2f_{\max} T > 1$.

(b) Consider an "X-band" radar with $\lambda_{cm} = 3$, 60-nm detection range, and $f_{\max} = 40$ KHz. From (7-60) deduce the highest possible range rate. Show that the radar would be categorized as low-, medium-, and high-PRF for $T = 10^{-3}$, 10^{-4}, and 10^{-5} sec, respectively.

7-28 How would very shallow antenna depression, hardly intercepting the horizon, affect the MBC spectrum? Would the effect be more pronounced for wet terrain or calm seas? Relate these questions, and the fourth power range law, to Ex. 7-26c and Fig. 7-7.

7-29 Compare the range rate between an airborne radar antenna and (1) the illuminated terrain below, versus (2) another aircraft with velocity perpendicular to the beam. Discuss MBC interference for a high-PRF radar. Could the second aircraft be detected by a low-PRF radar, if its range differed considerably from that of the illuminated terrain?

7-30 Use (7-64) and the definition of R_i to verify (7-63). Show that different range gates could employ different reference location vectors.

7-31 (a) Recall that, since $R_i \gg \lambda_{rf}$, the phase angle for the ith reflector is extremely sensitive to its location. Dynamic behavior of the phase angle, however, is not. Show this by writing the argument in (7-62) with

$$-(4\pi/\lambda_{rf})(\boldsymbol{R}_R \cdot \boldsymbol{R}_R - 2\boldsymbol{R}_R \cdot \boldsymbol{P}_i + \boldsymbol{P}_i \cdot \boldsymbol{P}_i)^{1/2}$$

with a time derivative $(4\pi/\lambda_{rf})\dot{\boldsymbol{R}}_R \cdot \boldsymbol{i}$, where $\boldsymbol{i}$ represents a unit vector along $\boldsymbol{P}_i - \boldsymbol{R}_R$. Would $\boldsymbol{i}$ differ markedly for two adjacent cells of Fig. 7-10?

(b) For a nominal aircraft position, show that nominal "midcell" values of d_i as well as $\boldsymbol{P}_i$ can be computed from the geometry in Fig. 7-10.

(c) Note that each spectral component of (7-67) has an arbitrary phase angle, implicit in the complex coefficient $\mathscr{C}_j$. How does this phase angle affect the image cell intensity discussed after (7-68)? Why is the extreme sensitivity of rf phase angle, mentioned in part (a), not a severe impediment to SAR imaging?

(d) Compare (7-60) versus (7-66). Does the latter represent a frequency shift beyond that associated with a reference image position?

7-32 (a) Use limit theory to determine the right of (7-74) when f_0 is zero. Express the resulting correlating function of (7-71) in terms of an unknown frequency shift as

$$\exp\{\imath 2\pi(f_i + 2\tilde{V}_0/\lambda_{rf})t\}$$

(b) On a display of range versus range rate, how could a cursor be expected to respond to an initial range rate error? Relate to the SAR update description in Section 7.1.3.

7-33 (a) Write the sine function as the difference of two exponential functions with imaginary arguments. Relate to the two spectral components in Fig. 7-11 and the factor $\frac{1}{2}$ on the left of (7-78).

(b) Apply (7-75) to components in (7-77), performing integration as necessary.

(c) Express an instantaneous phase error in terms of a double integration for (7-77).

(d) Relate the left of (7-78) to (7-76).

REFERENCES

7-1. Skolnik, M. I. (ed.), *Radar Handbook*. New York: McGraw-Hill, 1970.

7-2. Farrell, J. L., "Digital Mechanization for Single Beam Doppler Navigation," presented at ION Nat. Air Meeting, Atlanta, Georgia, Feb. 29–Mar. 2, 1972.

7-3. Reeves, R. M., and Blumgold, R. M., "Radar/Inertial Navigation Using a Forward Looking Multimode Radar," 1974 NAECON Record, Dayton, Ohio, pp. 483–490.

7-4. Harger, R. O., *Synthetic Aperture Radar Systems—Theory and Design*. New York: Academic Press, 1970.

7-5. Mims, J. H., and Farrell, J. L., "Synthetic Aperture Imaging with Maneuvers," *IEEE Trans.* **AES-8** (4), July 1972, pp. 410–418.

7-6. Sanders, L. L., and Fritch, V. J., "Instrument Landing Systems," *IEEE Trans.* **COM-21** (5), May 1973, pp. 435–454.

7-7. Pritchard, J. S., "The VOR/DME/TACAN System, Its Present State and Its Potential," *IEEE Trans.* **ANE-12** (1), March 1965, pp. 6–10.

7-8. VanEtten, J. P., and Zemlin, G. P., "Computers in Loran C/D and Omega Navigation and Guidance Systems," AGARDograph No. 158, London, Technical Editing and Reproduction Ltd., Feb. 1972, pp. 175–228.

7-9. Hedman, E. L., "A High-Accuracy Relationship between Geocentric Cartesian Coordinates and Geodetic Latitude and Altitude," *AIAA JSR* **7** (8), August 1970, pp. 993–995.

7-10. Klauder, J. R., Price, A. C., Darlington, S., and Albersheim, W. J., "The Theory and Design of Chirp Radars," *Bell System Tech. J.* **39** (4), July 1960, pp. 745–808.

7-11. Farrell, J. L., and Taylor, R. L., "Doppler Radar Clutter," *IEEE Trans.* **ANE-11** (3), Sept. 1964, pp. 162–172.

7-12. Friedlander, A. L., and Greenstein, L. J., "A Generalized Clutter Computation Procedure for Airborne Pulse Doppler Radars," *IEEE Trans.* **AES-6** (1), Jan. 1970, pp. 51–61.

7-13. Wong, J. L., Reed, I. S., and Kaprielian, Z. A., "A Model for the Radar Echo from a Random Collection of Rotating Dipole Scatterers," *IEEE Trans.* **AES-2** (2), March 1967, pp. 171–178.

7-14. Farrell, J. L., Mims, J. H., and Sorrell, A., "Effects of Navigation Errors in Maneuvering SAR," *IEEE Trans.* **AES-9** (5), Sept. 1973, pp. 758–776.

7-15. Farrell, J. L., Mims, J. H., and Sorrell, A., "Correction to 'Effects of Navigation Errors in Maneuvering SAR'," *IEEE Trans.* **AES-9** (6), Nov. 1973, pp. 964–965.

Chapter 8

Illustration of Navigation and Tracking Operations

8.0

A brief assessment will now be made for four phases of operation, i.e., initialization, en-route earth nav, airborne tracking, and point nav near destination. Detailed quantitative performance has been evaluated elsewhere for a variety of conditions and sensor combinations (e.g., Ref. 8-1). No attempt is made here to duplicate previous studies, but a general review can lend much insight to guide the design of specific systems. Only in this last chapter are simulation results explicitly presented, and even here the emphasis is on analytical insight. Familiarity with all preceding chapters is assumed in presenting this material.

A general system block diagram, already described in Section 5.4.1, is shown in Fig. 8-1. The goal is to provide up-to-date navigation data with accuracies comparable to, or better than, individual sensor capabilities. Because individual measurements are often indirect and single fixes are often incomplete, repeated observations with changing geometries are used to refine derivatives as well as position in all three directions. A means of combining the nonsimultaneous observations may be provided by a gimballed or strapdown IMU, a continuous doppler velocity measurement, or air data (with aircraft orientation added in either of the latter two schemes). All observables are associated with distances or angles, or their

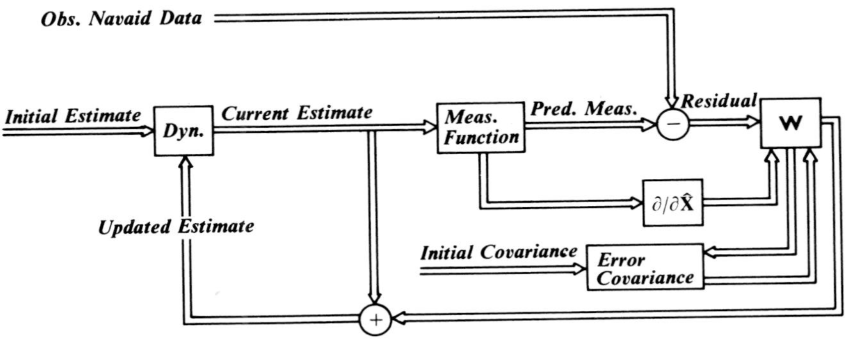

FIG. 8-1. Kalman estimation technique.

differences or rates, referenced to specified points or directions. Often a system will use only position updates, since all derivatives and INS orientation angles can be derived through kinematical coupling.

Specific equations for each block in Fig. 8-1 will depend upon the mode of operation. As explained in Section 5.4.2, the mode is influenced by the sources of measured data, as well as the states to be estimated. Sensor failure indicators, activated by abnormal residual sequences, can therefore be used to switch modes at any time. Another modal characteristic is the data window, i.e., the effective duration over which observations are averaged. This duration is heavily influenced by the primary source of velocity information (INS, doppler, or air data), due to the wide range of velocity accuracy levels. Data window reinitialization may be necessary, when kinematical error growth patterns are abruptly changed (e.g., Ex. 5-31d). Within any uninterrupted data window, the rate at which derivative uncertainties drive the position errors will determine the practical level of precision. With velocity noise at η_v (fps)^2/Hz and position updates spaced at ℓ sec, for example, any reduction of rms update error below $(0.1\eta_v \ell)^{1/2}$ ft could be futile.

8.1 INITIAL ALIGNMENT

In previous discussions the term "misalignment" often implied imperfect mechanical assembly, whereas "misorientation" meant departure between geographic and INS reference axes. The topic of initialization, however, uses the popular concept of "aligning" the two triads. Most of the applicable procedures rely on physical forces and rotations, rather than visible reference lines (Ex. 8-1). A subdivision of methods produces four different techniques, i.e., the so-called *self-alignment*, optical aiding, master INS drive, and in-flight alignment.

SELF-ALIGNMENT

On a stable support (e.g., an airstrip), the INS can be erected in a self-contained operation. As described in Section 3.2, leveling is achieved through servo-nulling the horizontal specific force components, and gyrocompassing provides the azimuth initialization; procedures are carried out physically or computationally, in accordance with the type of system. Gimballed platform operations can be expedited by a coarse/fine stage sequence, during which correction commands are fed to the platform servos and gyro torquers, respectively. Magnetic heading (Section 7.2.5), combined with magnetic variation, can be used in the coarse alignment stage.

Unlike many other nav operations, self-alignment can treat the INS instrument outputs (or their integrated effects) as observations in an estimation scheme. Whether taken directly from platform instruments or computationally derived from them, three observables (Y_{AN}, Y_{AE}, $Y_{\omega E}$) are easily related to earth nav states (ψ_N, ψ_E, ψ_A) by

$$Y_{AN} \doteq g\psi_E, \qquad Y_{AE} \doteq -g\psi_N, \qquad Y_{\omega E} \doteq -\omega_s(\psi_N \sin\lambda + \psi_A \cos\lambda) \tag{8-1}$$

so that platform orientation sensitivity is 1 mr per milli-g of north and east specific force uncertainty, and 1 mr per *milliearth rate unit* (MERU) of east angular rate uncertainty at the equator. An explanation can now be given for the usual rapidity of the leveling procedure. Consider a pair of integrating accelerometers, accurate to within $3 \times 10^{-5}g$, with resolution of $\frac{1}{32}$ fps. Leveling errors of order 0.2 mr are detectable after 20 sec, since the integrated effect of 1-g lift on a tilted accelerometer,

$$2(10)^{-4} \times 32.2 \text{ ft/sec}^2 \times 20 \text{ sec}$$

would far exceed the accelerometer measurement uncertainty. At the same time, a 1-mr azimuth misalignment would not be readily detectable, since an east gyro drift rate of only a few hundredths degrees/hour would exceed a measured sidereal rate projection of $10^{-3} \times 15(1.0027379)°/\text{hr}$.

Fortunately, self-alignment for ψ_A need not rely exclusively on gyro data; there is cross-axis coupling in (6-20), which, for known stationary geographic position, reduces to

$$\begin{bmatrix} \dot{\psi}_N \\ \dot{\psi}_E \\ \dot{\psi}_A \end{bmatrix} = \begin{bmatrix} 0 & -\omega_s \sin\lambda & 0 \\ \omega_s \sin\lambda & 0 & \omega_s \cos\lambda \\ 0 & -\omega_s \cos\lambda & 0 \end{bmatrix} \begin{bmatrix} \psi_N \\ \psi_E \\ \psi_A \end{bmatrix} + \mathbf{e}_\omega \tag{8-2}$$

Estimation of all three states ψ_N, ψ_E, ψ_A (Ex. 8-2) can rely primarily on

Y_{AN} and Y_{AE} after several minutes have passed.* For applications in which this delay is intolerable, one of the succeeding alternatives is commonly adopted.

OPTICAL AIDING

If two vectors $\hat{\boldsymbol{\lambda}}_1$, $\hat{\boldsymbol{\lambda}}_2$ expressed in a vehicle-based coordinate frame can be computed from airborne observations, and if the corresponding vectors $\boldsymbol{L}_1$, $\boldsymbol{L}_2$ are known in geographic coordinates, then the following relations apply to the apparent transformation $\hat{\mathbf{T}}_{G/V}$ from the vehicle-based axes to geographic coordinates:

$$\boldsymbol{L}_1 = \hat{\mathbf{T}}_{G/V}\hat{\boldsymbol{\lambda}}_1, \qquad \boldsymbol{L}_2 = \hat{\mathbf{T}}_{G/V}\hat{\boldsymbol{\lambda}}_2 \tag{8-3}$$

A similar relation naturally holds for the vector product $\hat{\boldsymbol{\lambda}}_1 \times \hat{\boldsymbol{\lambda}}_2$; by collection of all three relations into partitioned matrix form,

$$[\boldsymbol{L}_1 \mid \boldsymbol{L}_2 \mid \boldsymbol{L}_1 \times \boldsymbol{L}_2] = \hat{\mathbf{T}}_{G/V}[\hat{\boldsymbol{\lambda}}_1 \mid \hat{\boldsymbol{\lambda}}_2 \mid \hat{\boldsymbol{\lambda}}_1 \times \hat{\boldsymbol{\lambda}}_2] \tag{8-4}$$

and thus, if the original two vectors are noncollinear (in practice, preferably more than 45° apart; Ex. 8-3)

$$\hat{\mathbf{T}}_{G/V} = [\boldsymbol{L}_1 \mid \boldsymbol{L}_2 \mid \boldsymbol{L}_1 \times \boldsymbol{L}_2][\hat{\boldsymbol{\lambda}}_1 \mid \hat{\boldsymbol{\lambda}}_2 \mid \hat{\boldsymbol{\lambda}}_1 \times \hat{\boldsymbol{\lambda}}_2]^{-1} \tag{8-5}$$

There are various ways in which these relations can be applied to initial INS alignment. Reference 8-3 illustrates acquisition of strapdown IMU orientation using gravity and sighting vectors in (8-5), including a practical rationale for corrective computations during and after measurement in a rotating vehicle. Also, if $\hat{\boldsymbol{\lambda}}_1$ may be expressed in terms of azimuth and elevation angle measurements from platform-mounted optics, (8-3) provides the relation

$$|\boldsymbol{L}|\begin{bmatrix} \cos Y_{EL} \cos Y_{AZ} \\ \cos Y_{EL} \sin Y_{AZ} \\ -\sin Y_{EL} \end{bmatrix} = \begin{bmatrix} 1 & \psi_A & -\psi_E \\ -\psi_A & 1 & \psi_N \\ \psi_E & -\psi_N & 1 \end{bmatrix}\begin{bmatrix} L_x \\ L_y \\ L_z \end{bmatrix} \tag{8-6}$$

where any reference vector known in geographic coordinates can be used for $\boldsymbol{L}$. In particular, for a unit level line,

$$\begin{bmatrix} \cos Y_{AZ} - L_x \\ \sin Y_{AZ} - L_y \end{bmatrix} \doteq \begin{bmatrix} L_y \\ -L_x \end{bmatrix}\psi_A \tag{8-7}$$

which illustrates that this residual vector, normal to $\boldsymbol{L}$, is in effect a direct

* Further analysis (e.g., Ref. 8-2) shows that north and azimuth gyro drifts could also be estimated, although the latter can require hours for calibration by gyrocompassing. When aircraft can be maintained stationary since engine shutoff from preceding flight, a final azimuth angle reading can be stored for the next initialization. This "stored azimuth" technique offers a dramatic reduction in time required for alignment.

measure of ψ_A. For either in-flight or ground-based alignment, however, optical aids have operational limitations. Even if all field-of-view and accessibility problems are solved, the essential connection between platform reference and instrument IA directions inevitably involves mechanical assembly imperfections.

TRANSFER ALIGNMENT

Complete analysis of slaving one INS to another can involve a lengthy development (see Ref. 8-4), but a simplified description may be given as follows: Define a fixed displacement vector $\mathbf{d}_{N/S}$ from one IMU (the "master") to another (the "slave") on a rigid moving aircraft and assume that, for operational purposes, all master INS outputs are to be considered correct. Then the desired orientation of the slave IMU coincides with the master, and their velocities should differ by $\boldsymbol{\omega}_A \times \mathbf{d}_{N/S}$ (Ex. 8-4). Any departure from this desired difference can be modeled as an acceleration error in combination with slave misorientation, similar to (6-7) and (6-8); note that total acceleration error here includes effects of nonrigidity and imperfections in the kinematical velocity adjustment. Nonrigidity also interferes with angular rate matching, which would otherwise be a useful means of transfer alignment for strapdown systems.

IN-FLIGHT ALIGNMENT

By comparing changes in accumulated INS outputs versus successive fixes separated by known time intervals, $\boldsymbol{\psi}$ is forced to betray its presence. After leveling in cruise, horizontal specific forces from turns and speed changes can expedite azimuth correction (Ref. 8-5). In any case, in-flight alignment constitutes a perfect example of the overall nav data mixing scheme, discussed in the next section.

8.2 EARTH NAV DATA PROCESSING

This section discusses the transition from initialization through flight phases wherein navigation accuracy finally becomes commensurate with fix accuracies in three-dimensional space. Performance is described as a function of flight paths, instrument degradations, and geometry. Regardless of which instruments provide fixes and continuous velocity data (e.g., doppler radar plus HARS; gimballed platform or strapdown INS with wander azimuth mechanization), it is permissible to express position and velocity errors as resolved along geographic reference axes. The ensuing description considers north and east error components; altitude and vertical velocity are assumed to be adequately determined in a separate mechanization (Fig. 1-14).

TRANSIENTS AFTER INITIALIZATION

Position and velocity may be accurately known as takeoff, but INS orientation is often imprecise. One effect of misorientation, appearing shortly after takeoff, is interaction of the initial tilt with sustained 1-g lift, to produce horizontal position and velocity error transients. A second transient follows from drift effects. Insight regarding approximate error dynamics and corrections will now be provided, based on repeated reinitialization of equations from Section 3.4.2.

Suppose that a hastily conducted leveling procedure left an initial tilt of 3 mr at takeoff. From (3-48), velocity error would then grow to about 17 fps in 3 min. The dominating influence of tilt ($\tilde{V}^- \doteq \psi g t$) could then be corrected through a doppler measurement; a fix accurate to 2 fps could promptly reduce tilt error to a few tenths mr, since (Ex. 8-5)

$$\frac{\tilde{V}^+}{gt} \doteq \frac{2 \text{ ft/sec}}{32 \text{ ft/sec}^2 \times 180 \text{ sec}} \doteq 3.5(10)^{-4} \tag{8-8}$$

With range rate actually measured along skewed directions as in Fig. 8-2, the uncertainty remaining after the fix is amplified somewhat (e.g., 0.5 mr at 45°). Additional fixes would be desirable, to prevent eventual growth of velocity errors beyond a few fps.

A second transient effect arises from level gyro drifts, in combination with initial azimuth misorientation. Suppose that initial tilt effects are

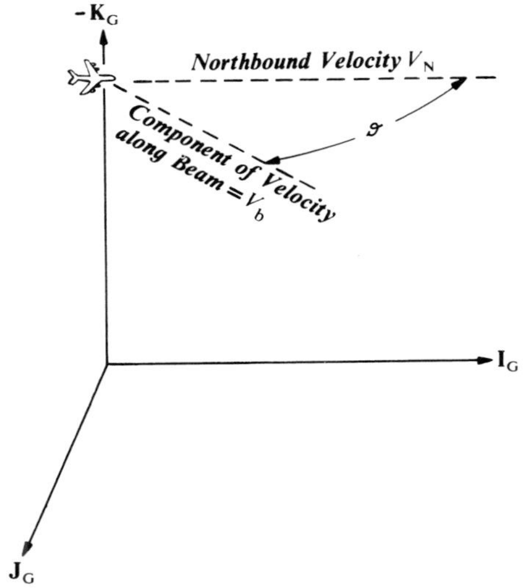

FIG. 8-2. Geometric influence on error remaining after doppler fix ($V_b = V_N \cos \vartheta$; $\tilde{V}_b = \tilde{V}_N \cos \vartheta$. Thus, $\tilde{V}_N = \tilde{V}_b \sec \vartheta$).

substantially removed, by doppler fixes during the first few minutes after takeoff, but a temporary cessation of fixes allows unattended propagation of $\psi_{A(0)}$. If a hasty initial alignment left an azimuth error of 0.5°, then (3-48) becomes dominated by a term of the form

$$\tilde{V}^- \doteq \tfrac{1}{2} g d t^2 \tag{8-9}$$

or, after 20 min of southbound flight at 600 fps, roughly 12 fps. By reasoning analogous to (8-8), another 2 fps doppler fix might then seem to leave azimuth uncertainty at roughly $1\frac{1}{2}$ mr. Actually it is the *combined* effect of drift rate about $\mathbf{I}_G$ and initial azimuth misorientation that would experience the six-fold reduction after 20 min; separate estimation could not yet be completed under the conditions described thus far. However, repeated intermittent fixes with different beam directions could enable an in-flight alignment and calibration of drift rates about level axes (Ex. 8-6). Applicable dynamics can conform to (6-25), truncated to nine states. The azimuth drift has been excluded thus far because of its more gradual influence, as illustrated in Section 3.4.2.

To supplement or replace velocity data for in-flight alignment, position fixes are quite effective. A tilt example paralleling (8-8) would reduce leveling error to (position fix uncertainty) ÷ ($\frac{1}{2}gt^2$) just after update, and the drift–azimuth correction is also exemplified through (3-53) instead of its derivative. Multiple fixes in all directions naturally provide further error reduction through statistical averaging. In any case, when each component of $\boldsymbol{\psi}$ is maintained at a value near the contribution from random drift accumulated between fixes, the transient period is over and a quasistatic error pattern governs further nav error growth. This concept is, of course, somewhat simplified, and subject to influence of irregular observation scheduling, geometry, sudden changes requiring reinitialization, etc.; still the generalization provides useful insight in many cases.

LIMITED ERROR EXCURSIONS

To maintain quasistatic accuracy levels continuously, enough measurements must be taken to provide observability in all directions during each data window. States that are not directly measurable can be determined via kinematical coupling, either through derivative relationships (e.g., Exs. 5-10a and 5-24a), or through cross-axis interaction, exemplified in (8-2). Required observations can be furnished by repeated usage of one sensor (e.g., re-examination of VOR station data at different points along the flight path, reception of multiple stations with the same time-shared receiver), or by redundant sensors. Redundancy is often motivated by availability considerations (e.g., loss of doppler at high altitude over calm seas, cloud hindrance of star tracking at low altitude) or backup in the event of sensor failure.

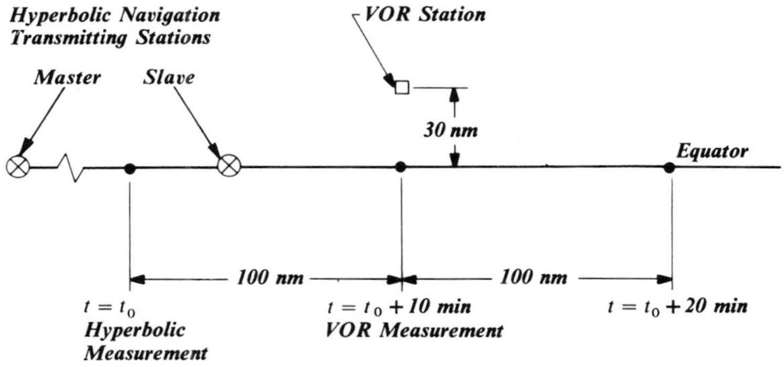

FIG. 8-3. Multiple earth-based observations.

In Fig. 8-3 suppose that the last reception from the master station occurs at time t_0, when ψ_N and $\tilde{V}_E$ have values of 0.2 mr and 5 fps, respectively. Let the corresponding slave station observation have a random relative timing error of 6 μsec (as in Ref. 8-1) but no bias. Since this measurement is obtained with the best possible geometry (Ex. 8-7), an initial longitude uncertainty $\tilde{X}_2 = 0.5 \overset{\frown}{\text{min}}$ (or $\tilde{x}_E = 0.5$ nm, at $t = t_0$) can be postulated. After 10 min of eastward cruise at 600 kt, with no doppler or position fixes, (3-44) and (3-51) give incremental errors comparable to initial uncertainties of 0.5-nm position, a few fps velocity, and a few tenths mr tilt. Inertial instrument bias of order $3 \times 10^{-5}g$ and a few hundredths degrees/hour would not yet be influential, and random drifts at that level would be still less effective. If VOR bearing could then be measured to within about a degree from the station shown, east position accuracy would be maintained at approximately $\frac{1}{2}$ nm. Refinements for velocity estimates could not yet be deduced through position data comparison, but there is no urgent need to do so until inertial instrument errors take effect. After ten more minutes the tilt would be appreciably influenced by the aforementioned gyro drift, while cumulative effects of tilt and velocity uncertainties would drive east velocity and position errors above their initial levels. Another position measurement of appropriate sensitivity (hyperbolic or satellite range difference, VOR, etc., with accuracies comparable to those just described) could reinitialize $\tilde{x}_E$ at about $\frac{1}{2}$ nm.

While restoring position accuracy, the estimator also tends to maintain the previous leveling and velocity accuracies, through kinematical coupling. An analogy can be made with (8-8), but the immediate example is more complex. Total position error at any time includes accumulated excursions from earlier tilt, velocity uncertainty, and drift; separate determination of each uses all measurements within a data window. More generally, there is additional coupling due to finite latitude and changing flight paths, with measurements typically sensitive to multiple states from different channels. Each fix then counteracts buildup of drift effects by reinitializing an estimated state, with the appropriate dynamic mode from Chapter 6.

SIMULATION EXAMPLE

A realistic simulation of various sensor combinations, described in Ref. 8-1, includes performance of a geographic-referenced platform updated by satellite observations. After coarse initialization (with errors typified at 1700 ft, 2 fps, 0.1°, and 2° in components of horizontal position, velocity, tilt, and azimuth misalignment, respectively), differences in range and doppler were inserted at 16-sec intervals throughout a flight path that included acceleration to 500 kt after takeoff, cruise at 45° heading, turns, and a dive–pullout sequence. In addition to rms random gyro drifts of 0.01°/hr with 30-min autocorrelation time, warmup drifts of 0.1°/hr with 2-min autocorrelation time, and $10^{-5}g$ level accelerometer noise with 5-min autocorrelation time, the simulation included unmodeled null offsets of 0.05°/hr and $10^{-4}g$ for the gyros and level accelerometers, respectively,

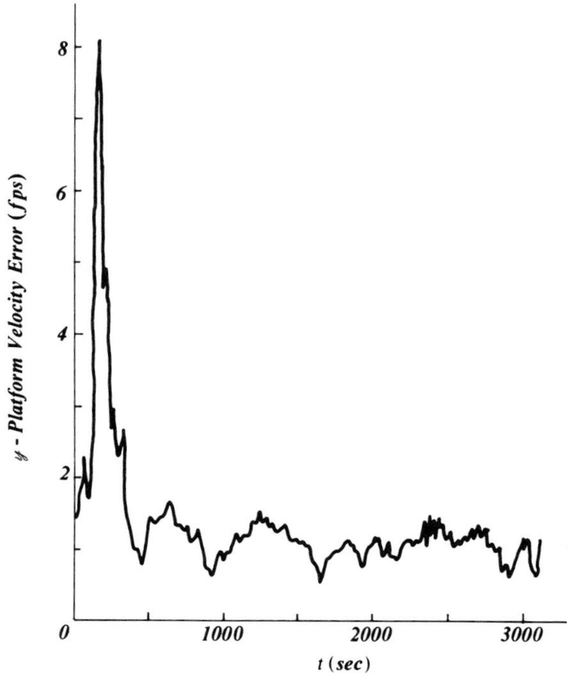

FIG. 8-4. Velocity error vs. time.

mass unbalance drifts (0.2°/hr/g), plus misalignments and scale factor errors at 1 min and 0.1%, respectively. Unmodeled satellite observation bias of 500 ft was also included, accompanied by assumed rms noise levels of 100 ft and 0.5 fps for differences in range and range rate.

Performance of the satellite-inertial system, as described in Ref. 8-1, can be summarized as follows: Position and velocity errors settle at rms values of roughly twice the total fix uncertainties. Unlike position errors, velocity estimates first undergo a transient (ψgt) due to tilt. One of the horizontal velocity error components is plotted in Fig. 8-4; the tilt transient peaks between 150 and 200 sec, when satellite geometry allows velocity to become observable (Ex. 8-8). Values eventually approached for tilt ($\sim$0.15 mr) and azimuth error ($<$2 mr) are comparable to examples considered previously here. Further attributes of the satellite-inertial system characterization included (1) separation of the vertical from horizontal navigation, as discussed in Sections 6.1 and 6.2, (2) augmentation states for gyro drift, as discussed in Section 6.4, (3) enhancement of azimuth observability by turn maneuvers, as discussed in Ref. 8-5, (4) degradation of azimuth gyro calibration by the dive–pullout sequence, caused by the unmodeled mass unbalance drift, and (5) an ensemble/theoretical position error ratio about equal to the bias/standard deviation factor of five, in the satellite range difference measurements.

8.3 AIR-TO-AIR TRACKING

In marked contrast to navigation, tracking determines a path for a *target*, from external measurements. The "target" may be an airliner observed by an air traffic controller, as a special case [e.g., $\mathbf{Q} \equiv \mathbf{0}$ in (6-43)] of the scenario to be discussed here. More generally, however, the position reference is a moving platform in a tracking vehicle, referred to as the *interceptor* in this section. Because tracking measurements are external, only a loose dynamic model (e.g., a band-limited acceleration vector, with time-varying autocorrelation properties) can be imposed on the estimated trajectory. At the same time, the target can hamper external tracker modeling by maneuvering unpredictably. Tracking with a simple dynamic model is then achieved by limiting data windows to short durations; instantaneous estimates are insensitive to observations taken several seconds earlier (Ex. 8-9).

Adoption of a Kalman estimator only begins to define an air-to-air tracking approach. Also included in a complete definition (Ex. 8-10) are the data rates, disposition of radar doppler information, means of antenna stabilization, and detailed computational procedures, as well as the number and types of states. Conventions chosen to represent state vectors illustrated here are Cartesian coordinates, with components expressed in the reference frame (e.g., geographic, wander azimuth) used for INS implementation. In addition to a natural interface with inertial data, this choice of convention facilitates isolation of the tracker from antenna servo errors, which have hindered earlier mechanizations (see Refs. 8-6 and 8-7 and Ex. 8-11). Further benefits include linear dynamics for accurate extrapolation through blind periods (Ex. 8-12), and an opportunity for analytical solution of applicable equations such as (5-48). By reducing the estimation equations to compact closed-form expressions, airborne computer software requirements are minimized.

TRACKING WITH A SLAVED ANTENNA

For single target tracking, beam *stabilization* involves control of antenna orientation, in order to maintain target illumination by the radar. Antenna-mounted gyros have been used in conventional stabilization designs but, if high attitude data rates are permissible (Ex. 8-13), a beam can be slaved computationally, without gyros. In *any* case, the approach considered here makes no use of gyro data for *tracking*; an unaugmented nine-state formulation includes Cartesian components of relative position $\boldsymbol{R}_S$, relative velocity $\dot{\boldsymbol{R}}_S$, and total target acceleration $\boldsymbol{Z}_T$. Estimator dynamics are based on a modification of (6-43), with band limiting of $\boldsymbol{Z}_T$ as in (5-61).

A system that incorporates the features just described is shown in Fig. 8-5. Initial conditions, plus target and interceptor maneuver sequences, define the state at all times. The estimator reacts to unpredictable target

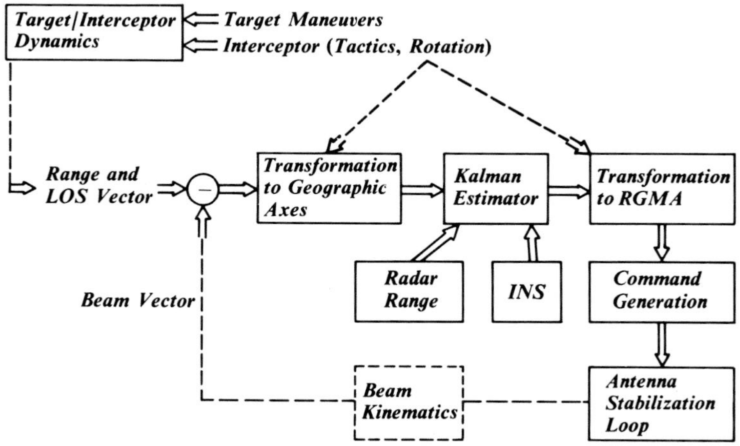

FIG. 8-5. Air-to-air tracking system.

maneuvers by prescribing the aforementioned band-limited acceleration model for data fitting. Radar range and LOS information, in conformance with (7-8) and (7-12), is transformed into weighted residuals expressed in a geographic frame for updating. Current estimates are used to compute LOS in radar gimbal mechanical axes (RGMA) for antenna stabilization.

Simulation of the system in Fig. 8-5 is reported in Ref. 8-6. A Kalman filter was inserted into the ENCOUNTER program (originally compiled by E. C. Quesinberry), which coordinates a sequence of target/interceptor dynamics, radar observation and processing, and antenna servo drive. The target velocity vector can be varied sequentially, in accordance with a maneuver plan having specified timing and g-levels. The interceptor can be given either its own independent maneuver sequence or a guidance course such as lead pursuit. A simplified aerodynamic force model can also be activated, with corresponding "wind axis-referenced" dynamic variables in conformance to accepted conventions and to the instantaneous orientation of the physical aircraft axes. These axes rotate in response to applied forces (which are dictated by the aforementioned maneuver sequence or guidance course), necessitating stabilization commands for the antenna servo. Also driving the servo are tracking outputs from the extended Kalman filter; antenna dynamics are modeled in detail, including digitized command response, torque limits, main beam pattern, and all gimbal rotation characteristics. Several sources of bias and random degradations are taken into account. These include scintillation effects on LOS ("glint" plus amplitude fluctuations) and on range measurements, data processing delays, mechanical response lags (e.g., due to inertia and friction in antenna drive mechanisms), gimbal pickoff resolution, radar resolution, misalignments, parameter uncertainties, range-dependent signal-to-noise ratios, INS tilt, drift, radar data blackouts, and computational truncation.

The subject of computation with limited word length goes beyond the scope of this book. Briefly, the mechanization described here is 16-bit fixed point—with over 95% of the computation performed in single precision. Comparison with 36-bit floating point computation showed word truncation effects to be small, relative to other error sources just listed. To achieve these results, over a six-octave range span, dynamic scaling and matrix fixing procedures were adopted.* For implementing the Kalman estimation equations, (5-50) was found adequate despite the computational superiority

* Every variable in a dynamically scaled algorithm can be expressed as a fraction of a prescribed maximum; scale factors are integral powers of two, which can be switched in accordance with any unit-consistent scheme when range octave boundaries are crossed. Matrix fixing can provide specified bandwidth and stability characteristics by preselecting elements for $\mathbf{P}$ in (5-44), while still accounting for changes in measurement geometry and accuracy via $\mathbf{H}$ and $\mathbf{R}$, respectively.

of (5-41). In concurrence with prefiltering concepts, integration and decrementing of **P** were not performed on every update. Data window length was controlled by process noise spectral densities (Ex. 8-14).

Simulation conditions in Ref. 8-6 include various parameter values, target/interceptor geometries, maneuvers, and ranges from less than 1 to nearly 80 nm. One situation that taxes the acquisition capability at short range is a large value for the quantity

$$\text{aspect angle} \triangleq \text{Arccos}\{-\boldsymbol{R}_S^{\mathrm{T}} \boldsymbol{V}_T/(|\boldsymbol{R}_S||\boldsymbol{V}_T|)\} \qquad (8\text{-}10)$$

so that there is a large directional deviation of total target velocity $\boldsymbol{V}_T$ from the range vector. For the case to be discussed here, initial conditions are 120° aspect, 180° interceptor heading, 6000-ft range, and zero relative bearing (Ex. 8-15). Both interceptor and target vehicles have speeds of 1000 fps. The target initiates a maneuver at the start of the run, reaching a nominal 3-g level at 0.8 sec, and the interceptor imitates the target maneuver after a 3-sec delay. In addition to the 1000-fps initial velocity error (Ex. 8-16), initial position errors corresponding to 1° in azimuth and elevation were introduced, as well as 500-ft initial range error in a direction chosen to aggravate the velocity uncertainty transient.

Estimation accuracy for the case just described is shown in Fig. 8-6. The interceptor roll maneuver (Ex. 8-17) at 3 sec produces no significant tracking degradation. Tracker isolation from antenna stabilization errors, as a marked improvement over early conventional designs, is also dramatically illustrated in Ref. 8-7. Other features worthy of attention are:

(1) *Prompt reduction of initial errors.* Reference 8-6, which employed a narrower tracking bandwidth during pull-in, also exhibited acceptable "lock-on" performance for *total* antenna/target deviations (i.e., combined effects of estimator uncertainty and imperfect stabilization).

(2) *Separate rate error reduction.* Velocity estimates can be conditionally adjusted from doppler information, as already postulated in Ex. 8-10d.

TRACK-WHILE-SCAN (TWS)

A radar antenna beam can be mechanically rotated or electronically steered, so that targets within a chosen sector in space are illuminated sequentially. Proper cataloging of detections received requires separate identification of individual echo sequences, i.e., each detection must be correctly associated with the detections from the same target on preceding scan frames. Correct association is hampered by several phenomena, including (1) motion of targets between frames, (2) flight of targets into or out of the scanned sector, (3) dispersion of target clusters that, for closing

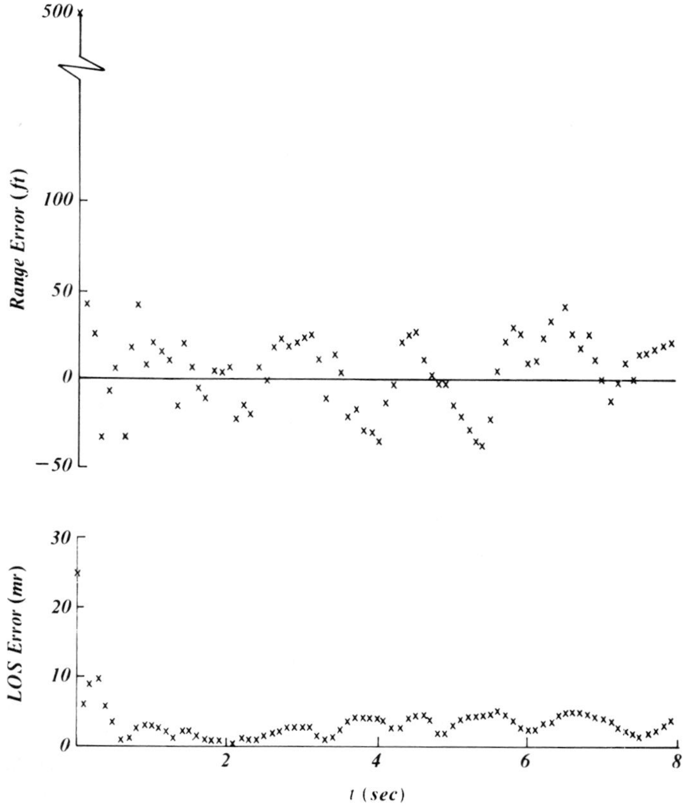

FIG. 8-6. Position error at 6 Kft range [*Notes*: (1) 1-mr misalignment in Φ, Θ, Ψ; (2) initial LOS error is root sum square of 1° (AZ) and 1° (EL); (3) antenna gimbal limit data cutoff occurs before end of run].

ranges, subtend increasing solid angles, (4) crossing target paths, (5) false detections, and (6) loss of detections through rf interference. In operation, the identification of individual target sequences is based on comparisons between successive measurements (e.g., range, range rate, LOS). Establishment and maintenance of correct track files will be straightforward only when detections are reasonably consistent and widely separated (e.g., when the time integral of relative velocity between frames is smaller than the distance between neighboring targets). The ensuing analysis will assume correct association of all detections and concentrate on the resulting estimation accuracy.

Consider a scanning radar beam with a uniform interval $t = 3$ sec between successive frames. Suppose that one of the targets repeatedly illuminated had an initial range (at the time $t = 0$ of the first detection) of 34 nm, after which the range closed at a constant 1200-kt rate for 72 sec (Ex. 8-18). Let a six-state Kalman filter provide three-dimensional estimates of position and velocity only (Ex. 8-19). A total of 24 frames would then be processed by the time $t = 72$ sec, when a 10-nm range is reached. At that distance, the data window could be allowed to span the 72-sec duration, as long as residuals conformed to the nominal straight path (Ex. 8-20). Tracking accuracy obtained at $t = 72$ will now be estimated on the basis of block processing relationships.

If one of the Cartesian position states were measured directly, with constant rms error σ, (5-17) would apply to that state and its derivative, with

$$\mathbf{H}_m = [1, 0], \qquad \mathbf{\Phi}_m = \begin{bmatrix} 1 & mt \\ 0 & 1 \end{bmatrix}, \qquad \mathbf{J}_m = [1, mt] \tag{8-11}$$

so that, for application of (5-26),

$$\mathbf{J} = \begin{bmatrix} 1 & t \\ 1 & 2t \\ 1 & 3t \\ \vdots & \vdots \\ 1 & \mathsf{M}t \end{bmatrix} \tag{8-12}$$

and

$$\mathbf{J}^{\mathrm{T}}\mathbf{J} = \begin{bmatrix} \mathsf{M} & t\mathsf{M}(\mathsf{M}+1)/2 \\ t\mathsf{M}(\mathsf{M}+1)/2 & t^2\mathsf{M}(\mathsf{M}+1)(2\mathsf{M}+1)/6 \end{bmatrix} \tag{8-13}$$

From (5-31), estimation error covariance after M observations is

$$\sigma^2(\mathbf{J}^{\mathrm{T}}\mathbf{J})^{-1} = \frac{2\sigma^2}{\mathsf{M}(\mathsf{M}-1)} \begin{bmatrix} 2\mathsf{M}+1 & -3/t \\ -3/t & 6/\{(\mathsf{M}+1)t^2\} \end{bmatrix} \tag{8-14}$$

and the interpretation is further enhanced by the following observations:

(1) All radar measurements are unambiguous for this analysis.
(2) With less than two measurements, a singularity is present.
(3) For $\mathsf{M} \gg 1$, rms uncertainty approaches $2\sigma/\mathsf{M}^{1/2}$ and

$$2\sqrt{3}\,\sigma/(t\mathsf{M}^{3/2}) = 2\sqrt{3}\,\sigma/(\mathsf{M}^{1/2}\mathsf{T}),$$

in position and velocity, respectively, where T represents elapsed time since track initiation.

(4) The constant-σ assumption used here is reasonable for estimation along range but, for cross-range directions, position measurement errors are better characterized by angular uncertainties multiplied by range.

Approximate performance can then be determined by placing bounds on rms error.

(5) Performance is easily bounded for irregular frame rates. If t can vary between 2 and 4 sec, for example, rms velocity uncertainty must lie between $(1/3)^{1/2}\sigma/\mathrm{T}$ and $(2/3)^{1/2}\sigma/\mathrm{T}$ in the modification of item (3).

8.4 POINT NAV PERFORMANCE AND OPERATION

Navigation with respect to a designated point is typified by data windows of a few minutes duration, with three-dimensional position and velocity determination as the primary objective. Often there is a concurrent effort to place the velocity in the vertical plane containing the aircraft and the reference point.* Present discussion is concerned with position and velocity estimation, irrespective of piloting considerations.

A flight path segment is shown in Fig. 8-7, along with a sensor-monitored area and its replica on a display. Cursor positioning controls,

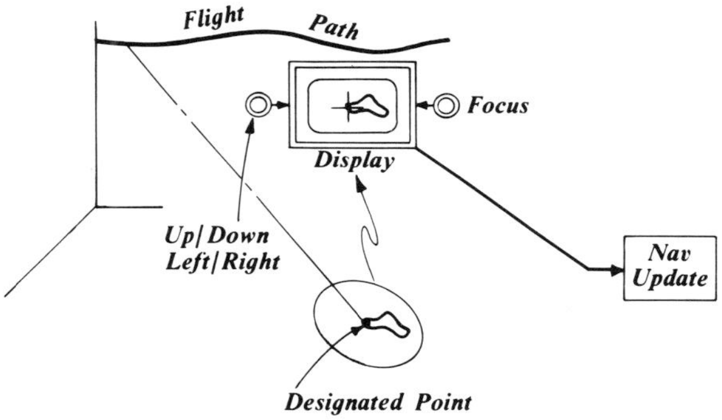

FIG. 8-7. Display operation.

as described in Section 6.5, provide residuals for discrete updating upon INSERT. Position residuals, accurate to within σ_m ft and spanning a data window T sec, can theoretically reduce velocity uncertainties to order σ_m/T fps. A restriction applicable to velocity error reduction, however, will now be recalled. With an initial velocity uncertainty far greater than σ_m/T, and all updates characterized by position relative to a fixed origin, the theoretical reduction of velocity uncertainty cannot be realized until T sec after point nav mode initiation.

* During a landing approach with crosswind present, for example, it is still desirable to maintain the localizer angle Y_L in Fig. 7-2 near zero. The aircraft is then flown such that crosswind is counterbalanced by either a heading offset or a rudder-induced lateral airspeed component.

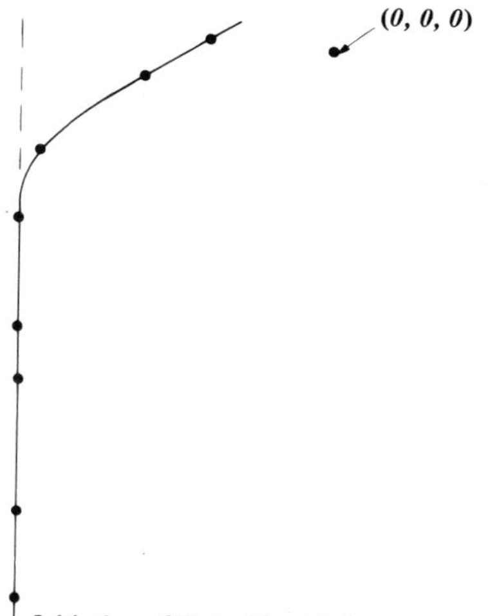

FIG. 8-8. Range observations for point nav mode (range measurements taken at marked dots).

In many applications, only partial position fixes are available. Geometry will then further restrict the velocity accuracy obtainable from sequential position observations. In Fig. 8-8, for example, suppose that only scalar range to the origin can update (6-33) at various points along the path. Before navigation estimates can be refined in three dimensions, the aircraft/origin displacement vector must rotate (preferably more than 30°) in both azimuth and elevation. A flight path segment near the origin is particularly effective for observability in the vertical direction. Range measurement scheduling is not critical, but near-uniform spacing is desirable.

Extension of the foregoing principles beyond a single stationary origin is straightforward. The case of a moving reference point constitutes a tracking operation, as in Sections 6.6 and 8.3, where velocities as well as displacements are relative. Multiple checkpoints of unknown location can also be used to update velocity; the equations in Section 6.5 are applicable, with a maximum of three position states introduced for each checkpoint. Partial knowledge of checkpoint location can alleviate the necessity for additional states (e.g., for horizontal navigation with radar range and azimuth fixes, to checkpoints unknown in elevation only).

INS updating from landing aids is a perfect example of point nav with data windows of a few minutes duration. The principles just discussed are applicable to azimuth, elevation, and distance observed with respect to a runway reference point (Ex. 8-21). Reference 8-8 describes the processing of localizer and INS data in a SceniCruiser test vehicle (probably the only

Greyhound bus in history ever equipped with that combination of sensors). Smooth convergence of position estimates toward the runway centerline was achieved.

EXERCISES

8-1 (a) Recall the definition of $(\mathbf{I}_P, \mathbf{J}_P, \mathbf{K}_P)$ for gimballed platforms given near the end of Section 3.3. Suppose that lines were etched outside a sealed IMU, and consider an attempt to point an etched line along a visually observed vertical reference. Would leveling errors exceeding the initial tilt in Table 3-2 seem likely to arise?

(b) After the leveling operation just described, consider an attempt to point another etched line along the sightline to an object ahead of the aircraft, at "known" bearing off the local meridian tangent. Discuss possible accuracy and field-of-view constraints.

(c) Recall the definition of strapdown reference axes, given in Section 3.3. Briefly discuss alignment by crude visual sightings.

8-2 (a) Are (8-1) and (8-2) in the proper form for estimation? If so, find equations of the corresponding general form in Chapter 5.

(b) Show that nonzero errors in $\mathbf{f}$ and $\mathbf{V}$ produce small modeling errors in (8-2).

(c) Show an augmented form of (8-2), using the form of (5-61) for drift rates.

(d) Recall the digital integrating accelerometer discussion in Section 3.3. Note that a pulsing rate of ($4g \div \frac{1}{8}$ fps) would be equal to ($1g \div \frac{1}{32}$ fps), and discuss the merits of rescaling accelerometer pulses during initial leveling.

8-3 (a) With $\boldsymbol{L}_1$ nearly parallel to $\boldsymbol{L}_2$, discuss possible difficulties in the computational inversion for (8-5). Generalize this conclusion for other angular separations between $\boldsymbol{L}_1$ and $\boldsymbol{L}_2$.

(b) Why are $\boldsymbol{L}_1$ and $\boldsymbol{L}_2$ not restricted to unit vectors?

(c) Relate the closing statement of Section 7.2.5 to the statement at the end of the subsection "Optical Aiding."

8-4 Write an expression similar to (7-79) for the position of a slave IMU. Differentiate and draw an analogy between the kinematical adjustment and the last term on the right of (4-102).

8-5 (a) Even with an accurate initial leveling operation, could warmup drifts introduce tilt transients amenable to short-term error analysis of Section 3.4.2?

(b) Verify the 17-fps figure in the text, using (3-48) and differentiation of (3-53).

(c) Verify (8-8) with estimation equations as follows: The first matrix on the right of (3-44) is a transition matrix, **H** is [1, 0], and inertial instrument errors are negligible for short durations. Let $\mathbf{P}_0$ be dominated by initial tilt effects, so that a 3-min extrapolation through (5-48) is dominated by the quantity

$$\mathbf{P}_1^- \doteq \begin{bmatrix} 1 & gt \\ -t/\mathscr{R} & 1 \end{bmatrix} \begin{bmatrix} 0 & 0 \\ 0 & 9(10)^{-6} \end{bmatrix} \begin{bmatrix} 1 & -t/\mathscr{R} \\ gt & 1 \end{bmatrix}$$

and (5-44) produces a weighting matrix

$$\mathbf{W} \doteq \left\{1 - \left(\frac{2}{3 \times 10^{-3} \times gt}\right)^2\right\} \begin{bmatrix} 1 \\ 1/(gt) \end{bmatrix}$$

Application of (5-50) then produces a covariance matrix with diagonal elements of approximately $(2)^2$ and $(3.5 \times 10^{-4})^2$.

8-6 (a) Azimuth alignment *could* be performed in flight, but not under every possible set of conditions. Show that fixed westward speed of $\mathscr{R}\omega_s \cos \lambda$ would render $\psi_{A(0)}$ ineffective, and therefore unobservable.

(b) Show that very low groundspeed in any direction would nearly isolate the drift component about $\mathbf{I}_G$, while combining the drift rate about $\mathbf{J}_G$ with $\psi_{A(0)}\omega_s \cos \lambda$. Compare with self-alignment after leveling.

8-7 (a) Hyperbolic measurements are most beneficial on the base line. Show that, for aircraft position between stations of Fig. 7-4, position uncertainty after a fix is half the range difference measurement error.

(b) Suppose that off the base line position uncertainty is amplified by the cosecant of half the angle subtended by the stations at the aircraft position. Draw an analogy with the discussion after (8-8), and after (7-30).

8-8 (a) Show that 0.1° tilt produces about 8-fps velocity error after roughly 160 sec. Show that the accompanying position error transient stays appreciably below 1700 ft.

(b) After correction of large velocity transients through optimal updating, it is not necessary to perform added computations to recover position errors accumulated before velocity became observable. Explain why, and extend this reasoning to tilt and azimuth error effects.

8-9 (a) Sketch an arbitrary curvilinear path and show that it can be subdivided into short contiguous arcs of varying radii. Show that if contiguous segments are restricted to straight lines, more segments are needed. Relate this conclusion to constant-velocity versus accelerating target models. Can allowable segment length be related to aerodynamic limitations of tracked objects?

(b) Instead of separate contiguous arcs, consider a track evolving from overlapping segments of a curvilinear path. Relate this concept to sequential processing of measurements, with an advancing data window as described near the start of Section 5.4.

8-10 (a) Given the equivalence between block and recursive optimal estimation, assume that (5-31) can characterize sequential estimation error covariance within a data window T. When (5-26) applies, show that rms estimation error varies directly with rms measurement error.

(b) With uniform direct position observations at equally spaced intervals, rms uncertainties in the block estimation case considered above would vary inversely as the square root of the number of measurements. State a proportionality relation between tracking radar data rate, data window, measurement errors, and tracking accuracies. Briefly discuss the three-dimensional implications.

(c) Consider the observed range and LOS between an interceptor and an engine on a target vehicle. Do these quantities provide a valid measure of target position?

(d) Recall how engines were mentioned briefly at the end of Appendix 7.B. Make a case for using range and LOS data, but excluding doppler measurements that are suspected of influence by engine modulation. If absence of modulation effects can be established, could doppler information be used to refine the *outputs* of a tracker fed by position data alone?

8-11 (a) Consider three rays emanating from a tracking radar. One ray points along a true target sightline vector $\boldsymbol{R}_S$, while a second ray is parallel to an estimated target sightline $\hat{\boldsymbol{R}}_S$ and the third ray represents an antenna axis $\mathbf{I}_b$. Show that, irrespective of beam orientation, cross-range position uncertainty can be expressed in terms of a vector product involving $\boldsymbol{R}_S$ and $\hat{\boldsymbol{R}}_S$.

(b) From (7-12) note that LOS observations are obtained in a form dependent upon beam orientation. A distinction is made here between tracking and stabilization; the latter involves correction of departures between target and beam orientation. Recall the statement following (7-12). Explain how a tracking and stabilization system can "render to Caesar the things that are Caesar's" by using adjusted observations for one purpose and undiminished signals for another.

8-12 (a) Recalling Ex. 7-5c, explain why angular measurements can be very nearly linear functions of the state, even with the INS reference frame described in the text.

(b) Show that the dynamic relation (6-43) is also linear, and, if (5-61) were used to characterize band-limited accelerations, the dynamics would remain linear.

(c) Consider an alternative tracker state formulation in terms of spherical coordinates. Sketch a pair of flight paths for "antiparallel fly-by," in which an interceptor and a target move in opposite directions along straight lines 1000 ft apart. With both vehicles flying at 1000 fps, sketch the time history of range R. Show that $\ddot{R}$ reaches a maximum of over $100g$!

(d) Suppose that mechanical gimbal constraints cause interruption of radar data before the sidelong position is reached, in the scenario just described, but another sensor illuminates the target after fly-by. In view of the large magnitudes and rapid variations in $\ddot{R}$, what are the prospects for extrapolation through an observationless interval? Compare with the feasibility of extrapolating separate Cartesian components of the relative (target/interceptor) position vector. How can accurate estimates for Z_T affect the extrapolation?

(e) When total target velocity V_T is perpendicular to the relative position vector, radar measurements are obscured by main beam clutter, defined in Appendixes 7.A and 7.B. Suppose that, at the same time, $\ddot{R}$ is large and subject to rapid variations. Even under less extreme conditions than depicted in part (c), would separate extrapolation of Cartesian position components be more reliable than range extrapolation?

8-13 (a) Imagine an interceptor experiencing high angular rates, causing target illumination by the radar beam to be disturbed. Can changes in the INS gimbal angles be used for generating beam pointing commands, to maintain target illumination without gyros? What problem would arise if the interval between successive commands were equal to the ratio of antenna beamwidth to the maximum angular rate?

(b) Must an antenna servo be capable of accommodating stabilization commands equal to the aircraft angular rates? Must the servo respond to changing commands when angular accelerations are present?

(c) Consider an antenna servo that maintains target illumination during interceptor rotation, but not without fractional-beamwidth pointing errors. Must tracking uncertainties be at least as large as servo pointing errors?

8-14 (a) Assume that, aside from target maneuvers, unpredicted drift in tracker extrapolation is characterized by a given rms distance in a given duration. Explain how spectral densities in (5-92) can be chosen in conformance to the assumption.

(b) Use Ex. 6-17a to describe the influence of maneuver characteristics on data window duration.

8-15 Recall from Fig. 1-9 that true bearing to an object is measured from $\mathbf{I}_G$, and relative bearing differs from true bearing by the heading angle. Show that, with no azimuth displacements between aircraft roll axis and velocity vectors, target heading for the case in the text is 240°.

8-16 (a) With no velocity measurements, and initial target velocities assumed equally likely in all directions, what should be the initial estimate for target velocity? For relative velocity? What is the corresponding error in the velocity estimate?

(b) Using the form of (7-9), express closing target acceleration along the range vector in terms of tracker state variables. Using the form of (7-35) and (7-36), write expressions for the target's true bearing and elevation above a level plane passing through the tracking antenna. Recalling Ex. 7-6b, define the LOS rate in terms of tracker states.

8-17 Use (4-97) to compute the roll angle corresponding to a 0.1-rad/sec turn rate at 1000 fps.

8-18 (a) Show that, from an initial range of 34 nm, a target closing at 1200 kt reaches a 10-nm range in 72 sec.

(b) *Dwell time*, i.e., the duration over which an individual target is illuminated by a scanning radar beam, is essentially the ratio of antenna beamwidth to the scan rate. Spectral width of the return from an isolated target, nominally twice the reciprocal of the dwell time, represents the limit of doppler resolution obtainable. For a 3-cm wavelength in (7-60), with a 3° beamwidth and 120°/sec scan rate, could two adjacent targets be detected separately if their range rates differ by 2 fps? By 8 fps?

(c) For any given frame rate, can larger volumes in space be scanned with faster or slower scan rates? In choosing a scan rate, does a compromise appear necessary for doppler resolution versus size of the scanned sector?

8-19 (a) Suppose that, midway between 34- and 10-nm range, a closing target initiates a $2g$ turn. Show that azimuth LOS changes by less than $\frac{1}{8}°$ within a 3-sec frame interval.

(b) Recall the velocity uncertainty discussion in Ex. 8-16a. At track initiation, are all effects of velocity uncertainties easily distinguishable from target acceleration effects?

8-20 (a) Consider a TWS operation, with a long-range target executing maneuvers that reverse every few seconds. Use the results of the preceding exercise to show that a six-state formulation with a 72-sec data window is still permissible between 34 and 10 nm, with velocity states representing components of an average velocity vector.

(b) If the reversing maneuvers just described were replaced by a turn, sustained for several 3-sec frame intervals, how would the residual editing test of Section 5.4.4 perform in a six-state formulation? Should the data window then be reduced? Relate the answer to Ex. 8-9.

(c) Postmultiply (8-14) by (8-13).

(d) Note that, for a fixed data window in (5-57), the shape of the effective attenuating characteristic for past data will vary with *relative* spectral densities in (5-56). The exact nature or amount of influence exerted by data from the remote past is far less important than the value of T, but the opportunity to shape the attenuation characteristic is worth noting. Show that, for any fixed T in correspondence to position measurements with dynamics of (5-92), the condition $\eta_v = 0$ provides the most gradual attenuation.

8-21 The Chester A. Budney International Airport, with runways 200 ft wide, can easily accommodate the localizer/INS updating system described in Ref. 8-8. Even with a 1-mr unmodeled platform tilt, azimuth errors are held to about one-eighth the runway width. With the same principles applied to the elevation channel, however, explain why dynamics suggested by (6-39) would be preferable to (6-33). In the full MLS/INS three-dimensional application, what would be the effect of unobservable DME bias?

REFERENCES

8-1. Danik, B., and Stow, R. C., "Integrated Hybrid-Inertial Aircraft Navigation Systems," presented at *AGARD Guidance and Control Symp.*, Delft, Netherlands, Sept. 23, 1969.

8-2. Bryson, A. E., "Simplified Design of Filters to Estimate Misalignment Angles and Gyro Drifts of a Stationary IMU," Stanford University Dept. Aero/Astro. Report 449, Dec. 1972.

8-3. Thompson, E. H., Farrell, J. L., and Knight, J. W., "Alignment Methods for Strapdown Inertial Systems," *AIAA JSR* **3** (9), Sept. 1966, pp. 1432–1434.

8-4. Baziw, J., and Leondes, C. T., "In-Flight Alignment and Calibration of Inertial Measuring Units—Part I: General Formulation, Part II: Experimental Results," *IEEE Trans.* **AES-8** (4), July 1972, pp. 439–465.

8-5. Bryson, A. E., "Rapid In-Flight Estimation of IMU Platform Misalignments Using External Position and Velocity Data," Final Report on Air Force Avionics Lab. Contract AFAL-TR-73-288, Sept. 1973, pp. 19–24.

8-6. Farrell, J. L., Quesinberry, E. C., Morgan, C. D., and Tom, M., "Dynamic Scaling for Air-to-Air Tracking," 1975 NAECON Record, Dayton, Ohio, pp. 157–162.

8-7. Clark, B. L., and Gaston, J. A., "On-Axis Pointing and the Maneuvering Target," 1975 NAECON Record, Dayton, Ohio, pp. 163–170.

8-8. Burrows, J. W., "Combined Inertial–ILS Aircraft Navigation Systems," *AIAA J. Aircraft* **8** (6), June 1971, pp. 439–443.

Appendix I

Nomenclature

Mathematical analysis of aircraft navigation involves far more variables than there are letters of the alphabet. A systematic and consistent notation is nevertheless afforded by the use of upper- and lowercase, Latin and Greek letters, upright, italicized, script, Fraktur, boldface, and sans serif type, subscripts, and arguments. Any of these devices can alter the meaning of a symbol. Superscripts, in contrast, are used here only to denote certain mathematical operations or modifiers applied to the basic variable. This section defines important notations (including units, dimensions, and/or coordinate frames where applicable) with meanings given for subscripts, superscripts, brackets, abbreviations, acronyms, and certain functional designations. Many standard functions [Arcsin (principal value), arctan (4-quadrant), ! (factorial), lim (limit), Σ (summation), etc.] are used without tabulation here, however, and no effort is made to catalog all specialized notation appearing in isolated parts of this book (such as appendices). Any potential uncertainties (e.g., whether to interpret parenthetical quantities as arguments) are easily resolved by context. The following definitions are applicable throughout this book.

I.1 BRACKETS

| | absolute value, root sum square of components, or determinant, as applied to scalars, vectors, or matrices, respectively

$[\cdot]_u$ 3 × 3 orthogonal transformation matrix corresponding to a rotation through (·) rad about the u axis (u replaceable by x, y, or z)

$\langle\ \rangle$ probabilistic mean; theoretical *ensemble* average

I.2 FUNCTIONS

Function	Applicable arguments	Indicated operation
av( )	scalar, vector, or matrix	*time* average
δ( )	scalar time	Dirac delta
exp( )	scalar or square matrix	exponential function
trace( )	square matrix	sum of diagonal elements
arctan(·/·)	ratio of two scalars	inverse tangent of the quotient, in quadrant dictated by algebraic signs given for the numerator and denominator

I.3 SYMBOLS

This section defines scalar, vector, matrix, and quaternion notation. Wherever appropriate, one possible set of units is prescribed which, though not intended to restrict the *text*, will allow application of all given *equations* with no conversion factors added. Dimensions for matrices and vectors are also given, wherever a consistent characterization is possible. Coordinate frames are indicated for vectors where applicable. Quantities are listed here with some essential subscripts, but brackets and superscripts are omitted from the alphabetical lists except for variables that never appear without them. Separate listings are also provided for superscripts and subscripts.

SCALARS

Symbol	Sections	Definition	Units
a_j, a_{yj}, a_{zj}	4.3, 4.3.1, 4.3.2, 4.B	vibratory specific force along nominal x, y, and z axis respectively, of jth inertial instrument	fps^2
a_E	3.4.1, 3.6, 3.A	semimajor axis of ellipsoid	ft
b_1, b_2, b_3; b_{L1}, b_{L2}, b_{L3}	2.3, 3.6	components of $\mathbf{b}$; of $\mathbf{b}_L$	dimensionless
b_1, b_2, b_3, b_4	2.3	components of $\boldsymbol{b}$	dimensionless
b_D	6.1	groundspeed bias	fps
b_Z	5.3.1	acceleration bias	fps^2

(continued)

Symbol	Sections	Definition	Units
b, b_j	2.3	scalar element in a quaternion, or in jth quaternion, respectively	dimensionless
b_L	3.6	scalar element in $\boldsymbol{b}_L$	dimensionless
c_{E1}, c_{E2}, c_{E3}	1.3.1	components of $\mathbf{c}_E$	dimensionless
c_{P1}, c_{P2}, c_{P3}	7.2.4	components of $\mathbf{c}_P$	dimensionless
c	7.2.2, 7.A, 7.A.1, 7.B, 7.C	speed of light	fps
d	3.4.2, 5.4.2, 8.2	constant drift rate	rad/sec
d_i	7.C	incremental range corresponding to ith image cell	ft
e_j	5.3.1	jth component of $\mathbf{e}$	see Note 1
e_{a1}, e_{a2}, e_{a3}	6.2, 6.3	components of $\mathbf{e}_a$	fps^2
e_{vj}	6.1	velocity noise in jth horizontal direction	fps
$e_{\omega 1}, e_{\omega 2}, e_{\omega 3}$	6.3	components of $\mathbf{e}_\omega$	rad/sec
e_E	3.4.1, 3.6, 3.A, 7.2.3	ellipsoid eccentricity	dimensionless
$\mathsf{e}_i, \mathsf{e}_i'$	7.B	radar energy received *at* and *from* ith reflector, respectively	arbitrary
f	7.A, 7.A.1, 7.B	carrier frequency	Hz
f_a	4.3, 4.3.2, 4.B	center frequency of vibratory specific force	Hz
f_i, f_j	7.C	incremental doppler shift corresponding to specified image cells	Hz
f_k	7.C	frequency in spectral expansion of motion compensation error	Hz
f_0	7.C	frequency of sinusoidal motion compensation error	Hz
f_w	4.3, 4.3.1, 4.4, 4.B	center frequency of oscillatory angular rate	Hz
f_i	7.B, 7.C	total doppler shift for a particular radar reflector	Hz
g	1.4, 1.5, 3.4.1, 3.4.2, 4.0, 4.1, 4.3, 4.3.1, 4.4, 4.A, 5.4.2, 6.2, 6.3, 6.A, 6.B, 8.1–8.3	acceleration due to gravity	fps^2
h	1.1, 1.3.3, 3.1, 3.4.1, 3.6, 5.1, 5.3.1, 6.3, 6.5, 7.2.1, 7.2.3	aircraft altitude (above sphere, ellipsoid, or local terrain, as dictated by context)	ft

(*continued*)

Symbol	Sections	Definition	Units
h_k	7.2, 7.2.1	altitude of *k*th station	ft
$\imath$	7.C	square root of (-1)	dimensionless
n_{a1}, n_{a2}, n_{a3}	3.4.2, 4.0, 4.3.2, 5.4.2, 6.B	components of $\mathbf{n}_a$	fps^2
$n_{\omega1}, n_{\omega2}, n_{\omega3}$	3.4.2, 4.3.1, 5.4.2, 6.B	components of $\mathbf{n}_\omega$	rad/sec
p, q, r	2.2.2	roll, pitch, and yaw rate, respectively; components of $\boldsymbol{\omega}_A$	rad/sec
$\mathscr{p}, \mathscr{q}, \mathscr{r}$	2.2.2, 4.A	components of $\boldsymbol{\zeta}$	rad/sec
r	3.6	difference of reference ellipsoid curvature radii	ft
s_j	6.5	*j*th component of **s**	see Note 1
$\mathfrak{s}$	4.2.1, 4.3.3	Laplace differentiation operator	sec^{-1}
t; t_0	see Note 2	time; reference time	sec
t_m	5.0, 5.1, 5.2.1, 5.2.2, 5.3.1, 5.3.2, 5.4.4, 6.1	time of *m*th discrete observation	sec
t	7.A.1	radar two-way transit time	sec
t	3.5	strapdown DDA iteration interval	sec
u	4.A	true airspeed (TAS)	fps
v_i	7.B	closing range rate between aircraft and a particular radar reflector	fps
υ	3.3, 3.5, 7.C	velocity resolution of integrating accelerometer	fps
$w_j, w_{\mathscr{y}j}, w_{\mathscr{z}j}$	4.3, 4.3.1, 4.3.4, 4.B	oscillatory angular rate about nominal $\mathscr{x}$, $\mathscr{y}$, and $\mathscr{z}$ axis, respectively, of *j*th inertial instrument	rad/sec
x; x_j	5.1, 5.2.1, 5.2.2, 5.3.1, 5.B, 6.1–6.4, 7.2.4	general state variable; *j*th component of **x**	see Note 1
$\tilde{x}_E$; $\tilde{x}_H$	3.4.2, 5.4.2, 6.A, 8.2	east position uncertainty; total horizontal position uncertainty	ft
x, y	7.2.2	aircraft position with respect to hyperbolic station reference	ft
$\mathscr{x}, \mathscr{y}, \mathscr{z}$	1.2, 2.1, 2.2, 2.2.1–2.2.3, 2.3, 2.4, 2.A.2, 3.5, 3.6, 3.A, 4.2.1, 4.2.2, 4.3, 4.3.1–	coordinate reference directional designations	not applicable

(*continued*)

Symbol	Sections	Definition	Units
	4.3.4, 4.4, 4.A, 6.3, 6.5, 6.6, 7.1.1–7.1.3, 7.2.2, 7.2.4		
y	5.2.2, 5.B	scalar observation free of measurement error	see Note 1
y_{aj}, z_{aj}	4.3.2	cross-axis specific force sensitivities due to misalignment of jth accelerometer	rad
$y_{\omega j}$, $z_{\omega j}$	4.3.1	cross-axis rate sensitivities due to misalignment of jth gyro	rad
z	5.0, 5.2.1, 5.2.2, 5.B, 6.1	scalar observation	see Note 1
A; A_{max}	3.1, 3.3, 3.5, 7.C	specific force; maximum value	fps^2
A_N, A_E, A_V	3.4.1, 3.4.2, 6.2, 6.3, 6.5	north, east, and downward components of specific force, respectively	fps^2
$\mathscr{A}$; $\mathscr{A}_j$	7.1.1, 7.1.2	radar antenna azimuth angle; value on jth observation	rad
$\mathfrak{a}$	7.1.3	optical sensor inner azimuth gimbal angle	rad
$\mathscr{A}_D$	2.4, 6.6	azimuth deflection of reflector	rad
B_j, B_{yj}, B_{zj}	4.3.1, 4.3.2	total specific force along nominal x, y, and z axis, respectively, of jth inertial instrument	fps^2
B	3.2, 4.2.1, 4.2.2, 4.3.3	viscous restraining torque constant of rate integrating gyro	slug-ft^2/sec
$\mathscr{B}$	5.5	noise bandwidth of velocity error excitation	Hz
C_j, C_{yj}, C_{zj}	4.3.1	steady specific force along nominal x, y, and z axis, respectively, of jth inertial instrument	fps^2
C_{ik}, D_{ik}	4.B	spectral component amplitudes for oscillatory angular rate waveform	rad/sec
D_{Cj}	7.1.4	indexed horizontal position coordinates of compass locator antenna	ft
D_L	7.1.4	localizer antenna coordinate	ft
D	3.4.2	time-varying effective drift rate	rad/sec

(*continued*)

Symbol	Sections	Definition	Units
$\mathscr{D}; \mathscr{D}_j$	7.1.1, 7.1.2	radar antenna depression angle; value on *j*th observation	rad
E_ω	4.3	spectral density of angular rate noise	$(\text{rad/sec})^2/\text{Hz} = \text{sec}^{-1}$
$\mathscr{E}$	7.1.3	optical sensor outer elevation gimbal angle	rad
$\mathscr{E}_D$	2.4, 6.6	elevation deflection of reflector	rad
$F_1, F_2, F_3;$ $\mathsf{F}_1, \mathsf{F}_2, \mathsf{F}_3$	4.A	components of **F** and ***F***, respectively	fps^2
F	4.3.3	transfer function in feedback path of platform servo	lb-ft/rad
G	3.1	gravitational acceleration	fps^2
G	4.3.3	gyro transfer function	dimensionless
H	3.2, 4.2.1, 4.2.2, 4.3.3	angular momentum of unperturbed gyro rotor	$\text{slug-ft}^2/\text{sec}$
$\mathscr{I}$	4.2.1	moment of inertia of gyro rotor about a diametral axis through its mass center	slug-ft^2
$\mathscr{I}_{Aj}$	2.A.2, 4.A	moment of inertia of aircraft about *j*th principal axis through its mass center	slug-ft^2
J	4.2.1, 4.2.2, 4.3.3	moment of inertia of gyro gimbal about OA line passing through its mass center	slug-ft^2
J	4.3.3	moment of inertia of gimballed platform about a principal axis through its mass center	slug-ft^2
$\mathscr{J}$	4.2.1	moment of inertia of gyro rotor about a polar axis through its mass center	slug-ft^2
K	4.3.4	number of active iterations by strapdown DDA	dimensionless
K	4.3.3	platform servo gain factor	sec
$\mathscr{K}$	7.2.2	nominal radius for hyperbolic navigation computations	ft
L	5.2.1, 5.A	weighted sum of squares of normalized true residuals	dimensionless
L	3.2	gyro torquing command	lb-ft
$\mathsf{L}_D, \mathsf{L}_E, \mathsf{L}_F$	4.3.3	disturbance, error, and feedback torque, respectively	lb-ft
$\mathscr{L}$	7.2.2	half-length of base line segment between hyperbolic navigation stations	ft
M	5.2.1, 5.2.2, 8.3	observation block size	dimensionless

(continued)

Symbol	Sections	Definition	Units
M_m	5.1, 5.2.1, 5.4.5	number of simultaneous scalar observations in *m*th data vector	dimensionless
N_1, N_2, N_3	2.4	components of $\boldsymbol{N}$	dimensionless
N	5.1, 5.2.1, 5.2.2, 5.4.5	number of state variables	dimensionless
$\mathrm{N_D}, \mathrm{N_S}$	1.2.1	number of days and additional seconds, respectively, elapsed since initialization of Ξ	dimensionless
$\mathcal{N}$	4.3.4	number of derivatives contained in numerical integration algorithm	dimensionless
R_E	1.1, 1.2.2, 1.4, 1.5, 3.4.1, 6.5, 7.2, 7.2.1	nominal earth radius	ft
R	7.A, 7.A.1, 7.B, 7.C	range	ft
$\mathrm{R_M}, \mathrm{R_P}$	1.1, 3.4.1, 3.6, 7.2.3	reference ellipsoid radii along meridian and equatorial directions, respectively	ft
$\mathcal{R}$	1.1, 1.2.2, 1.4, 3.4.1, 3.4.2, 3.B, 4.0, 6.1, 6.3, 6.B	nominal distance from earth center	ft
S_j	6.5, 6.6, 7.1, 7.1.1–7.1.4	*j*th component of **S**	see Note 1
S_{a1}, S_{w1}	4.3, 4.B	series proportional to components of **a** and **w**, respectively	fps^2 and rad/sec, respectively
S_{a2}, S_{w2}	4.3, 4.3.1	oscillatory series involving products of components of **a** and **w**, respectively	$(fps^2)^2$ and $(rad/sec)^2$, respectively
T	7.A.1, 7.B	interpulse period	sec
T	5.2.2, 5.3.1, 5.4, 5.4.2, 6.1, 6.2, 8.4	data window duration	sec
U_{I1}, U_{I2}, U_{I3}	2.0	components of $\mathbf{U}_I$	fps
V_H	1.3.2, 6.3, 7.1.1	groundspeed; total horizontal component of **V**	fps
V_{L1}, V_{L2}, V_{L3}	3.6	components of $\mathbf{V}_L$	fps
V_N, V_E, V_V	1.1, 1.2, 1.3.2, 1.3.3, 1.4, 3.4.1, 3.4.2, 3.6, 3.B, 4.0, 5.1, 5.3.1, 5.4.2, 6.1–6.3, 6.5, 6.B, 8.2	north, east, and downward components of geographic velocity, respectively	fps

(*continued*)

Symbol	Sections	Definition	Units
$\tilde{V}_b$, $\tilde{V}_0$	7.C	instantaneous value and amplitude of sinusoidal range rate uncertainty, respectively	fps
$\tilde{V}_k$	7.C	component amplitude in spectral expansion for range rate uncertainty	fps
W	3.5	integer exponent chosen for strapdown gyro calibration; $W = -\log_2 \phi$	dimensionless
W	3.4.2, 5.4.2, 6.3	Schuler frequency	rad/sec
$\mathscr{W}$	4.3.4	word length of strapdown DDA for attitude	bits
X_j	6.1, 6.3, 7.2, 7.2.1–7.2.4, 8.2	jth component of $\mathbf{X}$	see Note 1
Y	1.3.1, 6.1, 7.1.1–7.1.4, 7.2.1–7.2.4, 8.1, 8.4	observable; type of measurement indicated by subscript	see Note 1
Z_V	5.1, 5.3.1, 5.3.2	total downward acceleration	fps^2
$\tilde{Z}_b$	7.C	instantaneous acceleration error in motion compensation	fps^2
$\tilde{Z}_k$	7.C	spectral component of $\tilde{Z}_b$	fps^2
α	4.A	angle of attack	rad
α_t	4.A	trim angle of attack	rad
a	3.6	wander angle	rad
β	4.A	sideslip	rad
$\boldsymbol{\beta}$	4.3.2	coefficient for accelerometer error sensitive to motion indicated by axis subscripts	see Note 3
Y	4.2.2, 4.3.1	coefficient for gyro drift sensitive to motion indicated by axis subscripts	see Note 3
δ	4.3, 5.1	Dirac delta; see table of functions	sec^{-1}
ϵ; ϵ_j	5.0, 5.2.1, 5.3.2, 5.4.2	measurement error; jth component of $\boldsymbol{\epsilon}$	see Note 1
ϵ_A, ϵ_E	6.6, 7.1.2	azimuth and elevation pointing error measurements, respectively	rad
ζ	1.3.2, 4.A, 6.1, 6.5, 7.1.1	ground track angle	rad
$\Delta\zeta$	6.5	change in ζ since mode initialization	rad
$\boldsymbol{\zeta}$	4.3.3	gimballed platform servo damping coefficient	dimensionless

(*continued*)

Symbol	Sections	Definition	Units
η_a	5.5	acceleration noise spectral density	$(fps^2)^2$/Hz
η_g	6.2	acceleration noise spectral density	g^2/Hz
η_v	5.5, 6.1, 8.0	velocity noise spectral density	$(fps)^2$/Hz
$\boldsymbol{\eta}$	7.2.1, 7.2.5	local eastward magnetic variation	rad
θ	2.1, 2.2.1, 2.2.3, 2.3	angle (general)	rad
$d\theta$	2.2.2	infinitesimal angular increment	rad
$\boldsymbol{\theta}$; $\boldsymbol{\theta}_j$	3.2, 4.2, 4.2.1, 4.2.2	gyro gimbal displacement about OA relative to case; value for jth gyro	rad
ϑ, ϑ_i	7.C, 8.2	angle between velocity vector and a sighting vector	rad
κ_j, κ_{yj}, κ_{zj}	4.3.1	quasi-static angular rate about nominal x, y, and z axis, respectively, of jth inertial instrument	rad/sec
λ	1.1, 1.2.2, 1.3.1, 1.3.2, 1.4, 2.1, 3.2, 3.4.1, 3.6, 3.B, 6.1, 6.3, 6.5, 8.1	geodetic latitude	rad
λ_j, λ_k	7.2, 7.2.1, 7.2.2	geodetic latitude of jth and kth stations	rad
$\boldsymbol{\lambda}_{cm}$, $\boldsymbol{\lambda}_{rf}$	7.A–7.C	radar wavelength	cm and ft, respectively
$\boldsymbol{\mu}_{aj}$	4.3.2	scale factor error of jth accelerometer	dimensionless
$\boldsymbol{\mu}_{\omega j}$	4.3.1	scale factor error of jth gyro	dimensionless
μ_1, μ_2, μ_3; ν_1, ν_2, ν_3	3.5	discrete rotation increments for aircraft and reference frame, respectively, with possible values $(0, \pm\phi)$	rad
$\boldsymbol{\nu}_{aj}$	4.3.2	motion-independent bias of jth accelerometer	fps^2
$\boldsymbol{\nu}_{\omega j}$	4.3.1	motion-independent drift bias of jth gyro	rad/sec
ξ_1, ξ_2, ξ_3	4.3.4	components of $\boldsymbol{\xi}$	rad
ρ_{ij}	4.3.4, 4.B	normalized cross-correlation coefficient between oscillatory motion about ith and jth axes	dimensionless

(*continued*)

Symbol	Sections	Definition	Units
σ	5.2.1, 6.1, 8.3, 8.4	standard deviation of measurement error	see Note 1
σ_W	6.1	rms effective gust velocity	fps
τ	5.3.1, 6.4	bias time constant	sec
τ_{W1}, τ_{W2}	6.1	time constant for effective velocity disturbance in north and east directions, respectively	sec
$\boldsymbol{\tau}_{aj}, \boldsymbol{\tau}_{\omega j}$	4.3.1, 4.3.2, 4.3.4	differential delay (relative to slowest inertial instrument) of jth accelerometer and jth gyro, respectively	sec
$\boldsymbol{\tau}_G$	7.A.1, 7.C	pulse width	sec
φ	1.1, 1.2.2, 1.3.1, 1.3.2, 1.4, 2.1, 3.2, 3.4.1, 3.6, 6.1, 6.5	longitude	rad
φ_j, φ_k	7.2, 7.2.1, 7.2.2	longitude of jth and kth stations	rad
ϕ	3.3, 3.5, 4.3.1, 4.3.4	strapdown gyro resolution	rad
$\boldsymbol{\phi}; \boldsymbol{\phi}_j$	3.2, 4.2, 4.2.1, 4.2.2	rotation experienced by gyro about IA; value for jth gyro	rad
$\boldsymbol{\varphi}$	4.3.3	gimballed platform servo error	rad
ψ_N, ψ_E, ψ_A	1.2.3, 1.4, 2.1, 3.4.2, 3.B, 4.3, 5.4.2, 6.2–6.4, 6.B, 7.2.4, 7.2.5, 8.1, 8.2	components of $\boldsymbol{\psi}$	rad
$\tilde{\boldsymbol{\psi}}$	7.C	dynamic rf phasing error	rad
ω_{ba}, ω_{be}	7.1.2	azimuth and elevation angular rates, respectively, of beam coordinate frame	rad/sec
$\omega_j, \omega_{yj}, \omega_{zj}$	4.2.2, 4.3, 4.3.1–4.3.3	total angular rate about nominal x, y, and z axis, respectively, of jth inertial instrument	rad/sec
ω_{max}	3.5	maximum angular rate of vehicle	rad/sec
ω_s	1.2.1, 1.2.2, 3.1, 3.2, 3.4.1, 3.6, 3.B, 6.3, 8.1	sidereal rate of earth rotation	rad/sec

(*continued*)

Symbol	Sections	Definition	Units
$\omega_{A1}, \omega_{A2}, \omega_{A3}$	2.2.2, 2.3, 2.A.2, 4.3.2	components of $\boldsymbol{\omega}_A$	rad/sec
ω_n	4.3.3	gimballed platform servo natural frequency	rad/sec
$\varpi_1, \varpi_2, \varpi_3$	3.2, 3.4.1, 3.4.2, 3.6, 3.B, 6.3	components of $\boldsymbol{\varpi}$	rad/sec
Δ_{ij}	3.5	change in direction cosine produced by strapdown gyro pulse or reference increment	dimensionless
Θ	1.2.3, 2.1, 2.2.1, 2.2.2, 3.3, 4.4, 4.A, 6.6	aircraft pitch angle	rad
Λ	5.5, 6.3	eigenvalue	see Note 1
Ξ	1.2.1, 1.3.1, 2.1, 3.6, 7.2.4	hour angle of the vernal equinox	rad
Φ	1.2.3, 2.1, 2.2.1, 2.2.2, 3.3, 4.A, 6.6	aircraft roll angle	rad
Ψ	1.2.3, 1.3.2, 2.1, 2.2.1, 2.2.2, 3.2, 3.3, 4.4, 4.A, 6.6, 7.1.1, 7.1.2, 7.2.5	aircraft heading angle	rad
Ω	4.2.1, 4.2.2	gyro rotor speed	rad/sec

COLUMN VECTORS

For 3 × 1 Cartesian vectors, a parenthetical "index" below refers to the coordinate frame in which the symbol is defined. For all other vectors, the index without parentheses denotes the dimensionality.

Symbol	Index	Sections	Definition	Units
a	(see Note 3)	4.3, 4.3.1, 4.B	vibratory specific force experienced by inertial instruments	fps^2
b	(see Note 4)	2.3, 3.6	vector partition of four-parameter rotation operator	dimensionless
b	4	2.3, 2.A.1	complete set of rotation parameters with respect to inertial reference	dimensionless
c	(see Note 4)	1.2.1, 2.3, 7.2.4	star sightline vector	dimensionless

(*continued*)

Symbol	Index	Sections	Definition	Units
$\mathbf{d}$	(A)	2.1	vector between two points fixed on rigid aircraft	ft
$\mathbf{d}_{C/N}$	(A)	4.A	fixed displacement from aircraft mass center to IMU reference point	ft
$\mathbf{d}_{N/R}$	(A)	7.C	fixed displacement from IMU reference point to radar antenna phase center	ft
$\mathbf{d}_{N/S}$	(A)	8.1	fixed displacement between reference points in master and slave IMU's	ft
$\mathbf{e}$	N	4.0, 5.0, 5.1, 5.2.2, 5.3.2, 5.4, 5.4.1, 5.4.5, 6.1–6.3, 6.5, 6.6	white noise forcing function for state dynamics	see Note 1
$\mathbf{e}_a$	(G)	4.0, 4.3, 4.4, 4.B, 6.3	acceleration noise	fps^2
$\mathbf{e}_\omega$	(G)	4.0, 4.3, 4.3.1, 4.3.3, 4.3.4, 4.4, 4.B, 6.2, 6.3, 8.1	random rotational drift rate	rad/sec
$\boldsymbol{\varepsilon}$	N	6.4	random forcing function for augmented state dynamics, modeled as white noise	see Note 1
$\mathbf{f}$	(G)	3.4.2, 4.0, 4.4, 6.3, 6.A, 6.B	computed acceleration adjustment	fps^2
$\mathbf{g}$	(G, L, 0)	3.4.1, 3.4.2, 3.6, 3.A, 4.0, 4.4, 4.A, 6.3, 6.A, 6.B	gravity vector in any level frame	fps^2
h	4	3.6	complete set of wander azimuth rotation parameters with respect to earth reference	dimensionless
i	(A)	7.B	unit sightline vector from aircraft to a particular radar reflector	dimensionless
$\mathbf{n}_a$	(see Note 3)	3.4.2, 4.0, 4.4, 4.B, 6.B	general representation for acceleration error	fps^2
$\mathbf{n}_\omega$	(see Note 3)	3.4.2, 3.B, 4.0, 4.3.4, 4.4, 4.B	general representation for rotational drift rate	rad/sec
$\mathbf{q}$	(0)	6.5	incremental velocity since nav mode initiation	fps

(*continued*)

Symbol	Index	Sections	Definition	Units
$\mathbf{r}_j$	(A)	4.3.2	displacement of jth accelerometer from IMU reference point	ft
s	N	6.5, 6.6, 7.0	uncertainty in state **S**	see Note 1
w	(see Note 3)	4.3, 4.3.3, 4.3.4, 4.4, 4.B	oscillatory angular rate experienced by inertial instruments	rad/sec
x	N	4.0, 5.0, 5.1, 5.2.1, 5.2.2, 5.3.2, 5.4.1, 5.4.3, 5.A, 6.1, 6.3, 6.5, 7.0	state of linear system or uncertainty in state **X** of nonlinear system	see Note 1
y	M	5.1, 5.2.2	data vector free of measurement error	see Note 1
z	M or M	5.1, 5.2.1, 5.2.2, 5.3.2, 5.4.1, 5.A	data vector	see Note 1
A	(see Note 4)	1.1, 3.1, 3.4.1, 3.4.2, 3.5, 3.6, 3.A, 4.1, 4.3, 4.4, 4.B, 6.A, 6.B	specific force	fps²
B	(see Note 3)	4.3, 4.3.1	total specific force experienced by inertial instruments	fps²
C	(see Note 3)	4.3, 4.3.1	static specific force experienced by inertial instruments	fps²
D, ***D***	3	2.4	incident and reflected vectors, respectively, as expressed in a mirror reference frame	arbitrary
E	3	2.2.3, 2.3, 2.A.1	principal eigenvector of 3×3 orthogonal matrix, as expressed in either of the two frames involved in the transformation	dimensionless
F	(A)	4.4, 4.A, 4.B	specific force experienced at IMU reference point	fps²
F	(A)	4.A	specific force experienced at aircraft mass center	fps²
G	(see Note 4)	1.1, 3.1, 3.4.1, 3.6, 3.A, 4.1, 6.3, 6.A	gravitational acceleration	fps²
H	(see Note 4)	2.A.2	angular momentum	slug-ft²/sec
$\boldsymbol{I}_j$	(0)	7.1.1, 7.1.2	unit vector along beam at time of jth doppler observation	dimensionless

(*continued*)

Symbol	Index	Sections	Definition	Units
I, J, K	3	see Note 2	unit vectors denoting reference directions in three-dimensional space	dimensionless
L	(see Note 4)	2.A.2, 3.2, 4.2.1, 4.3.3	torque	lb-ft
$\boldsymbol{L}_1$, $\boldsymbol{L}_2$	(G)	8.1	reference vectors known in geographic coordinates	arbitrary
$\boldsymbol{M}$, $\boldsymbol{N}$	3	2.4	unit vectors parallel and perpendicular to reflecting plane, respectively, expressed in a mirror reference frame	dimensionless
$\langle \mathbf{N}_a \rangle$	(G)	4.0, 4.4, 4.B, 6.4	total effective acceleration bias, including rectification effects	fps^2
$\langle \mathbf{N}_\omega \rangle$	(G)	4.0, 4.4, 4.B, 6.4	total effective angular drift rate bias, including rectification effects	rad/sec
0	general	4.0, 5.1, 5.2.2, 6.6, 8.3	null vector	arbitrary
$\mathbf{P}_j$, $\mathbf{P}_k$	(E)	7.2, 7.2.1–7.2.3	satellite or earth station positions	ft
$\boldsymbol{P}_\mathrm{E}$	(E)	7.2.3	aircraft location in geocentric reference	ft
Q	(0)	6.5, 6.6, 8.3	position change since nav mode initiation	ft
R	(see Note 4)	2.2.2, 3.1, 3.4.1, 3.A, 4.1, 6.3, 6.5, 6.A, 7.2.1, 7.2.2	geographic position of aircraft or of airborne IMU	ft
$\boldsymbol{R}$	(see Note 4)	1.1, 5.5, 7.1, 7.1.1, 7.1.2, 7.C, 8.3	point-referenced position; definition completed by subscript	ft
S	*N*	6.5, 7.0	point-referenced state	see Note 1
$\mathbf{T}_0$	(G)	6.5	earth-referenced position of target point	ft
U	(see Note 4)	2.0, 2.2.2, 3.1, 3.2, 3.6	aircraft velocity with respect to earth center	fps
V	(see Note 4)	1.1, 2.0, 3.1, 3.2, 3.4.2, 3.6, 4.4, 4.A, 4.B, 6.3, 6.5, 6.A, 6.B, 7.B, 7.C	geographic velocity experienced by aircraft or by airborne IMU	fps
$\boldsymbol{V}_\mathrm{A}$	(A)	4.A	airspeed vector	fps
$\boldsymbol{V}_\mathrm{W}$	(A)	4.A	wind velocity	fps
$\boldsymbol{V}_\mathrm{T}$	(see Note 4)	1.1, 6.6, 8.3	target velocity	fps

(*continued*)

Symbol	Index	Sections	Definition	Units
$\mathbf{W}$	N	6.1	set of estimation weights for a scalar measurement	see Note 1
$\mathbf{X}$	N	5.4.1, 5.4.3, 6.1, 6.3, 7.0	earth-referenced state; full state with nonlinear dynamics	see Note 1
$\mathbf{Y}$	M or $\mathbf{M}$	5.4.1	noise-free data vector in nonlinear system	see Note 1
Z_T	(see Note 4)	1.1, 6.6, 8.3	total acceleration of target	fps^2
$\boldsymbol{\beta}$	variable	5.3.2	bias vector	see Note 1
$\boldsymbol{\epsilon}$	M or $\mathbf{M}$	5.1, 5.2.1, 5.2.2, 5.3.2, 5.4, 5.4.1, 5.4.3, 5.4.5, 5.B	data vector observation error	see Note 1
$\boldsymbol{\zeta}$	(A)	2.2.2, 4.A	aircraft angular rate with respect to platform	rad/sec
$\mathbf{d\theta}$	(A)	2.2.2	infinitesimal rotation	rad
$\boldsymbol{\kappa}$	(see Note 3)	4.3, 4.3.1	quasi-static angular rate experienced by inertial instruments	rad/sec
$\hat{\boldsymbol{\lambda}}_1$, $\hat{\boldsymbol{\lambda}}_2$	(vehicle)	8.1	reference vectors as computed from airborne observations	same as $\boldsymbol{L}_1$, $\boldsymbol{L}_2$, respectively
$\boldsymbol{\xi}$	(A)	4.3.4, 4.4	strapdown system attitude uncertainty with respect to inertial reference	rad
$\boldsymbol{\rho}_E$, $\boldsymbol{\rho}_I$	(I)	4.1	position of earth and of accelerometer triad, respectively, in an inertial reference frame	ft
$\boldsymbol{\psi}$	(G)	2.1, 3.3, 3.4.2, 3.B, 4.0, 4.3.3, 4.4, 4.B, 6.2, 6.3, 6.5, 6.B, 7.2.4, 8.1, 8.2	inertial system attitude uncertainty with respect to geographic reference	rad
$\boldsymbol{\psi}_G$	(G)	4.4	contribution to $\boldsymbol{\psi}$ from uncertainty in orientation of geographic frame itself	rad
$\boldsymbol{\omega}$	(see Note 3)	4.3	directly measureable (absolute) angular rate	rad/sec
$\boldsymbol{\omega}_A$	(A)	2.2.2, 2.3, 2.A.1, 2.A.2, 4.3, 4.3.2–4.3.4, 4.4, 4.A, 8.1	absolute angular rate of aircraft	rad/sec
$\boldsymbol{\omega}_B$	(see Note 4)	3.B, 4.3.4	apparent angular rate	rad/sec

(*continued*)

Symbol	Index	Sections	Definition	Units
$\boldsymbol{\omega}_E$, $\boldsymbol{\omega}_{sL}$, $\boldsymbol{\omega}_s$	(E), (L), (G), respectively	3.1, 3.4.1, 3.6, 3.A, 6.A	sidereal rate of earth rotation	rad/sec
$\boldsymbol{\omega}_G$	(G)	2.2.2, 3.2	angular rate of geographic frame	rad/sec
$\boldsymbol{\omega}_L$	(L)	3.6	angular rate of wander azimuth frame	rad/sec
$\boldsymbol{\omega}_P$	(P)	2.2.2, 3.B, 4.3.3, 4.3.4	platform angular rate	rad/sec
$\boldsymbol{\omega}_R$	(R)	4.2.1, 4.2.2	angular rate of gyro rotor	rad/sec
$\boldsymbol{\varpi}$	(ref. frame)	2.2.2, 3.2, 3.4.1, 3.4.2, 3.5, 3.6, 3.B, 4.0, 4.3, 4.4, 4.B, 6.3, 6.B	angular rate of reference frame (equated to $\boldsymbol{\omega}_G$ in text)	rad/sec
$\mathbf{1}_j$	K (see definition)	3.6, 7.1.1, 7.1.3, 7.1.4, 7.2.4	jth column of $K \times K$ identity matrix	dimensionless

MATRICES

The "index" column above is no longer applicable, but array dimensions are given as conventionally defined. Units are not tabulated below, since most matrices have mixed or dimensionless elements, easily determined from applicable text or definitions. For those few matrices allowing a consistent characterization of elements, units are included in the definition below.

Symbol	Dimensions	Sections	Definition
A	$N \times N$	4.0, 5.0, 5.1, 5.2.2, 5.3.2, 5.4.1, 5.5, 6.1–6.5	dynamics matrix; $\partial \dot{\mathbf{X}}/\partial \mathbf{X}$
$\mathcal{A}$	$N \times N$	6.1, 6.4	augmented dynamics matrix; $\partial \dot{\mathbf{X}}^*/\partial \mathbf{X}^*$
E	$N \times N$	5.1, 5.2.2, 5.3.2, 5.4, 5.4.2, 5.4.3, 5.5, 6.1, 6.2	spectral density for dynamic excitation noise
F	3×2	6.5	proportionality between current position error vector and initial platform tilts in point nav mode (fps^2)
G	3×3	6.5	proportionality between current geographic distance uncertainty and initial platform misorientation (fps^2)

(*continued*)

Symbol	Dimensions	Sections	Definition
$\mathbf{H}$	$M \times N$	5.0, 5.1, 5.2.1, 5.2.2, 5.3.1, 5.3.2, 5.4.1–5.4.3, 6.1, 8.3	proportionality between measurements and current state
$\mathscr{H}$	$1 \times N$	5.2.2, 5.3.2, 5.5, 6.1	$\mathbf{H}$-matrix for scalar measurement
$\mathbf{I}$	square	2.2.2, 2.3, 2.4, 2.A.1, 3.3, 3.4.2, 3.5, 4.0, 4.4, 5.2.1, 5.2.2, 5.5, 5.A, 6.4–6.6, 6.A, 6.B, 7.2.4	identity matrix (dimensionless)
$\mathscr{I}_{\mathrm{A}}$	3×3	2.A.2	centroidal principal inertia matrix for aircraft (slug-ft^2)
$\mathscr{I}_{\mathrm{R}}$	3×3	4.2.1	centroidal principal inertia matrix for gyro rotor (slug-ft^2)
$\mathbf{J}$	$\mathrm{M} \times N$	5.2.1, 5.3.2, 5.A, 5.B, 8.3	proportionality between measurements and initial state
$\mathbf{O}$	variable	4.0, 4.4, 4.B, 5.3.2, 5.5, 6.1, 6.2, 6.4–6.6	null matrix
$\mathbf{P}$	$N \times N$	5.2.2, 5.3.2, 5.4, 5.4.1–5.4.3, 5.5, 6.1, 6.5, 8.3	state uncertainty covariance matrix
$\mathscr{Q}$	$N \times N$	5.2.2	integrated effect of $\mathbf{E}$
$\mathbf{R}$	$M \times M$	5.1, 5.2.2, 5.3.2, 5.4, 5.4.1, 5.4.3, 6.1, 8.3	measurement uncertainty covariance matrix
$\mathbf{S}$	$\mathrm{M} \times N$	5.2.1, 5.A, 5.B	proportionality between estimate and residuals
$\mathbf{T}$	3×3	2.0, 2.1, 2.2.1–2.2.3, 2.3, 2.A.1, 2.A.2, 3.1–3.3, 3.5, 3.6, 3.B, 4.2.1, 4.3.3, 4.3.4, 4.4, 4.A, 4.B, 6.3, 6.6, 6.A, 7.1.1–7.1.4, 7.2.1, 7.2.2, 7.C, 8.1	orthogonal transformation between two right-hand coordinate frames (dimensionless)
$\mathscr{T}$	3×3	2.4, 6.6	orthogonal transformation obtained with pairs of reflectors (dimensionless)
$\mathbf{U}$	$\mathrm{M} \times \mathrm{M}$	5.2.1, 5.A, 5.B	inverse square root of $\mathbf{R}$
$\mathbf{W}$	$N \times M$	5.2.2, 5.4.1–5.4.3	set of estimation weights for a data vector
$\mathbf{Z}$	$M \times M$	5.2.2	covariance matrix of predicted residuals
$\mathscr{L}$	9×9	6.5	azimuth transformation for position and velocity states (dimensionless)

Symbol	Dimensions	Sections	Definition
$\boldsymbol{\mu}$	3 × 3	3.5	skew-symmetric gyro increment matrix (dimensionless)
$\boldsymbol{\nu}$	3 × 3	3.5	skew-symmetric reference rotation increment matrix (dimensionless)
$\boldsymbol{\tau}$	square	5.3.2, 6.4	diagonal matrix of time constants (sec)
$\boldsymbol{\Gamma}$	3 × 3	2.4	transformation between a right- and a left-hand coordinate frame (dimensionless)
$\boldsymbol{\Phi}$	$N \times N$	5.1, 5.2.1, 5.2.2, 5.3.1, 5.3.2, 5.5, 6.5, 8.3	state transition matrix between two discrete instants of time (e.g., t_{m-1} and t)
$\boldsymbol{\Psi}$	4 × 4	2.3, 2.A.1	skew symmetric matrix of angular rates in four-parameter differential equation (rad/sec)
$\boldsymbol{\Omega}_A$	3 × 3	2.2.2, 2.A.1, 2.A.2, 3.5, 4.3.4, 4.A	$(-\boldsymbol{\omega}_A \times\ )$ operator (rad/sec)
$\boldsymbol{\Omega}_G$	3 × 3	2.2.2, 3.4.1, 3.5	$(-\boldsymbol{\omega}_G \times\ )$ operator (rad/sec)
$\boldsymbol{\Omega}_P$	3 × 3	3.B	$(-\boldsymbol{\omega}_P \times\ )$ operator (rad/sec)

QUATERNIONS

Section 2.3 uses uppercase sans serif italic boldface type to define the rotation operator $\boldsymbol{B}$, the null quaternion $\boldsymbol{O}$, and three quaternions ($\boldsymbol{C}_A$, $\boldsymbol{C}_I$, $\boldsymbol{U}$), which correspond to the vectors $\mathbf{c}_A$, $\mathbf{c}_I$, $\frac{1}{2}\boldsymbol{\omega}_A$, respectively.

NOTES

1. States, observables, and all related functions contain quantities of varying units and dimensions.
2. Scalar time and reference axis vectors enter most of the analyses contained in this book.
3. Forces, angular rates, and degradations experienced by inertial instruments are referenced to geographic coordinates for gimballed platforms, and to aircraft axes for strapdown systems. Coefficients for errors sensitive to these motions, defined in Sections 4.3.1 and 4.3.2, have standard units (ft, sec, rad).
4. Applicable coordinate frames for many vectors are stated in text, often with a subscript indicator (e.g., $\mathbf{c}_A$, $\mathbf{c}_E$, $\mathbf{c}_I$, $\mathbf{c}_P$). Subscripts in the list below can also indicate where dynamic quantities are acting (e.g., $\mathbf{H}_A$, $\mathbf{L}_P$, Z_T). Vectors **A**, **G**, **R**, **V** without subscripts are expressed in the geographic frame.

I.4 SUBSCRIPTS AND SUPERSCRIPTS

A few subscripts used in this book carry connotations following from definitions already given, e.g., a parenthetical (t_0) referring to previously defined reference time, vector or matrix dimensional subscripts $N \times 1$, $M \times N$, etc. Others are used in a general sense, such as (i, j, k) denoting the ith radar reflector, the jth vector component or inertial instrument, the kth station, etc.; these are also used in combination (e.g., to represent an element of a matrix $\mathbf{T}$ by T_{ij}). Still others have definitions dictated by context (e.g., numerals 1, 2, 3 may correspond to vector components along reference **I**, **J**, **K** axes, but are not restricted to that meaning). Certain designations are worth tabulating here, however, since they convey meanings that are not self-evident. The subscripts listed below may appear alone or in combination with others, in or out of parentheses.

COORDINATE AXIS SUBSCRIPTS

In connection with reference directions (**I**, **J**, **K**) or transformation matrices $\mathbf{T}$, coordinate frames are designated as follows:

A	aircraft
A1	axes displaced from the aircraft frame by the roll angle
A2	axes displaced from the aircraft frame by the roll and pitch angles
b	beam
B	apparent directions of axes, slightly misoriented
E	earth
g	gyro gimbal
G	geographic
I	inertial
L	locally level, wander azimuth
M	magnetic
P	platform
R	gyro rotor
0	"target" coordinate frame

DESIGNATION OF OBSERVABLES

As subscripts of Y or $\mathbf{Y}$, the following measurements are indicated:

a	inner azimuth gimbal angle
A	outer azimuth gimbal angle
AE	east specific force (self-align mode)
AN	north specific force (self-align mode)
AZ	azimuth angle from platform-mounted optics
B	marker beacon
C	compass locator
d	doppler range rate, with beam directed toward a designated point
D	inner elevation gimbal depression angle
e	outer elevation gimbal angle
EL	elevation angle from platform-mounted optics
G	glide path angle
h	altitude
H	hyperbolic range difference

j	scanned beam doppler range rate, associated with a specific beam orientation
L	localizer angle
P	distance from DME
R	range
S	satellite range difference
V	VOR bearing
ωa	azimuth LOS rate
ωe	elevation LOS rate
ωE	angular rate about east axis (self-align mode)

ADDITIONAL SUBSCRIPTS WITH REPEATED USAGE

Notations already defined are often used with quantities other than coordinate axes and transformations (e.g., R_E; $\mathbf{L}_g$; $\boldsymbol{\omega}_G$; $\mathbf{U}_I$; $\mathbf{V}_L$) or observables (e.g., D_L; D_{C1}; D_{C2}). The numeral 0 pertains to initial or reference quantities (often in connection with the point navigation reference origin or frame), while a T subscript refers to a *moving* target point. Subscripts of the form m or $m-1$ denote a particular discrete observation. Some variables associated with motion are subscripted by ω or $\underline{w}$ for rotation, or $\underline{a}$ for translation; a subscript $\underline{r}$ denotes relative or rms value. Directional subscripts include N, E, V for north, east, and vertical (downward) translation, respectively; N, E, A for rotation about north, east, and azimuth axes, respectively; and (x, y, z) for coordinate reference designations. When the latter appear as separate or multiple subscripts of scalars, they refer to inertial instrument axes.

SUPERSCRIPTS

The conventional notation $(\ )^{-1}$ is used for the reciprocal of a scalar, and for the inverse of a matrix or quaternion operator. Other standard conventions include dot operators $[(\dot{\ }), (\ddot{\ })]$ for differentiation with respect to time, and angles occasionally expressed in degrees $[(\ )^\circ]$. Additional superscripts used in this book are tabulated below:

$(\ )^T$	transpose of a matrix (or of a vector treated as a matrix)
$(\ )^{\#}$	generalized inverse
$(\hat{\ })$	apparent value; observed, estimated, or computed from observations and estimates
$(\hat{\hat{\ }})$	optimal estimate obtained from smoothing
$(\tilde{\ })$	uncertainty; error in estimate or observation
$(\ )^-$	pertaining to a time immediately before an observation
$(\ )^+$	pertaining to a time immediately after an observation
$(\ )^*$	augmented quantity
$(\breve{\ })$	pertaining to a 90° phase-shifted waveform

I.5 ABBREVIATIONS AND ACRONYMS

The following terms may appear in text or as subscripts, with or without parentheses:

ADF	automatic direction finding
AZ	azimuth
brg	bearing
cm	centimeter
CW	continuous-wave
dB	decibels
DDA	digital differential analyzer
DEC	declination
DFT	discrete Fourier transform
DME	distance measuring equipment
E	east
EL	elevation
E/O	electro-optical
FFT	fast Fourier transform
FM	frequency modulation
fps	feet per second
fps^2	feet per second per second
ft	feet
g	unit of acceleration ($\doteq 32.2$ fps^2)
HARS	heading and attitude reference system
hr	hour
Hz	hertz
IA	input axis (of gyro or accelerometer)
ILS	instrument landing system
im	imaginary component of complex quantity
IMU	inertial measuring unit
in.-Hg	inches of mercury (pressure)
INS	inertial navigation system
IR	infrared
Kft	kilofeet
KHz	kilohertz
kt	knots (nautical miles per hour)
LORAN	LOng RAnge Navigation
LOS	line of sight
mag	magnetic
max	maximum
MBC	main beam clutter
meas	measured
MERU	milli-earth rate unit
milli-g	$10^{-3}g$
min	minute of time
$\widehat{\text{min}}$	minute of arc
MLS	microwave landing system
MMR	multimode radar
mr	milliradian
msec	millisecond
N	north
nav	navigation
nm	nautical mile
OA	output axis (of gyro or accelerometer)
platf	pertaining to a gimballed platform
PRF	pulse repetition frequency
rad	radian
radar	RAdio Detection And Ranging
re	real component of complex quantity
Ref	reference
rf	radio frequency
RGMA	radar gimbal mechanical axes
rms	root mean square
R-nav	area navigation
S	south
SA	spin axis (of gyro)
SAR	synthetic aperture radar
SDF	single-degree-of-freedom (adjective)
sec	second of time
$\widehat{\text{sec}}$	second of arc
SHA	sidereal hour angle
SLR	side-look radar
SRA	spin reference axis (of gyro)
TACAN	TACtical Air Navigation
TAS	true airspeed
TV	television
TWS	track-while-scan
UHF	Ultra High Frequency
UT	Universal Time; Greenwich civil time
VHF	Very High Frequency
VOR	VHF OmniRange
W	west
μsec	10^{-6} second; microsecond
2DF	two-degree-of-freedom (adjective)

Appendix II

Applicable Matrix Theory

Matrices offer a convenient shorthand for summations and simultaneous linear equations. As used here, their properties are also exploited for insight into analytical developments. This appendix, a restrictive collection of well-known principles, contains most of the general matrix theory necessary for this book. Although a few selected topics in matrix analysis appear in individual chapters, text presentation generally presupposes knowledge of the material to follow.

II.1 DEFINITIONS AND CONVENTIONS

Vectors and matrices are written in boldface, with the latter appearing in sans serif type. A boldface script notation $\mathcal{A}$ is used to represent a few matrices. The order of subscripts labeling the ij element of a matrix denotes an entry in the ith row and the jth column; these are reversed by the transpose operator (superscript T). To express an intermediate level of detail without writing every scalar element, a matrix is sometimes decomposed into partitioned submatrices. All of these conventions are illustrated by the vector (or the $N \times 1$ matrix) **A**, the $N \times N$ square matrix A, the larger $K \times L$ matrix $\mathcal{A}$, and the $M \times N$ rectangular matrix B:

$$\mathbf{A} = \begin{bmatrix} A_1 \\ A_2 \\ \vdots \\ A_N \end{bmatrix}, \qquad \mathbf{A}^{\mathrm{T}} = [A_1, A_2, \ldots, A_N] \tag{II-1}$$

$$\mathbf{A} = \begin{bmatrix} A_{11} & A_{12} & \cdots & A_{1N} \\ A_{21} & & & \\ \vdots & & & \vdots \\ A_{N1} & & \cdots & A_{NN} \end{bmatrix}; \qquad \mathbf{A}^{\mathrm{T}} = \begin{bmatrix} A_{11} & A_{21} & \cdots & A_{N1} \\ A_{12} & & & \\ \vdots & & & \vdots \\ A_{1N} & & \cdots & A_{NN} \end{bmatrix} \tag{II-2}$$

$$\mathcal{A} = \left[\begin{array}{c:c} \mathbf{A}_{N \times N} & \mathbf{Q}_{N \times (L-N)} \\ \hdashline \mathbf{R}_{(K-N) \times N} & \mathbf{S}_{(K-N) \times (L-N)} \end{array}\right], \qquad K > N, \quad L > N \tag{II-3}$$

$$\mathbf{B} = \begin{bmatrix} B_{11} & B_{12} & \cdots & B_{1N} \\ B_{21} & & & \\ \vdots & & & \vdots \\ B_{M1} & & \cdots & B_{MN} \end{bmatrix}; \qquad \mathbf{B}^{\mathrm{T}} = \begin{bmatrix} B_{11} & B_{21} & \cdots & B_{M1} \\ B_{12} & & & \\ \vdots & & & \vdots \\ B_{1N} & & \cdots & B_{MN} \end{bmatrix} \tag{II-4}$$

With the column matrix representation for vectors in (II-1), the scalar product (also called the dot product or inner product) of two $N \times 1$ vectors ($\mathbf{U}$, $\mathbf{V}$) can be written

$$\mathbf{V}^{\mathrm{T}}\mathbf{U} = \mathbf{U}^{\mathrm{T}}\mathbf{V} = U_1 V_1 + U_2 V_2 + \cdots + U_N V_N \tag{II-5}$$

In (II-2) the elements A_{11}, A_{22}, ..., A_{NN} comprise the principal *diagonal*; this holds for any square matrix.

As applied to vectors, the absolute value sign $|\,|$ denotes the root sum square of vector components. As applied to matrices, the notation $|\,|$ and $(\ )^{-1}$ denote determinant and unique inversion, respectively, e.g., if

$$\mathbf{TU} = \mathbf{V} \tag{II-6}$$

then

$$\mathbf{U} = \mathbf{T}^{-1}\mathbf{V}, \qquad |\mathbf{T}| \neq 0, \quad \mathbf{T}^{-1}\mathbf{T} = \mathbf{T}\mathbf{T}^{-1} = \mathbf{I} \tag{II-7}$$

where $\mathbf{I}_{N \times N}$ denotes the $N \times N$ identity matrix (i.e., all diagonal elements are unity and all off-diagonal elements are zero; thus for any N-dimensional vector $\mathbf{V}$, $\mathbf{I}_{N \times N}\mathbf{V} \equiv \mathbf{V}$). Similarly, $\mathbf{O}_{M \times N}$ represents the $M \times N$ *null* matrix, consisting entirely of zeros. Subscripts for identity and null matrices are sometimes omitted where context permits.

A rectangular $M \times N$ matrix is square if $M = N$; in general, the *rank* (i.e., greatest number of linearly independent rows or columns) of a matrix can be any integer from 1 to M if $M \leq N$, so that a matrix of rank M is said to have *maximum rank*. At the other extreme, a *unit rank* matrix is exemplified by a vector $\mathbf{V}$ and also by the *outer product*,

$$\mathbf{V} = \mathbf{V}\mathbf{V}^{\mathrm{T}} = \begin{bmatrix} V_1^2 & V_1 V_2 & \cdots & V_1 V_N \\ V_1 V_2 & & & \\ \vdots & & & \vdots \\ V_1 V_N & & \cdots & V_N^2 \end{bmatrix} \tag{II-8}$$

which, when postmultiplied by any vector **U** with the same dimension as **V**, produces a vector parallel to **V** [recall (II-5)]:

$$\mathbf{V}\mathbf{U} = (\mathbf{V}\mathbf{V}^{\mathrm{T}})\mathbf{U} = \mathbf{V}(\mathbf{V}^{\mathrm{T}}\mathbf{U}) = (\mathbf{V}^{\mathrm{T}}\mathbf{U})\mathbf{V} \tag{II-9}$$

This latter exercise involves operations that will now be further scrutinized.

II.2 MATRIX OPERATIONS

For matrix addition, the ij element of a sum $\mathbf{C} + \mathbf{D}$ is simply the sum of the ij elements of $\mathbf{C}$ and $\mathbf{D}$. This operation has meaning only for matrices of equal dimension, e.g., when $\mathbf{C}$ is $M \times N$, then both $\mathbf{D}$ and $\mathbf{C} + \mathbf{D}$ are also $M \times N$. Matrix addition is associative,

$$(\mathbf{C} + \mathbf{D}) + \mathbf{E} = \mathbf{C} + (\mathbf{D} + \mathbf{E}) \tag{II-10}$$

and commutative,

$$\mathbf{D} + \mathbf{E} = \mathbf{E} + \mathbf{D} \tag{II-11}$$

To multiply a matrix by a scalar, each element of that matrix is multiplied by the scalar. Application of this rule with the scalar (-1) easily establishes validity of the above rules for matrix subtraction as well as addition.

For *conformable* matrices a product can be defined; e.g., $\mathbf{BA}$ or $\mathbf{A}^{\mathrm{T}}\mathbf{B}^{\mathrm{T}}$ can be formed from (II-2) and (II-4), where $\mathbf{AB}$ would be meaningless. Thus, matrix multiplication is *not* commutative in general,

$$\mathbf{AB} \neq \mathbf{BA} \tag{II-12}$$

and a $K \times L$ matrix can be postmultiplied by any $M \times N$ matrix only if $L = M$. The specific rule for forming a product can be expressed as follows: Let $\mathbf{A}$ and $\mathbf{B}^{\mathrm{T}}$ of (II-2) and (II-4) be partitioned into columns,

$$\mathbf{A} = [\mathbf{A}_1, \mathbf{A}_2, \ldots, \mathbf{A}_N], \qquad \mathbf{B}^{\mathrm{T}} = [\mathbf{B}_1, \mathbf{B}_2, \ldots, \mathbf{B}_M] \tag{II-13}$$

so that the scalar product $\mathbf{B}_i^{\mathrm{T}}\mathbf{A}_j$ defines the ij element of $\mathbf{BA}$. This definition can be used to show that the transpose of a product is the product of transposed matrices with the order of multiplication reversed,

$$(\mathbf{BA})^{\mathrm{T}} = \mathbf{A}^{\mathrm{T}}\mathbf{B}^{\mathrm{T}} \tag{II-14}$$

A similar rule holds for inverses of square matrices, provided all determinants exist,

$$(\mathbf{BA})^{-1} = \mathbf{A}^{-1}\mathbf{B}^{-1}, \qquad |\mathbf{A}||\mathbf{B}| \neq 0 \tag{II-15}$$

Matrix multiplication is associative,

$$\mathbf{F}(\mathbf{G}\mathbf{H}) = (\mathbf{F}\mathbf{G})\mathbf{H} \tag{II-16}$$

and distributive over addition,

$$(\mathbf{F} + \mathbf{G})\mathbf{H} = \mathbf{F}\mathbf{H} + \mathbf{G}\mathbf{H} \tag{II-17}$$

Matrix derivatives (or integrals) can be obtained by differentiating (or integrating) each element. When vector elements are variables of differentiation, the operator $\partial/\partial\mathbf{X}$ denotes partial differentiation with respect to each component of $\mathbf{X}$. For a scalar variable of differentiation X, the derivative of a product can be expressed as

$$\frac{d}{dX}(\mathbf{F}\mathbf{G}) = \mathbf{F}\frac{d\mathbf{G}}{dX} + \frac{d\mathbf{F}}{dX}\mathbf{G} \tag{II-18}$$

In this book, the only exception to the above notation arises when the variable of differentiation is time. The reason for the departure is a need for two distinct time derivative operators, which generally differ as applied to vectors. The operator d/dt denotes intrinsic variation as *experienced* in a coordinate frame of fixed orientation, although projections of the derivative may be *resolved* along any axes after differentiation. A second operator, denoting separate time differentiation for each scalar element of a vector $\mathbf{V}$ or matrix $\mathbf{M}$ is denoted by a dot ($\dot{}$):

$$\dot{\mathbf{V}} = \begin{bmatrix} \dot{V}_1 \\ \dot{V}_2 \\ \vdots \\ \dot{V}_N \end{bmatrix}, \qquad \dot{\mathbf{M}} = \begin{bmatrix} \dot{M}_{11} & \dot{M}_{12} & \cdots & \dot{M}_{1N} \\ \dot{M}_{21} & & & \\ \vdots & & & \vdots \\ \dot{M}_{N1} & & \cdots & \dot{M}_{NN} \end{bmatrix} \tag{II-19}$$

It is to this latter time differentiation operator that the above product derivative rule applies.

The remainder of this section is dominated by nonsingular square matrices, i.e., matrices having finite nonzero determinant. An operation of particular importance here is *diagonalization*, which will now be addressed from a theoretical viewpoint.* Consider the possibility that the $N \times 1$ vector $\mathbf{U}$, after premultiplication by the $N \times N$ matrix $\mathbf{M}$, produces a vector proportional to $\mathbf{U}$ (e.g., when $N = 3$ the original and transformed vectors are parallel in three-dimensional space):

$$\mathbf{M}\mathbf{U} = \Lambda\mathbf{U} \tag{II-20}$$

* Theoretical properties investigated here do not necessarily reflect the means of computing the quantities of interest. It suffices to note here that accurate algorithms are available for obtaining these quantities.

or, for a specific scalar *eigenvalue* Λ and its corresponding *eigenvector* $\mathbf{U}$,

$$(\mathbf{M} - \Lambda \mathbf{I})\mathbf{U} = \mathbf{0} \tag{II-21}$$

Eigenvalues are found as roots of the equation

$$|\mathbf{M} - \Lambda \mathbf{I}| = 0 \tag{II-22}$$

which, as an Nth order algebraic equation, has N roots. At this point a general exposition would raise the issue of repeated roots versus distinct eigenvalues, requiring expanded development along those lines. The discussion immediately below, however, is concerned with two specific types of (real, square, nonsingular) matrices, i.e., an *orthogonal* matrix $\mathbf{T}$ with positive determinant $(+1)$, and a *symmetric* matrix $\mathbf{P}$ with no vanishing eigenvalues. Only distinct roots will be considered initially, and modifications for repeated eigenvalues will follow.

The orthogonal matrix $\mathbf{T}$ obeys the relation

$$\mathbf{T}^{\mathrm{T}}\mathbf{T} = \mathbf{T}\mathbf{T}^{\mathrm{T}} = \mathbf{I} \tag{II-23}$$

so that, if partitioned as in (II-13), the sum of squares of elements in each column would be unity, while the scalar product involving any two different columns would vanish (a similar statement holds for rows). This definition, in combination with the foregoing rule for matrix multiplication, provides the rule that products of orthogonal matrices are always orthogonal. Solution of (II-22) under conditions of orthogonality produces roots of unit magnitude (whether real or complex). There is always a positive real root $(+1)$, which is the principal eigenvalue of interest here. When this root is not repeated, it is always possible to choose $N - 1$ independent nonvanishing rows of the matrix

$$\mathbf{Z} = \mathbf{T} - \mathbf{I} \tag{II-24}$$

and solve for the principal eigenvector (denoted $\mathbf{E}$) as follows: From (II-21) and (II-24), $\mathbf{E}$ is orthogonal to every vector in the column-partitioned matrix

$$[\mathbf{Z}_1, \mathbf{Z}_2, \ldots, \mathbf{Z}_N] \triangleq \mathbf{Z}^{\mathrm{T}} \tag{II-25}$$

There are then $N - 1$ independent linear equations in N unknowns. To complete the definition it is permissible to choose any convenient vector magnitude without violating (II-20). Let

$$\mathbf{E}^{\mathrm{T}}\mathbf{E} = 1 \tag{II-26}$$

As a familiar example, consider the case $N = 3$ with two nonvanishing nonparallel 3×1 vectors $(\mathbf{Z}_a, \mathbf{Z}_b)$. The unit vector perpendicular to both of these is a normalized cross product,

$$\mathbf{E} = \pm \mathbf{Z}_a \times \mathbf{Z}_b / |\mathbf{Z}_a \times \mathbf{Z}_b| \tag{II-27}$$

More generally, the $N \times 1$ vector orthogonal to $N - 1$ linearly independent vectors $\mathbf{Z}_A, \mathbf{Z}_B, \ldots, \mathbf{Z}_{N-1}$ (of N components each) has a direction given by the determinant of a matrix containing unit reference vectors in the top row, and transposed $\mathbf{Z}$-vectors in every other row:

$$\mathbf{E}_N' = \left\| \begin{bmatrix} \mathbf{1}_1 \quad \mathbf{1}_2 \quad \cdots \quad \mathbf{1}_N \\ \hdashline \mathbf{Z}_A^{\mathrm{T}} \\ \mathbf{Z}_B^{\mathrm{T}} \\ \vdots \\ \mathbf{Z}_{N-1}^{\mathrm{T}} \end{bmatrix} \right\| \tag{II-28}$$

(where $\mathbf{1}_j$ is associated with the jth column of the $N \times N$ identity matrix), and division by the root sum square of the elements can provide normalization as in (II-27).

Symmetric matrices can be defined by the relation

$$\mathbf{P} = \mathbf{P}^{\mathrm{T}} \tag{II-29}$$

which includes the following examples:

(1) a *diagonal* matrix, i.e., a matrix in which every element off the principal diagonal is zero;

(2) a 3×3 *inertia* tensor, in which the diagonal and off-diagonal elements represent moments and products of inertia, respectively, for a given physical structure and a chosen set of orthogonal reference axes;

(3) a probabilistic *covariance* matrix, of the form

$$\mathbf{P} = \begin{bmatrix} \sigma_1^2 & \rho_{12}\sigma_1\sigma_2 & \cdots & \rho_{1N}\sigma_1\sigma_N \\ \rho_{12}\sigma_1\sigma_2 & \sigma_2^2 & & \\ \vdots & & \ddots & \vdots \\ \rho_{1N}\sigma_1\sigma_N & & \cdots & \sigma_N^2 \end{bmatrix}, \qquad |\rho_{ij}| < 1 \tag{II-30}$$

Symmetric matrices can always be characterized by orthogonal sets of eigenvectors, and their eigenvalues are always real. If, in addition, the eigenvalues are all positive, the symmetric matrix is said to be *positive definite*, because the *quadratic form* obtained with any nonzero vector $\mathbf{U}$ obeys the relation

$$\mathbf{U}^{\mathrm{T}}\mathbf{P}\mathbf{U} > 0, \qquad \mathbf{U} \neq \mathbf{0} \tag{II-31}$$

In particular, when $\mathbf{U}$ is a unit eigenvector $\mathbf{T}_n$ of $\mathbf{P}$,

$$\mathbf{T}_n^{\mathrm{T}}\mathbf{P}\mathbf{T}_n = \Lambda_n, \qquad 1 \leq n \leq N \tag{II-32}$$

which follows from (II-20), premultiplied by the transpose of the unit eigenvector. When all equations of this form are collected, the result can be written in the form

$$\mathbf{T}^{\mathrm{T}}\mathbf{P}\mathbf{T} = \mathbf{\Lambda} \tag{II-33}$$

where $\mathbf{\Lambda}$ is diagonal while, in analogy with (II-13),

$$\mathbf{T} = [\mathbf{T}_1, \mathbf{T}_2, \ldots, \mathbf{T}_N] \tag{II-34}$$

and, if (II-33) were expanded into scalars, each row of the product matrix would conform to (II-32). This formulation utilizes the aforementioned orthogonality of eigenvectors for symmetric matrices; $\mathbf{T}$ obeys (II-23). Consequently $\mathbf{P}$ is the result of a *similarity transformation* performed on $\mathbf{\Lambda}$,

$$\mathbf{P} = \mathbf{T\Lambda T}^{-1} \tag{II-35}$$

and $\mathbf{\Lambda}$ is the result of a similarity transformation (i.e., obtained through premultiplication by a nonsingular matrix and postmultiplication by its inverse) performed on $\mathbf{P}$:

$$\mathbf{\Lambda} = \mathbf{UPU}^{-1}, \qquad \mathbf{U} = \mathbf{T}^{\mathrm{T}} \tag{II-36}$$

Significantly, both the *trace* (i.e., sum of diagonal elements) and the determinant of a matrix are unchanged by a similarity transformation. It follows that the trace and the determinant are identified with the sum and product of eigenvalues, respectively. Also, the determinant of a product is the product of determinants.

The aforementioned inertia and covariance matrices are positive definite, with readily interpretable physical properties. For the inertia matrix of an arbitrarily shaped rigid structure, eigenvectors correspond to principal axes* and eigenvalues represent principal moments of inertia about these axes. The covariance matrix example is conveniently illustrated here by a zero mean random vector $\mathbf{x}$ (i.e., a vector whose every component is a random variable of zero mean), with less than 100% correlation between any two components. In this case

$$\mathbf{P} = \langle \mathbf{x}\mathbf{x}^{\mathrm{T}} \rangle \tag{II-37}$$

where the angular brackets denote probabilistic mean. Formation of the outer product, and of each *variance* σ_n^2 and each *normalized cross correlation coefficient* ρ_{nk},

$$\sigma_n^2 = \langle x_n^2 \rangle, \qquad \rho_{nk} = \langle x_n x_k \rangle / (\sigma_n \sigma_k) \tag{II-38}$$

immediately produces (II-30). The eigenvectors of $\mathbf{P}$ then form *uncorrelated* random variates χ_n from linear combinations of components x_i, since

$$\chi = \mathbf{T}^{\mathrm{T}}\mathbf{x}, \qquad \langle \chi\chi^{\mathrm{T}} \rangle = \mathbf{T}^{\mathrm{T}}\mathbf{PT} \tag{II-39}$$

and, by (II-33), this latter product must be diagonal ($\langle \chi_n \chi_k \rangle = 0,\ n \neq k$).

* In a rigid body with uniform mass distribution, an axis of symmetry is a principal axis. When there are no axes of symmetry, the principal axes are unique.

Eigenvalues of $\mathbf{P}$ in this example are mean squared values of the uncorrelated variates.

To determine the eigenvector corresponding to the nth distinct eigenvalue Λ_n of a symmetric matrix, (II-24) can be replaced by

$$\mathbf{Z} = \mathbf{P} - \Lambda_n \mathbf{I} \tag{II-40}$$

so that the right of (II-27) or (II-28) could be applied with (II-25), in three-dimensional or N-dimensional space, respectively.

The last topic remaining in this section concerns repeated eigenvalues of positive definite symmetric and orthogonal matrices. For purposes of this book, only simple additions to the above development are needed. Consider, for example, a double root in (II-22) for the inertia tensor of a right circular cylinder. Any two directions normal to the cylinder axis can then be chosen for eigenvectors. For a sphere, any triad can represent principal axis directions. More generally, when there are fewer than $N - 1$ linearly independent $\mathbf{Z}$-vectors, the analyst can impose added conditions to constitute a unique set of relationships.

The only orthogonal matrix diagonalization of interest in this book concerns 3×3 transformations between coordinate frames, and only the principal eigenvector $\mathbf{E}$ is of interest. Equation (II-27) is then applicable except in trivial cases (e.g., rotation through 2π rad), for which no special eigenvector direction need be determined.

II.3 MATRICES IN LINEAR DYNAMICS ANALYSIS

Any set of simultaneous independent linear differential equations of any order can be re-expressed as a collection of first-order linear differential equations. With the homogeneous case considered first, a dynamical system can then be represented by the *vector differential equation*

$$\dot{\mathbf{x}} = \mathbf{A}\mathbf{x} \tag{II-41}$$

Since $\mathbf{x}$ and $\dot{\mathbf{x}}$ must have equal dimension ($N \times 1$), the dynamics matrix $\mathbf{A}$ must be square ($N \times N$). In general $\mathbf{A}$ may be singular and/or time-varying. When it is constant, however, there is a closed-form solution of the type

$$\mathbf{x} = \exp\{\mathbf{A}(t - t_0)\}\mathbf{x}_0, \qquad \dot{\mathbf{A}} = \mathbf{O} \tag{II-42}$$

where, just as in the scalar case,

$$\exp\{\mathbf{A}(t - t_0)\} \triangleq \mathbf{I} + \mathbf{A}(t + t_0) + \frac{1}{2!}\mathbf{A}^2(t - t_0)^2 + \cdots \tag{II-43}$$

and a diagonalization provides further insight as follows: Define $\mathbf{B}$ by the relation

$$\mathbf{B}^{-1}\mathbf{A}\mathbf{B} = \mathbf{\Lambda} \tag{II-44}$$

where $\mathbf{\Lambda}$ contains the eigenvalues* of $\mathbf{A}$. In general, these represent "complex frequencies," characterizing natural frequencies of oscillation and/or time constants applicable to the dynamics of interest. This can be seen more clearly through the transformation

$$\mathbf{x} = \mathbf{B}\mathbf{w}, \qquad \mathbf{w} = \mathbf{B}^{-1}\mathbf{x} \tag{II-45}$$

so that

$$\dot{\mathbf{w}} = \mathbf{B}^{-1}\dot{\mathbf{x}} = \mathbf{B}^{-1}\mathbf{A}\mathbf{x} = \mathbf{\Lambda}\mathbf{w} \tag{II-46}$$

and therefore

$$\mathbf{w} = \exp\{\mathbf{\Lambda}(t - t_0)\}\mathbf{w}_0 \tag{II-47}$$

or

$$\mathbf{x} = \mathbf{B}\exp\{\mathbf{\Lambda}(t - t_0)\}\mathbf{B}^{-1}\mathbf{x}_0 \tag{II-48}$$

Thus, the dynamics of $\mathbf{x}$ can be found by a straightforward transformation; to illustrate the simplicity of this concept, consider the case of no repeated roots in the eigenvalue equation for $\mathbf{A}$ (so that, as just noted, $\mathbf{\Lambda}$ will be diagonal). Then (II-47) is identical to a set of uncoupled scalar equations,

$$w_n = \exp\{\Lambda_n(t - t_0)\}w_{n(t_0)}, \qquad 1 \le n \le N \tag{II-49}$$

where complex roots are expressible as

$$\Lambda_n = \Lambda_{n(\text{re})} + i\Lambda_{n(\text{im})} \tag{II-50}$$

and

$$\exp\{\Lambda_n(t - t_0)\} = \exp\{\Lambda_{n(\text{re})}(t - t_0)\}\,\text{cis}\{\Lambda_{n(\text{im})}(t - t_0)\} \tag{II-51}$$
$$\text{cis}(\cdot) = \cos(\cdot) + i\sin(\cdot)$$

Any conjugate root pairs will interact with the (in general, complex) elements of $\mathbf{B}$ to produce a real result in (II-48). Since

$$\mathbf{A}\mathbf{B} = \mathbf{B}\mathbf{\Lambda} \tag{II-52}$$

it follows that eigenvectors of $\mathbf{A}$ are formed by columns of $\mathbf{B}$, which influence relative magnitude and phasing in the spectral distribution for each

* For example, $\mathbf{\Lambda}$ is diagonal if all eigenvalues of $\mathbf{A}$ are distinct. The only restriction to be placed on the definition of $\mathbf{B}$ is nonsingularity. This latter condition can always be satisfied for the linear system (II-41).

component of $\mathbf{x}$. To emphasize the significance of the word *relative*, note that a diagonal matrix $\mathbf{D}$, effectively rescaling the columns of $\mathbf{B}$ through post-multiplication, could define another diagonalizing matrix in the above example; since diagonal matrices always commute,

$$(\mathbf{BD})\mathbf{\Lambda}(\mathbf{BD})^{-1} = \mathbf{B\Lambda B}^{-1} = \mathbf{A} \tag{II-53}$$

Therefore, in analogy with the normalization in the preceding section, $\mathbf{B}$ can consist of unit magnitude column vectors and the smallest imaginary component magnitude in each column can be zero.

Stability of the linear dynamical system (II-41) is expressed in terms of asymptotic behavior resulting from nonzero initial conditions $\mathbf{x}_0$. From the exponential nature of the solution (II-48), it is clear that instability results when the real part of any eigenvalue is positive. By way of contrast, if all eigenvalues have negative real parts, the system eventually approaches a null state ($\mathbf{x} = \mathbf{0}$) for any finite initial displacement. Multiple roots with zero real parts may or may not produce instability, but such roots cannot cause instability if they are not repeated.*

The fact that so much insight can be gained in the above case of constant $\mathbf{A}$ is not surprising. Ordinary linear differential equations, producing real and/or complex eigenvalues in conjugate pairs, are indicative of simple dynamic trajectories such as harmonic oscillators, coupled or uncoupled, with or without damping (or growth). A somewhat less restrictive case results from a dynamics matrix that varies but commutes with its time integral. In that case (II-42) is replaced by

$$\mathbf{x} = \exp\left\{\int_{t_0}^{t} \mathbf{A}\, d\tau\right\}\mathbf{x}_0, \qquad \mathbf{A}\left[\int \mathbf{A}\, dt\right] = \left[\int \mathbf{A}\, dt\right]\mathbf{A} \tag{II-54}$$

but the commutability condition is too restrictive for many cases. A more general time-varying dynamics matrix can be accommodated by the use of a *transition matrix*, defined as the solution to the *matrix differential equation*

$$\dot{\mathbf{\Phi}} = \mathbf{A\Phi} \tag{II-55}$$

subject to the initial conditions

$$\mathbf{\Phi}_0 = \mathbf{I} \tag{II-56}$$

* Stable solutions with purely imaginary or null ($0 + \imath 0$) roots in general contain terms that, although bounded, do not approach zero. A case of repeated null roots without instability is exemplified by inserting a null matrix into (II-41). As a counterexample, consider the system

$$\begin{bmatrix} \dot{x}_1 \\ \dot{x}_2 \end{bmatrix} = \begin{bmatrix} 0 & 1 \\ 0 & 0 \end{bmatrix} \begin{bmatrix} x_1 \\ x_2 \end{bmatrix}$$

which, when given a nonzero initial value for x_2, produces a response in x_1 that grows with time.

The solution in the more general case is then simply

$$\mathbf{x} = \mathbf{\Phi}\mathbf{x}_0 \tag{II-57}$$

which, of course, reduces to (II-42) or (II-54) if the stated restrictions are imposed. This solution is easily verified by differentiation, using (II-55).

As a further generalization, consider an arbitrary forcing function $\mathbf{u}$ added to the linear system. The time-varying nature of all quantities will be emphasized by writing vectors and matrices with time arguments:

$$\dot{\mathbf{x}}(t) = \mathbf{A}(t)\mathbf{x}(t) + \mathbf{u}(t) \tag{II-58}$$

The solution in this case is

$$\mathbf{x}(t) = \mathbf{\Phi}(t, t_0)\mathbf{x}_0 + \int_{t_0}^{t} \mathbf{\Phi}(t, \tau)\mathbf{u}(\tau)\, d\tau \tag{II-59}$$

because this satisfies $\mathbf{x}(t_0) = \mathbf{x}_0$ and, by differentiation,

$$\dot{\mathbf{x}}(t) = \mathbf{A}(t)\mathbf{\Phi}(t, t_0)\mathbf{x}_0 + \mathbf{\Phi}(t, t)\mathbf{u}(t) + \int_{t_0}^{t} \mathbf{A}(t)\mathbf{\Phi}(t, \tau)\mathbf{u}(\tau)\, d\tau \tag{II-60}$$

Since the matrix $\mathbf{A}(t)$ in the integrand is independent of the variable τ, this readily reduces to (II-58). An application of interest here is the case in which $\mathbf{x}_0$ and $\mathbf{u}$ are uncorrelated zero mean random vectors, the latter in conformance with the *white noise* property,

$$\langle \mathbf{u}(\tau)\mathbf{u}^{\mathrm{T}}(\tau') \rangle = \mathbf{F}(\tau)\, \delta(\tau - \tau') \tag{II-61}$$

where $\mathbf{F}(\tau)$ contains spectral densities of the excitation and δ represents the Dirac delta function. From (II-37) and (II-59), under the conditions stated, with the product of integrals conveniently written as a double integral,

$$\mathbf{P}(t) = \mathbf{\Phi}(t, t_0)\mathbf{P}_0\, \mathbf{\Phi}^{\mathrm{T}}(t, t_0) + \left\langle \int_{t_0}^{t} \int_{t_0}^{t} \mathbf{\Phi}(t, \tau)\mathbf{u}(\tau)\mathbf{u}^{\mathrm{T}}(\tau')\mathbf{\Phi}^{\mathrm{T}}(t, \tau')\, d\tau'\, d\tau \right\rangle \tag{II-62}$$

or, in view of (II-61) and the deterministic nature of the transition matrix, placement of the averaging operator inside the integrand and subsequent application of the Dirac delta integration property yields

$$\mathbf{P}(t) = \mathbf{\Phi}(t, t_0)\mathbf{P}_0\, \mathbf{\Phi}^{\mathrm{T}}(t, t_0) + \int_{t_0}^{t} \mathbf{\Phi}(t, \tau)\mathbf{F}(\tau)\mathbf{\Phi}^{\mathrm{T}}(t, \tau)\, d\tau \tag{II-63}$$

This last equation is the basis for covariance extrapolation between discrete fixes, in estimation algorithms used throughout this book.

Appendix III

Navigation Functions and Data Flow

A fundamental step in defining an approach to nav system integration is to list the various operations and the data required by each function. Following is one possible breakdown of these functions, expressed in a general form that can be modified for any specific application. Data flow into, out of, and within the navigation system (i.e., from one part of the nav system to another) are discussed.

III.1 INITIALIZATION OF EARTH NAV QUANTITIES*

Whether on a stable (e.g., airstrip) or a moving (e.g., aircraft carrier) support, position initialization consists of straightforward insertion of accurately known coordinates, available from an external source. Velocity initialization is also reasonably straightforward, e.g., in the stable case, the north, east, and vertical components of velocity can be nulled; and in the moving reference case, the initial velocity vector is that of the supporting structure, augmented by the kinematical effect of rotations (e.g., ship roll). Platform alignment, in the case of the stable support, typically consists of

* For brevity of this description, these operations are described for a particular INS configuration (a north slaved, Schuler tuned platform). The basic concepts, however, remain applicable for other INS implementations.

torquing the north and east gyros until a satisfactory null is obtained from the level accelerometers and torquing the azimuth gyro in accordance with gyrocompassing or external (e.g., optical) alignment procedures. On moving structures, however, supplementary operations are required to prevent degradation from perturbing forces (e.g., wind or irregular angular oscillation of an aircraft carrier). These operations may take the form of velocity matching from a master reference, transfer of an external orientation reference via radio frequency (rf) or wire communication, manual means, etc.

III.2 POINT NAV INITIALIZATION

In this operation, all navigation quantities are re-referenced with respect to a new origin (e.g., designated target) in a new coordinate frame (e.g., locally level but independent of north direction). Typically this involves only an amount of data flow commensurate with a single position fix (see Section III.6), while the velocity information is simply transformed into the new coordinate frame, and the mechanized angular orientation reference is left unchanged.

III.3 PLATFORM ORIENTATION

A north slaved local vertical reference is maintained by driving the platform servos from the gyros (which are sensitive to angular rate with respect to inertial space) and by torquing the gyros in accordance with the nav computations (which indicate the rotation of the local coordinate frame relative to the inertial reference; Section III.4). The data rate for the first of these would be quite high in a digital mechanization (dictated by the ratio of the maximum aircraft angular rate to the quantization increment); thus platform gimbals may be controlled by analog servos. The data rate for the latter is the navigation computer iteration rate, as described below.

III.4 PERIODIC INCREMENTATION

At regular intervals (e.g., 50/sec) the velocity increments in the local coordinate frame are counted for velocity incrementing in each channel, supplemented by standard navigation equation correction terms and kinematical latitude/longitude incrementation with respect to the reference geoid. Corresponding torquing commands are determined for north slaving and Schuler tuning.

III.5 EARTH NAV UPDATE

Each time a navaid reading is accepted, the sensor information is fed to the data mixer (e.g., a Kalman filtering computation) to adjust the current estimate of latitude, longitude, altitude; north, ground east, and vertical velocity; and angular orientation. This is a low-data-rate operation (<1 Hz); sensors with high data rates are prefiltered prior to data mixing.

III.6 POINT NAV UPDATE

From the standpoint of data flow, this operation is the same as described in Section III.5; only the reference origin and coordinate orientation have been changed.

III.7 FLIGHT CONTROL

For fuel management, control of ailerons, elevators, rudder, etc., information is received from local inertial sensors as well as altitude, dive and drift angles, attitude, and air data; the air-to-air mode uses radar data (range, range rate, LOS, and LOS rate) also for this function (e.g., in a lead pursuit course). Although the 0 dB crossover frequencies are low (<5 Hz) for the control surfaces, the information rates must be sufficiently high to produce negligible phase shift at about 2 Hz. Typically, this calls for information rates of 30 Hz from the local inertial sensors (and the radar data, when in the air-to-air mode) and lower rates for the other information sources just listed for this item.

III.8 BALLISTICS COMPUTATION

Prior to airdrop the airspeed is vectorially subtracted from groundspeed, to yield estimated wind. After computation of ballistics with respect to the air mass, the integrated effect of the wind is reinserted to produce the aircraft-to-impact-point displacement vector; this is compared with the aircraft-to-target displacement vector so that timing and steering commands can be obtained from the resulting miss vector (impact-point-to-target). For an essentially continuous display and smooth command sequence generation, the impact computations must be computed at a rate on the order of 25 Hz. For displays that call for roll stabilization (whether for indication of roll on a synthetic display or for computed impact transformation into aircraft coordinates on a real display), pitch and roll data can be required at rates on the order of 25 to 50 Hz, respectively.

III.9 RADAR IMAGING

For systems that employ SAR, two additional functions must be included; i.e., motion compensation and automatic cursor drive. The first is by far the faster of the two operations, requiring data from the integrating accelerometers as the velocity increment pulses are generated. For example, at 0.032-fps velocity resolution, the data rate would be 1 KHz per g. Since this far exceeds any nav computer iteration rate, a cyclic accumulation would be required between successive nav computer increments. The north, ground east, and vertical velocity (V_N, V_E, V_V, respectively) are also required at the nav computer iteration rate. When the antenna phase center is displaced from the IMU, this displacement must also be transformed into reference coordinates through the aircraft orientation matrix and added to the inertial system output. In order that the antenna phase center position be updated at the time of every pulse repetition frequency (PRF) pulse for motion compensation, the data rate for this latter computation may be as high as the PRF, or the ratio of angular rate to pickoff resolution, whichever is less.

Automatic cursor drive simply calls for up-to-date point navigation information at a rate sufficient for smooth visual operation (e.g., 15 Hz).

III.10 AIR-TO-AIR TRACKING

The air-to-air mode calls for vector velocity, angular orientation, and radar data (range and LOS) at rates typically on the order of 30 Hz for all operations other than antenna steering (see Section III.12) without antenna-mounted gyros.

III.11 CLUTTER TRACKING

In conventional systems, this is a closed-loop operation whereby a frequency servo automatically maintains the center of the main beam clutter (MBC) spectrum close to the center frequency of a notch filter. Quite conceivably, however, an advanced mechanization might employ nav information directly in performing this function. The required data rate would be ten times the tracking bandwidth.

III.12 ANTENNA STEERING

Although this function is included in other operations (e.g., SAR imaging, ground target tracking, etc.), it is listed here as a separate function because of the high data rates often implied by it. In air-to-air operation, the aircraft orientation is sampled at typical rates of 100 Hz (search) or 200 Hz (track), while the radar information is sampled at rates on the order of 30 Hz.

CONCEPTUAL APPROACH TO NAVIGATION DATA FLOW

This section defines a premechanization system concept, identifying the source and destination for various nav signals to be transmitted at widely differing rates. An effort was made to maintain generality in this discussion. The present description does, however, establish a basis whereby a configuration can be defined for a given application, once all system provisions (available navaids, external information sources, etc.) have been specified.

All navaid measurements are regarded as prefiltered low-data-rate entries into a Kalman estimator, and the nav outputs will automatically be expressed with respect to the appropriate coordinate reference, as dictated by the current mode of operation. Simple functions that can be computed directly from these navigation outputs (e.g., dive angle and drift angle, which are completely defined by aircraft attitude, heading, and velocity vector) need not necessarily appear explicitly; also, wherever attitude information is required at high rates (e.g., 100 Hz), it is assumed for the present discussion that computing provisions are included wherever needed. With these ground rules, Table III-1 has been constructed to exemplify data flow into, out of, and within the navigation system. While different ground rules can certainly be postulated and the tabulation could also be affected by mode switching or function changes, the table does provide a realistic example of data flow in an integrated system approach. It is noted that certain points in the system act as both source and destination for signals; these include the following:

(a) *the navigation computer*, which accepts the integrating accelerometer pulses and provides outputs to the Kalman filter and the gyro torquers,

(b) *the Kalman filter*, which accepts inputs from the navigation computer and the navaids, while providing outputs to other points in the system,

(c) *the clutter tracker*, which provides doppler information as an output, while (in an advanced mechanization) accepting updated navigation information to maintain the MBC center frequency at a notch-filter center, and

(d) *navaids*, which, while furnishing update information, may use INS velocity for slaving under rapidly changing geometry.

In interpreting the table, the Kalman filter output should be regarded as earth-referenced data (latitude, longitude, altitude above mean sea level, north, ground east, and vertical velocity) or point-referenced data (position with respect to a designated point and velocity expressed in a locally level frame that may be independent of north direction) as appropriate for each

TABLE III-1

TYPICAL DATA RATES FROM SOURCE TO DESTINATION (Hz)

Source / Destination		IMU pulses			Nav. computer								
	Ext.	Gyro	Accel.	Gimbal pickoffs	Posit.	Vel.	Torq. comm.	Local iner.	Air Data	Nav-aids	Kalm. fltr.	Radar	Clut. trkr.
Gyro torquers	—	—	—	—	—	—	50	—	—	—	—	—	—
Platform gimbal servos	—	Analog	—	Analog	—	—	—	—	—	—	—	—	—
Navigation computer	—	—	3000	—	—	—	—	—	—	—	<1	—	—
Kalman filter	<1	—	—	—	50	50	—	—	—	<1	—	50	<1
Antenna servos	—	—	—	200	50	—	—	—	—	—	50	—	—
Flight control computer	—	—	—	10	10	10	—	30	10	—	—	—	—
Clutter tracker	—	—	—	200	—	25	—	—	—	—	—	—	—
Ballistics computer	—	—	—	50	25	25	—	—	25	—	—	—	—
SAR motion compensation	—	—	1000	1000	50	50	—	—	—	—	—	—	—
SAR cursor	—	—	—	—	50	50	—	—	—	—	—	—	—
Navaid aiding signals	—	—	—	—	—	50	—	—	—	—	—	—	—

specific function. It is seen that certain functions (e.g., platform orientation,* antenna servo drive, and SAR motion compensation) contain data inputs at very high speeds, and other functions are involved with flight safety; separate provisions will sometimes be preferred for those functions.

* To circumvent excessive data rate requirements, this function may be performed by analog servos.

Index

Note to reader: (1) See Appendix I for mathematical index. (2) Abbreviations are used as headings only when the abbreviation is the most common form in usage. When this occurs, the reader can refer to p. 319.

D

E

F

G

S

T

U

V

W

A 6
B 7
C 8
D 9
E 0
F 1
G 2
H 3
I 4
J 5

9

Farrell, James L.

Integrated aircraft navigation

DATE DUE
